Engineering Steels and High Entropy-Alloys

Edited by Ashutosh Sharma,
Zoia Duriagina, Sanjeev Kumar

Published in London, United Kingdom

IntechOpen

Supporting open minds since 2005

Engineering Steels and High Entropy-Alloys
http://dx.doi.org/10.5772/intechopen.84991
Edited by Ashutosh Sharma, Zoia Duriagina, Sanjeev Kumar

Contributors
Sanjeev Kumar, Lavinia Cosmina Ardelean, Lucien Reclaru, Alexandru Florian Grecu, Catalin Adrian Miu, Yong A Zhang, Ya Song Li, Farid Akhtar, Sajid Ali Alvi, Hanzhu Zhang, Victor Geanta, Ionelia Voiculescu, Fahim Khan, Hossain M.M.A Rashed, Yeeli Kelvii Kwok, Tatsuhiko Aizawa, Tomomi Shiratori, Takafumi Komatsu, Petrică Vizureanu, Andrei Victor Sandu, Miklós Tisza, Valentina Loganina, Yerkebulan Mazhitov, Anas Ahmad Siddiqui, Avanish Dubey, Zoia Duriagina, Ashutosh Sharma, Taras Kovbasyuk, Volodymyr Kulyk, Andriy Trostianchyn, Tetiana Tepla

First published in London, United Kingdom, 2020 by IntechOpen
IntechOpen is the global imprint of INTECHOPEN LIMITED, registered in England and Wales, registration number: 11086078, 7th floor, 10 Lower Thames Street, London, EC3R 6AF, United Kingdom
Printed in Croatia

British Library Cataloguing-in-Publication Data
A catalogue record for this book is available from the British Library

Additional hard and PDF copies can be obtained from orders@intechopen.com

Engineering Steels and High Entropy-Alloys
Edited by Ashutosh Sharma, Zoia Duriagina, Sanjeev Kumar
p. cm.
Print ISBN 978-1-78985-947-8
Online ISBN 978-1-78985-948-5
eBook (PDF) ISBN 978-1-83880-556-2

For EU product safety concerns:
IN TECH d.o.o., Prolaz Marije Krucifikse Kozulić 3, 51000 Rijeka, Croatia, info@intechopen.com or visit our website at intechopen.com.

Meet the editors

Ashutosh Sharma is currently working in the Department of Energy Systems Research, Ajou University, Suwon, South Korea. He earned his Ph.D. degree in Metallurgical and Materials Engineering from the Indian Institute of Technology (IIT) Kharagpur, which is a premier institute of India. His research interests include electrochemical deposition, lead-free soldering and brazing, additive manufacturing, high entropy alloys, gas sensors, and transparent nanoheaters. Dr. Sharma is a life member of various scientific and professional bodies. In a very short time, he has contributed more than 100 international journals, 14 patents, 3 book chapters, and 1 authored book so far. Recently, he was awarded the Extraction and Processing Division (EPD) Award by The Minerals, Metals and Materials (TMS, USA) in 2016 in recognition of his outstanding research in the area of nonferrous materials processing.

Zoia Duriagina , Sc.D., Prof., Head of the Department of Applied Materials Science and Materials Engineering, NULP, Ukraine. She is a professor, dr.hab. of the Lublin Catholic University, Poland. Her scientific and research interests include: stainless steels and functional alloys, surface engineering: application of functional coatings, creating new thermoelectric materials, and investigation the properties of Ti-powders alloys. She has authored and co-authored 268 scientific publications, including 52 articles in journals indexed in Scopus and Web of Science (Author ID: 6507291021, http://orcid.org/0000-0002-2585-3849, h-index:6), 8 scientific monographs, 4 textbooks, and 16 patents. Awards received included the Golden Medal of Scientific Society of Materials Science and Diploma of International expert in the field of materials science, China. Sh is a member of the Academy of Higher Education of Ukraine and a member of the World Academy of Materials and Manufacturing Engineering.

Dr. Sanjeev Kumar has been working as a senior postdoc researcher in the Christian Doppler Laboratory for Interfaces and Precipitation Engineering at IMST, TU Wien, Austria, since 2018. He also worked as a postdoc researcher from 2017 to 2018 on an FWF project at the Institute of Materials Science, Forming and Joining, TU Graz, Austria. He obtained his MTech (2012) and PhD (2017) degrees in the field of Welding Metallurgy from IIT Roorkee, India, and his BTech (2009) degree in Mechanical Engineering from AKGEC GZB UPTU, Lucknow, India. His primary research interest is physical metallurgy, e.g. physical simulation studies for welding, thermomechanical processing of materials, new alloy design, microstructure-properties correlation, heat treatments, kinetic precipitation of materials, and OpenPhase field modeling, etc. He has published 29 papers in reputed international journals, conferences, and books.

Contents

Preface

It gives us immense pleasure to have this opportunity of editing this book on "Engineering Steels and High Entropy-Alloys". With the advancement of new materials, there is no doubt today that "Engineered Steels and High Entropy-Alloys" (HEAs) are the friendliest group of structural and instrumental materials that meet the needs of almost all industries. Besides, these are the materials that will be popular forever as they can be completely recycled. According to the World Steel Association, there are over 3,500 different grades of steel, each with unique physical, chemical, and environmental properties. It is well known that steel is composed of Fe and C, although it is the amount of C, as well as the extent of impurities and/or alloying elements that contribute to the properties of each steel grade. Indeed, the use of various technological methods allows us to purposefully change the microstructure, phase composition, and functional properties of steel products. Moreover, employing surface engineering methods, it is possible to create coatings that possess the properties of new materials.

Engineering steels are typically wrought steels engineered for mechanical, specialized, and allied technological applications requiring critical and often stringent levels of strength, ductility, toughness, and fatigue resistance. In special cases, steel may be required to exist at extremely high/low temperatures, corrosive, or harsh environments. Applications of these engineered high strength steel grades, often called Ultra High Strength Steels, are gripping current production of car bodies all around the world owing to their unique characteristics that not only allow lightweight but stronger parts than conventional steel counterparts. The composition, manufacturing route, and heat treatment procedures are carefully chosen to meet the customers' needs.

In recent times, the traditional alloy design approach has been shifted towards multi-component alloys known as HEA, thanks to the works of Prof. Brian Cantor, Prof. Jien-Wei Yeh, and Prof. S. Ranganathan. The traditional alloys such as steel, bronze, superalloys, etc. consist of one principal element with other elemental additions to enhance their microstructural and mechanical performance. In contrast, HEAs contains multiple principal elements in equimolar/near equimolar fractions to have a high configurational entropy that results in solid solution phases. The high entropy approach includes not only alloys but also ceramics and is regarded as a novel alloy design approach for various alloys including steels, superalloys, coatings, and ceramics, which possess excellent mechanical, thermal, and chemical stability, excellent creep resistance, oxidation/corrosion, and tribological and magnetic properties.

This breakthrough in HEA came after almost a century of research and development that resulted in a highly mature alloy design approach. It is expected that HEAs can satisfactorily replace traditional materials and design in advanced engineering applications. The ongoing research results in high entropy materials are very exciting covering a wide range of possibilities in areas such as jet engines, nuclear and power plants, marine structures, hard and thermal barrier coatings, and tools design.

We have a great honor to thank all the co-editors, contributing authors, co-authors, and publication staff from IntechOpen for their efforts in preparing this book. We hope that the new experimental methodology given in the book will be the foundation for the development of new materials and technologies in materials science. This book will be useful to researchers, scientists, engineers, graduate and undergraduate level students across multi-disciplines. We hope that readers enjoy this book and that it will serve as an aid to create new exciting materials with unique properties.

Dr. Ashutosh Sharma
Research Professor,
Department of Energy Systems Research,
Ajou University,
Suwon, South Korea

Department of Materials Science and Engineering,
Ajou University,
Suwon, South Korea

Zoia Duriagina,
Professor,
Dr.Sc, PhD, Dr.Hab.Eng. Head of Department of Applied Materials,
Science and Materials Engineering of Lviv,
Polytechnic National University,
Ukraine

Prof. of the John Paul II Catholic University of Lublin,
Poland

Dr. Sanjeev Kumar
Christian Doppler Laboratory for Interfaces
and Precipitation Engineering, CDL-IPE,
TU Wien,
Vienna, Austria

Section 1

Fundamentals of Engineering Steels

Chapter 1

Integrated Manufacturing of Fine-Grained Stainless Steels for Industries and Medicals

Tatsuhiko Aizawa, Tomomi Shiratori and Takafumi Komatsu

Abstract

Austenitic stainless steels have been widely utilized in industries, infrastructures, housing structures, kitchen components, and medical tools. Higher hardness and strength as well as more improvement of wear and corrosion toughness are often required in the industrial and medical applications. Fine-grained stainless steel (FGSS) provides a solution to increase the strength without loss of ductility and toughness. Deeper research and development in manufacturing of FGSS is required to make full use of its properties toward its applications in industries and medicals. First, its mechanical properties and microstructure is introduced as a basic knowledge of FGSS with comparison to the normal stainless steels. Mechanical and laser machinability of FGSS is stated and discussed to finish the products in seconds. Its performance in metal forming and diffusion bonding is explained to explore its applications in third. Its surface treatment and tooling is discussed to describe the grain-size effect on the low temperature plasma nitriding and to demonstrate its effectiveness in die-making in forth. Finally, every aspect in manufacturing of FGSS sheets and solids is summarized as a conclusion.

Keywords: stainless steels, fine grain, milling, laser machining, metal forming, diffusion bonding, surface treatment, tooling, medical applications

1. Introduction

Hall-Petch relationship was found in 1951 [1] and 1953 [2]. Since this finding, the grain-size refinement has become one of the key development topics to improve properties of metals and alloys. High-pressure torsion (HPT) is one of the early technologies to refine microstructure for aluminum or copper [3]. A disk of these material is confined in a cylinder and subjected to torsion by rotating punch under high vertical pressure. As the strains increased during torsions in a disk, the grain size of these materials becomes smaller; then rounding by four times increases its hardness together with homogeneity of grain size [4]. However, the method is only applicable for relatively soft metallic alloys. It is difficult to apply to hard metals such as carbon steels or stainless steels. In addition, since the strains during a torsion is different in the radial direction of disk, the grain size remains larger than that in outer diameter; the rotation is not enough. Hence, this method might be suitable to create hard surface around the edge of products like a miller disk in the office copy machine; its application to bulk work materials is limited by cost and production process. The continuous production process, like rolling, stamping, or cutting by lathe in the production area, need proper and stable bulk materials. For the part productions of medical, IT, or automobile industries, stable shapes of bulk material are important to produce precise parts with high accuracy and lower cost. Especially, minimally invasive medical treatment needs smaller devices and higher strength. Thus, wire and coil shapes are preferable for these parts industries. Often, the improvements of material need changing the chemical composition, but it usually needs approval from medical regulations in each country. The approval waist a lot of investments and time. Thus, the technology of improvement without changing the chemical composition was evitable. Fine-grained stainless steel (FGSS) is highlighted as the most reliable work materials to fabricate the miniature automotive parts as well as the fine medical components.

In the present chapter, various manufacturing methods are employed to investigate each workability of FGSS materials; e.g., mechanical and lase machining, metal forming, diffusion bonding, and surface treatment after explanation on their mechanical properties and microstructure. The superiority of FGSS to this integrated manufacturing is a key to utilize FGSS to fabricate the miniature mechanical parts and medical fine components.

2. Mechanical properties and microstructure of FGSS

There are only two ways to produce FGSS in the production place in 2019. One method is oval-square rolling, which has been developing by Torizuka and Muramatsu since 1998 as a suitable means to produce the bar and wire [5]. **Figure 1** shows the change of work cross section during oval-square rolling process. Steel or stainless steel coil with ϕ6.0 mm is used as a starting work material. Grains are decreased by heating and severe plastic deformation with use of oval and square shaping rolls. The high strains at the center of coil drive dynamic recrystallization from 20 to less than 3 μm across the all cross section. This process continues till coil diameter reaches ϕ3.0 mm.

Figure 2 compares the IPF (inverse pole figure) of AISI316L by Electron Back Scattered Diffraction Pattern (EBSD) before and after this oval-square rolling. There are wide range of grain size before the process and these are austenite (**Figure 2(A)**), then the average grain size becomes 1.3 μm (**Figure 2(B)**).

The tensile strength of initial work ranges from 520 to 580 MPa. This is increased up to 980–1150 MPa by the grain reduction by 1/10 just after the Hole-Petch relationship. To be noticed, this strengthening also accompanies with work hardening and processing-induced transformation to martensite. This difference in strengthening mechanism between work hardening and grain-size reduction must be distinguished carefully.

Inverse phase transformation method is also cost effective method to produce the thin FGSS sheets [6, 7]. The grain-size reduction is driven by the inverse phase transformation from martensite by work hardening to austenite phase. Austenitic stainless steels of type AISI304 were employed as a work. The work materials were manufactured in a heat batch processing. **Table 1** shows the chemical composition of the material.

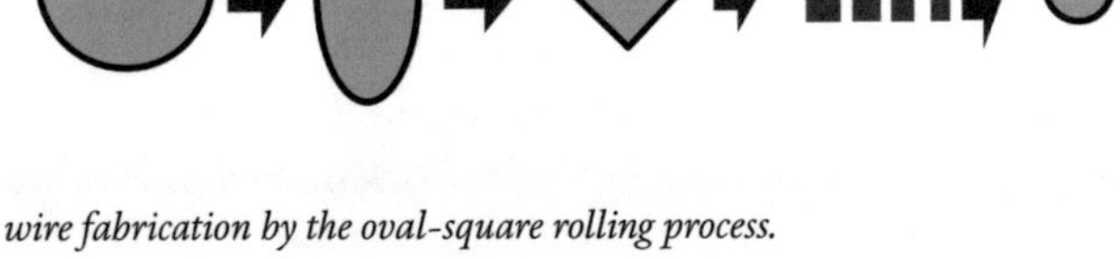

Figure 1.
Illustration on the wire fabrication by the oval-square rolling process.

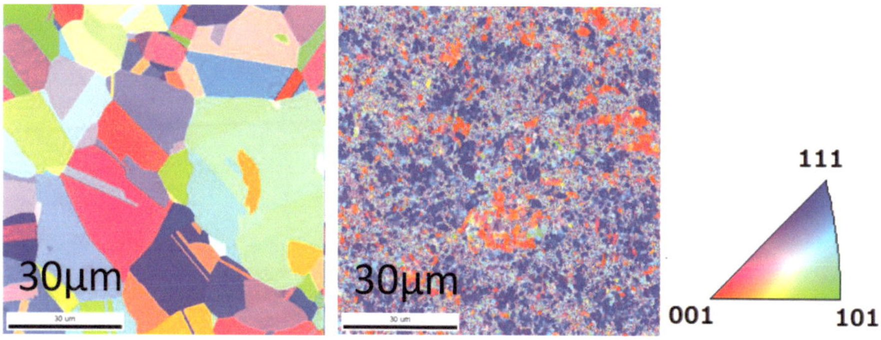

(A) (B)

Figure 2.
Picture of grain condition by EBSD. (A) Left: before oval-square rolling. The average grain size is 14.7 μm and (B) right: after the process. It becomes 1.3 μm.

	Composition %
Carbon (C)	0.06
Silicone (Si)	0.4
Manganese (Mn)	1.09
Phosphorus (P)	0.03
Sulfur (S)	0.004
Nickel (Ni)	8.03
Chromium (Cr)	18.02

Table 1.
Chemical composition of the specimen.

The normal grain stainless steel was formed to reduce the plate thickness from 3 to 0.2 mm in rolling with heat treatment. **Figure 3(A)** shows the microstructure, which is observed by EBSD. θ is the crystal orientation angle; and α' and γ are the crystal phase. The marten-sites structures are observed and the grain sizes are ranged from 2 to 20 μm, where the average grain size is 9.10 μm.

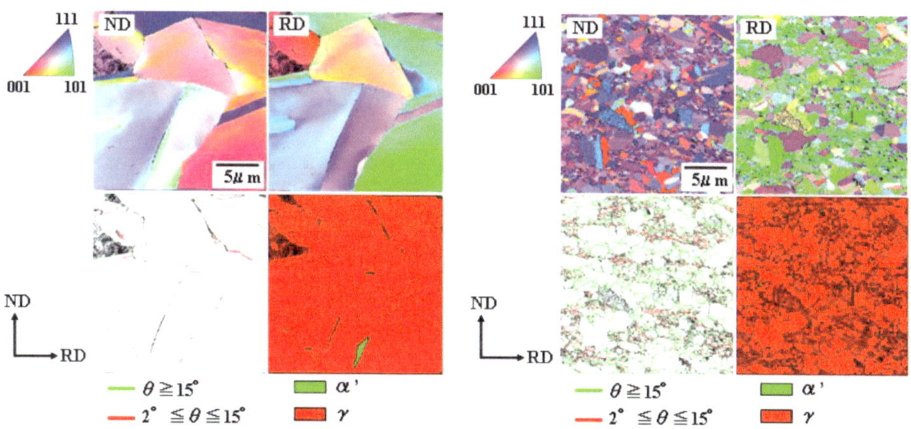

(A) (B)

Figure 3.
Picture of grain condition by EBSD. (A) Left: normal grain stainless steel. The average grain size is 9.8 μm and (B) right: FGSS by reverse phase transformation. The average grain size is 1.52 μm.

		Normal grain steel	Ultra fine-grained steel
Vickers Hardness [HV]		260	260
Tensile strength [MPa]	RD	870	919
	ND	858	880
Elongation [%]	RD	51.1	42.5
	ND	57.5	46.4
Ave. grain size [μm]		9.10	1.52

Table 2.
Mechanical properties.

The fine-grained stainless steel was formed with repeating plastic deformation and reverse phase transformation. **Figure 3(B)** shows the microstructure, where the average grain size is 1.52 μm.

Although the grain size was different among specimens, their ultimate tensile strength became around 900 MPa, irrespective of the grain size.

Table 2 shows the mechanical properties of the normal stainless steel and FGSS. Here, the mechanical properties are similar to each other. The hardness and tensile strength is controlled by work handing for normal grain, and by decreasing the grain size by reverse phase transportation for FGSS. These different hardening methods for stainless steel should be distinguished in terms of microprocessing.

As a summary, FGSS has refined microstructure with the average grain size of 1.5 μm and higher ultimate strength than 900 MPa. Its hardening process is governed by the reverse phase transformation in different from the work hardening in the normally grained AISI304.

3. Mechanical machining of FGSS

Several researchers have discussed the effect of grain size and boundaries in micro/nano processing such as cutting, piercing, forming, and so on. As the removal depth becomes smaller than the grain size, which is usually more than 20 μm, it is difficult to ignore the crystalline direction, size, and distributions in such microprocessing.

Simoneau et al. discussed the grain-boundary effect in medium carbon steels, and he found the different grain deformations in cutting chips [8]. Lee et al. discussed material-induced vibration, which is caused by the changing crystallography of the material substrate [9, 10]. They also analyzed the variation in microcutting forces in diamond tool turnings of crystalline materials, based on a microplasticity model and spectrum analysis technique [11]. Furukawa and Moronuki discussed the grain-boundary effect during the cutting of large grained steel. They found that when the grain boundary is crossed by the tool, the cutting force increases [12]. Komatsu have also discussed the effects of grain size on two-dimensional microcutting [13–15]. When the grain size is less than 3 μm, the cutting tool vibration is reduced, and the quality of the groove is improved. This is because small grained materials are regarded as homogeneous materials in terms of hardness.

However, study of the effect of grain size on other processes and product functions, based on materials of various grain sizes, are limited because the size of bulk material with FGSS has been too small for the studies.

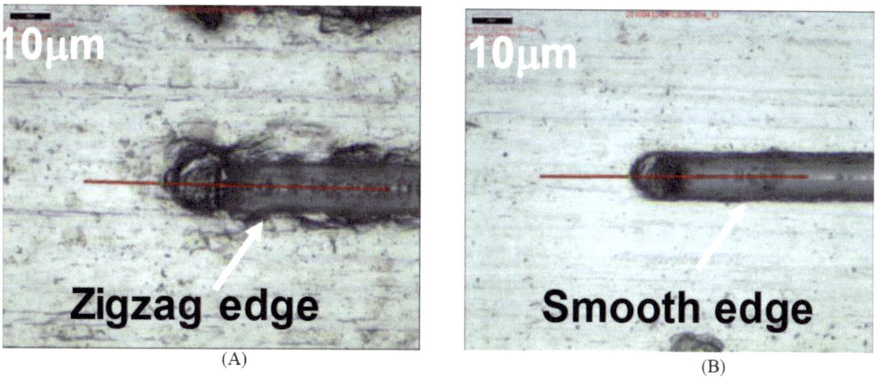

Figure 4.
Picture after microcutting by diamond tool. (A) Left: groove condition on normal grain stainless steel. The average grain size is 9.8 μm and (B) right: Groove condition on FGSS by reverse phase transformation. The average grain size is 1.52 μm. FGSS can get smooth edge when cutting in microscale.

This chapter will present the cutting process of FGSS. Two different grain sizes were employed, as is mentioned before, to discuss the effect of the grain size on the cutting process in the microscale.

Figure 4 shows the surface finishes by diamond tool in microcutting. The groove width and depth was about 8 μm. It should be noted that the edge quality is incredibly improved as compared between **Figure 4(A)** and **(B)**. This feature had benefits for fine parts or medical equipment.

The effect of the grain size was discussed with measuring the cutting forces and their oscillation. The stability in the cutting process was associated with the oscillation of the cutting force in **Figure 5**.

When large grain material was cut, the tool was widely vibrated as shown in **Figure 5(A)**, because the deformation reached to the grain boundaries and it needed force more. On the other hand, smaller grain had narrow length of between grain boundaries, it constantly effected on the cutting force [13, 14] in **Figure 5(B)**. Komatsu discovered some more features in microcutting this FGSS. Higher shear-plane angles, cutting FGSS less than 5 μm, reduce the cutting force even the tensile strength is increased [12]. This features reduced the burr. Ball end-mill cutting examined for these material and the burr size become from 1/5 to 1/3 [15]. Finally, the residual strains were also reduced by using FGSS.

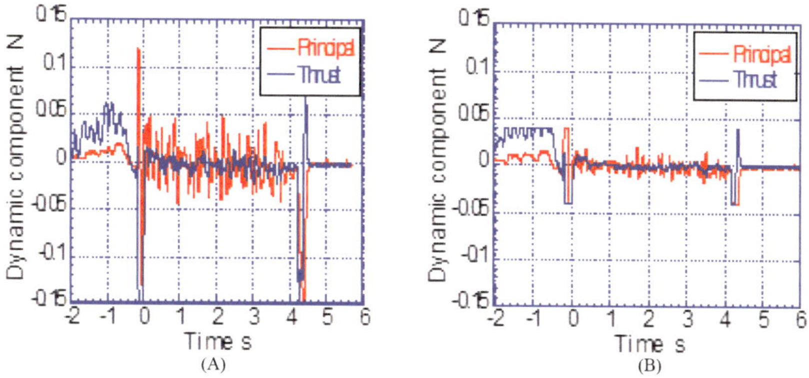

Figure 5.
Differentials of vibration results from principal and thrust cutting force. (A) Left: the average grain size is 9.8 μm and (B) right: the average grain size is 1.52 μm.

In summary, mechanical machinability of FGSS is much improved by fine-grained microstructure; smooth cutting edges, reduction of in situ vibrations and burr size.

4. Laser machining of FGSS

Laser machining grew up as one of useful means to produce miniature parts. This laser processing has two different aspects; one is thermal process like welding, and the other is mechanical process. When producing accurate parts for medical or precision parts, thermal effects on the works surface became a main driving force in the CO_2-laser machining. When the laser wave length became shorter or when using the pico- or femtosecond laser, the thermal effects are ignored because materials are removed before transferring the temperature [16]. Siegelet et al. investigated the picosecond pulse ablation behavior and industrial relevance in 2009. He noted that the accurate surface and structures were introduced and the thermal effect was ignorable [17].

Since the picosecond laser reduces the thermal influence on cutting, the grain-size effect on the laser machining changes by its mechanical process. Using FGSS is one of the methods to apply to the enhanced parts of medical, because it has higher tensile strength and hardness without changing any chemical composition. Komatsu used the picosecond laser to investigate the grain-size effect on the laser machining of AISI 304 with the thickness of 1.5 mm [18]. **Figure 6(A)–(C)** shows the cross-sectional images after laser drilling AISI304 sheet with different grain sizes by 30 times. The laser machining system (PANASONIC AP-3220) was employed; the laser output was 500 mW, the wavelength, 1053 nm, the pulse duration, 15–25 ps, and the repetitive frequency, 5 kHz. The spot size was constant by 30 μm.

There are significant differences of processing depth in different grain size. When processing smaller grains such as FGSS (1 or 2 μm), the depth of groove was from 0.1 to 0.12 mm. On the other hand, there are from 20 to 40% deeper groove in large grain's material. After 30 times laser processing, the material nearly cut. **Figure 7** describes the relationship between the one-shot machined depth and the grain size. Deeper groove is formed by machining the large grain AISI304, while shallower groove becomes unstable in case of FGSS. This might come not only from the hardness of material but also from the absorptivity, which related to surface roughness and structure [18]. The smaller grain size is preferable to micro deep laser cutting.

In summary, laser machinability of FGSS is characterized by precise cutting of relatively shallow microgrooves.

(A) (B) (C)

Figure 6.
SEM cross-sectional image after laser processing machining the AISI304 with different grain size (d) by 30 times in the same condition. (A) Left: D = 1.1 μm, (B) center: D = 2.0 μm, and (C) right: D = 9.2 μm.

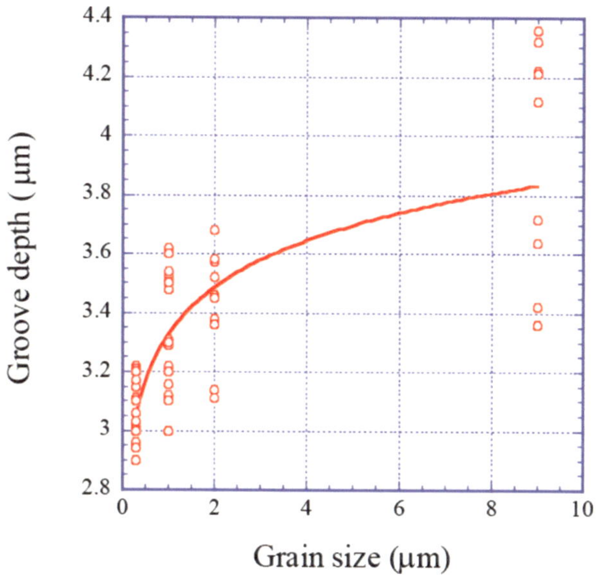

Figure 7.
Relationship between the grain size and groove depth after one-shot laser processing.

5. Metal forming of FGSS

5.1 Deformation and transformation in micropunching of FGSS304

As well known, size effects were observed during metal forming [19–21]. The size effect also occurs in punching process. In this chapter, introducing the difference in deformation and transformation of austenitic stainless steel AISI304 when micropunching with grain size changing. AISI304 stainless steel causes strain-induced transformation during plasticity process [22]. The strain-induced transformation of AISI304 austenitic stainless steel was classified as non-diffusive shear transformation, and transformation occurs in grain-size units [23]. Therefore, shear deformation in micropunching became complicate.

5.2 The stability of sheared surface length when changing a grain size

Figure 8 shows four IPFs of AISI304 stainless steels with different grain sizes, which were analyzed by EBSD [24]. Grain size (Gs) 7.5 μm was standard grain size of AISI304. The grain size from Gs 3.0 to Gs 1.0 μm were specially minimized AISI304 that were not changed chemical composition. The characteristics of materials were shown in **Table 3** [24]. From this table, as grain size decreasing, 0.2 % proof stress and tensile strength increasing and elongation became decreasing. The Hall-Petch relationship works effectively in these grain-size conditions.

Figure 9 shows the constitution of sheared surface and cross-sectional SEM images of punched microhole [24]. The punching condition was following; material thickness 100 μm, punch diameter 80 μm, die diameter 87 μm, clearance between punch and die 3.5 μm, punching speed 4.2 mm/s. The sheared surfaces were consisted by shear droop, burnished surface, and fractured surface, respectively. The burrs were not shown in this figure. **Figure 10** shows the ratio of sheared surface that calculated from five samples [24]. From this figure, as grain size decreasing, the shear droop ratio became small and fractured surface ratio became large. However,

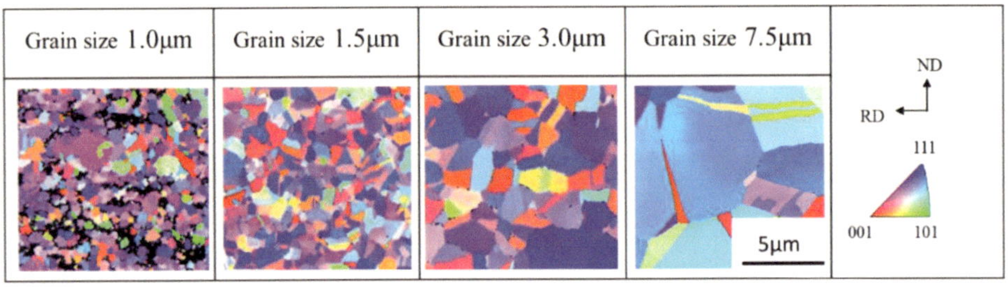

Figure 8.
EBSD inverse pole figure of four grain size austenitic stainless steel AISI304 including FGSS304.

Grain size [μm]	**1.0**	**1.5**	**3.0**	**7.5**
Tensile strength [N/mm^2]	1181	875	845	803
0.2% proof stress [N/mm^2]	1000	599	504	433
Elongation [%]	6.0	45.8	49.5	56.6
Hardening exponent [−]	—	0.31	0.37	0.43
Vickers hardness [HV]	350	261	227	191

Table 3.
Mechanical characteristics of four grain size austenitic stainless steel AISI304 including FGSS304.

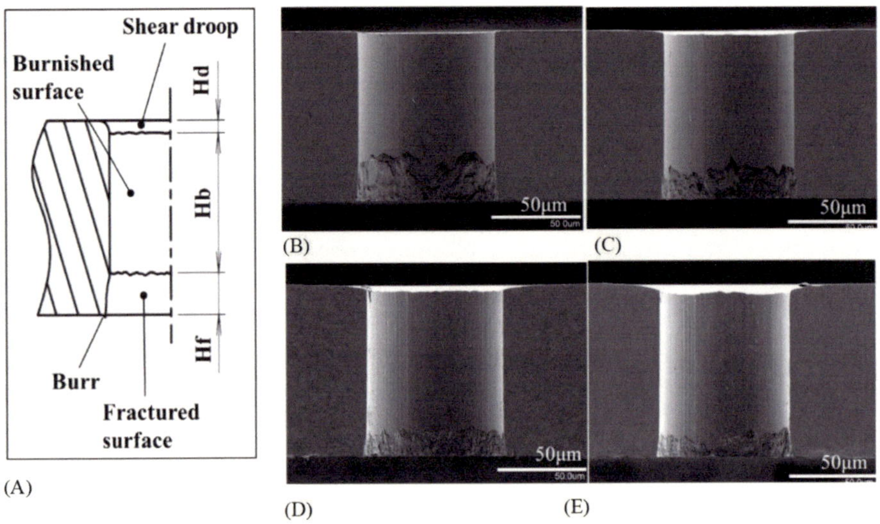

Figure 9.
Constitution of sheared surface and SEM images of cross section of micropunched hole. (A) Constitution of sheared surface, (B) Grain size 1.0μm, (C) Grain size 1.5μm, (D) Grain size 3.0μm, (E) Grain size 7.5μm.

from **Figure 11**, the standard deviation of burnished surface ratio took a different trend [24]. The grain size 3.0 μm took a smallest standard deviation. It was found that the stability of the sheared surface length changing with grain size.

5.3 Process affected zone in micropunched hole

Why the stability of burnished surface length changed with grain size? **Figure 12** shows the EBSD analysis results of punched samples with three different varying grain sizes of 1.5, 3.0, and 7.5 μm, respectively. The punching conditions were following; the material thickness is 100 μm, the punch diameter, 80 μm, the die diameter, 85 μm,

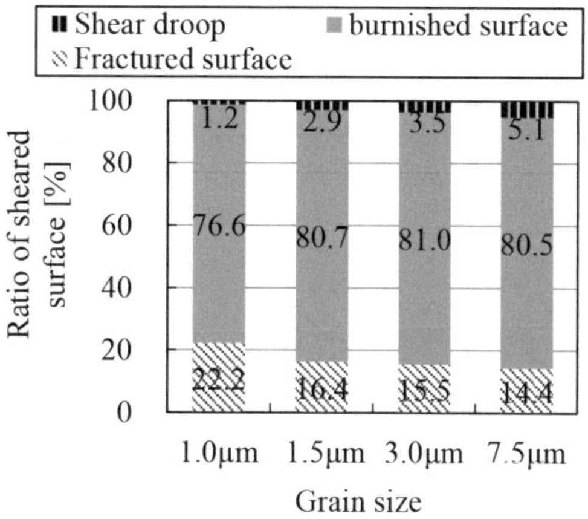

Figure 10.
Ratio of the sheared surface.

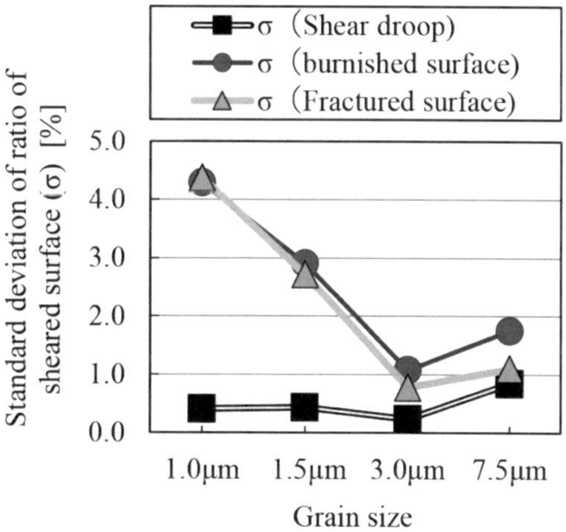

Figure 11.
Standard deviation of ratio of sheared surface.

the clearance between punch and die, 2.5 μm, and the punching speed, 4.2 mm/s. In **Figure 12**, IQ + phase map shows the grain phase and the IQ + KAM (Kernel average misorientation) map shows the grain misorientation angle. The grain misorientation angle was known to correspond to equivalent plastic strain [25].

From **Figure 12(A)–(C)**, the strain-induced martensitic α' phase was remained along with the punched hole. From the grain size 7.5 μm phase map results (**Figure 12(C)**), the border line between α' phase and austenitic γ phase looks like hackly. This trend is caused by grain unit transformation. On the other hand, from the grain size 1.5 μm phase map results (**Figure 12(A)**), the border line between α' phase and γ phase looks like stable. From the **Figure 12(D)–(F)**, process-affected areas were existed along with punched hole, respectively. Especially, the highly 5° misorientation angle color red has spread along with the wall of punched hole at each samples.

5.4 Deformation and transformation mechanism in micropunching of FGSS304

Figure 13 shows the total frequency of α'-phase and total amount of misorientation angle that calculated from the sample shown in **Figure 12** and the other two samples. From **Figure 13(A)**, Z minimizes at the grain size of 3 μm. From **Figure 13(B)**,

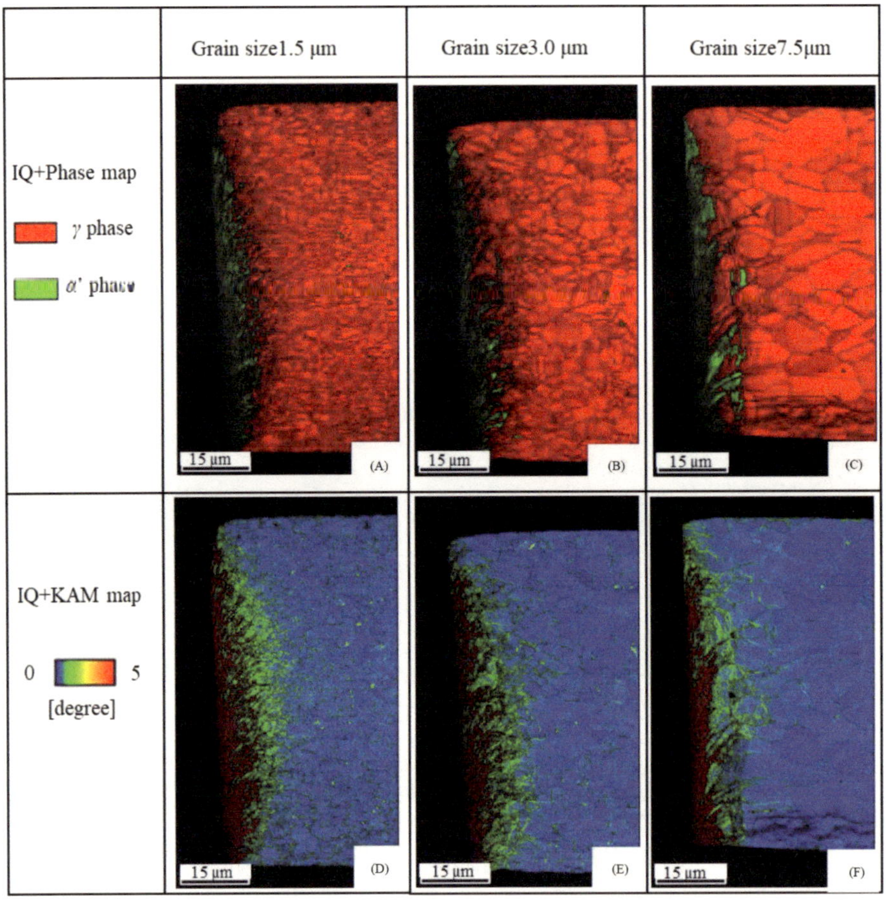

Figure 12.
EBSD analysis results of cross section of punched hole.

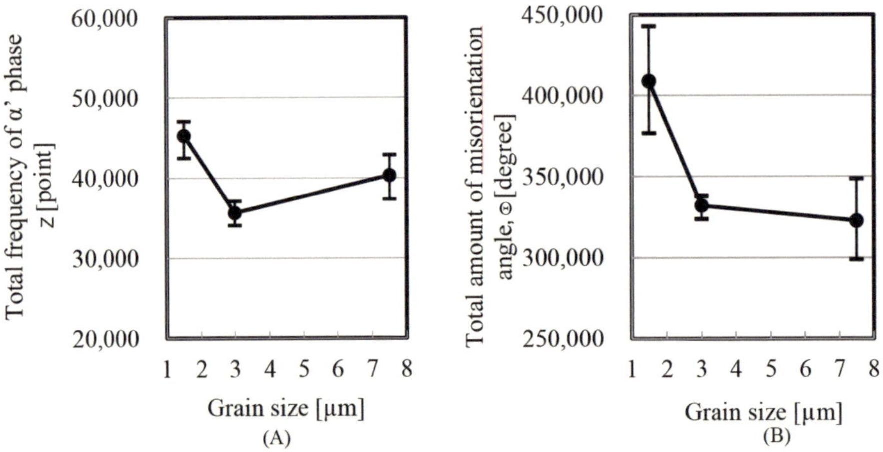

Figure 13.
Total frequency of α′-phase and total amount of misorientation angle in the cross section of punched hole. (A) Total frequency of α′ phase and (B) total amount of misorientation angle.

Θ maximizes at the grain size of 1.5 μm; Θ1.5 μm is more than Θ3.0 μm by 24 %. When choosing the grain size 1.5 μm, small grain size has limited work hardening ability. So that, strain-induced transformation works actively and the frequency of

α' phase more increase than the other large grains as shown in **Figure 13(A)**. In our experiment, full-α' phase material has only 1.2% elongation and 1877 MPa tensile strength. This smallest elongation and higher tensile strength make fracture surface longer than the other large grains as shown in **Figure 10**. When choosing the grain size of 7.5 μm, the stability of sheared surface length become unstable as shown in **Figure 11**. These characteristics is related to the effect of distribution variation in strain-induced martensitic transformation [26, 27]. If choice the grain size 3.0 μm, the work hardening ability is maintained during micropunching and effect of distribution variation in strain-induced martensitic transformation decrease with the grain size decreasing. Therefore, the deformation and the transformation characteristics are optimized and the stability of sheared surface length is minimized as shown in **Figure 11**. For the stable punching, selecting a grain size 3.0 μm is considered to be effective at the AISI304 micropunching.

5.5 Summary

Fine-grained microstructure in FGSS changes the deformation and phase transformation characteristics in the punching process. In particular, the sheared surface length is more stabilized to improve the punched-out product quality. The punching process can be advanced by using FGSS in practice.

6. Diffusion bonding of FGSS

In the solid-phase diffusion bonding process, the holding temperature (T_H) plays a role to govern the microstructure. When T_H is higher than the recrystallization temperature, the crystal grain easily grows to be coarse one within the bonding time and reduces the strength. Hence, T_H must be lowered as possible to maintain high strength in the mechanical characteristics. In particular, this task in the diffusion bonding of AISI304 must also solve an issue to remove the passive film at low temperature. In the conventional process, T_H becomes higher than 1100 K [28] to eliminate the oxide film on the stainless steel and to accelerate the bonding.

In recent years, new materials have been developed. **Figure 14** shows the newly developed full-martensitic stainless steel (hereafter called WC). This material can accelerate the diffusion bonding at low temperature by introducing a large amount of strain into austenitic stainless steel before bonding. The process of introducing distortion into the material was carried out as follows and as shown in **Figure 14** [29]. First, blocks of AISI304 were cut into 40 × 40 × 20 mm samples (**Figure 15(A)**). The samples were then compressed, cut, and rolled into a 10-mm-thick sheet, so that the equivalent strain in the compression direction was 90% at 573 K (less than

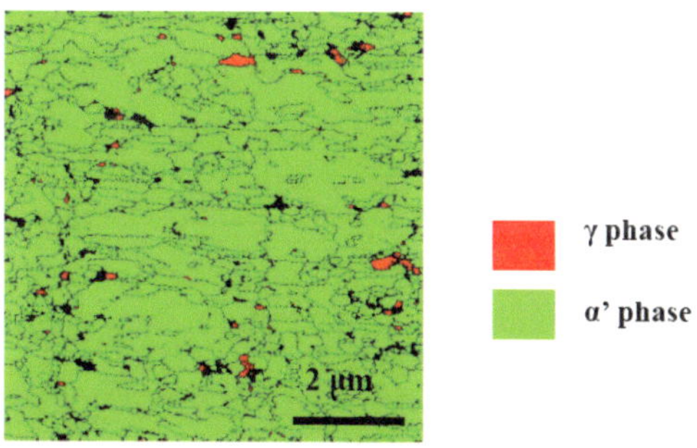

Figure 14.
Developed full-martensitic stainless steel AISI304 (WC).

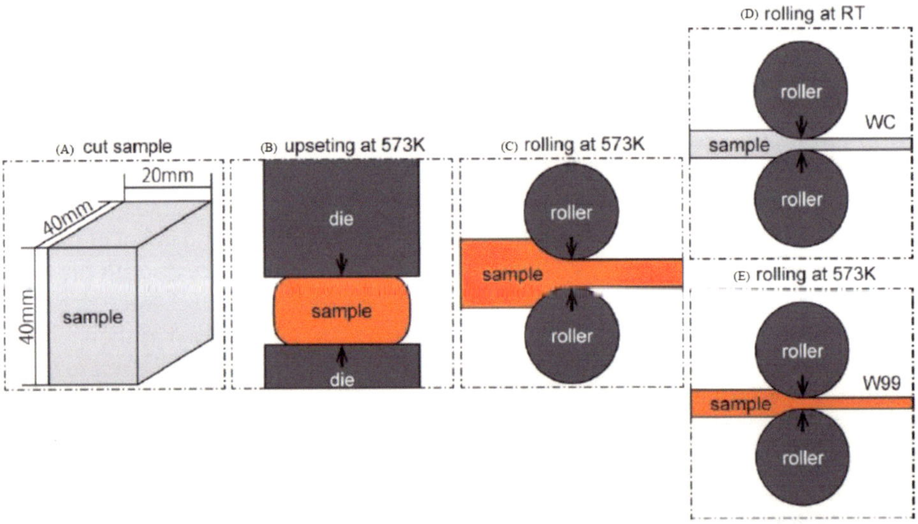

Figure 15.
Schematics of distortion-introduction process. Arrows show compression direction.

the recrystallization temperature) (**Figure 15(B)** and **(C)**). These warm-rolled samples were then rolled to 1-mm thickness at room temperature to transform the microstructure into martensite from austenite (**Figure 15(D)**). To evaluate the effects of the presence or absence of martensite on diffusion bonding, a sample was warm-rolled at 573 K (thus containing deformed austenite) with a reduction of 99% (hereafter called sample W99), as depicted in **Figure 15(E)**. For the cross-tensile test, all the bonding specimens were cut from the 1-mm-thick sheet, and then their bonding areas were mirror polished. The bonding experiment was executed at punching diameter 5 mm and high vacuumed.

Figure 16 shows the effects of strain-induced martensitic and strain for fracture load of bonded AISI304 seats on bonding temperature [30]. WC/WC bonded sample only at 973 K has a fracture load of 2 kN, while T_H must be increased up to 1073 K to attain the same fracture load in case of SOL/SOL bonded sample. That is, WC/WC

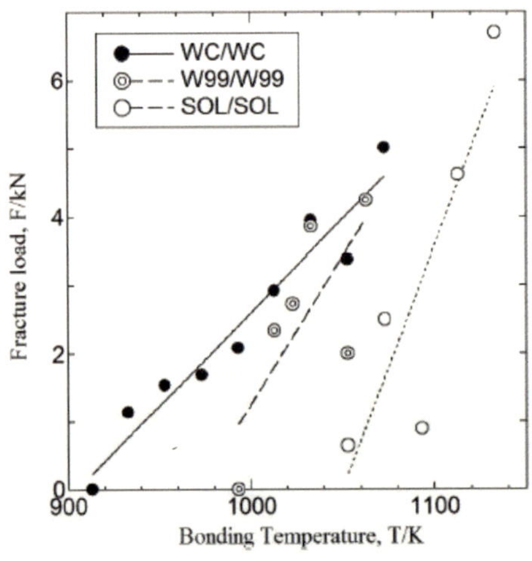

Figure 16.
Effect of strain-induced martensite and strain for fracture load of bonded SUS304 seats on boning temperature.

full-martensitic stainless steel bonding has a capacity to decrease T_H by 100 K against conventional SOL/SOL sample. Furthermore, this T_H for WC/WC bonded sample to attain 2 kN is reduced down to 20,123 K by using the fine-grained W99/W99 boding. This difference in T_H by 50 K between WC/WC and W99/W99 suggests that the martensitic phase works effectively when diffusion bonding stainless steel.

In summary, the holding temperature in diffusion bonding is lowered by using FGSS sheets to successfully fabricate the bonded parts without grain size coarsening. This selection of FGSS sheets for joining is suitable to produce the precision parts with high strength.

7. Surface treatment of FGSS

In standard heat and surface treatments, the stainless steel parts and tools are subjected to high-holding temperature [31]. The first issue of engineering is a crystalline coarsening by the grain growth. As shown in **Figure 17**, FGSS is easy to be coarsened by heat treatment for 1.8 ks at 1073 K [32]. Even below 1000 K, the thermal distortion is issued to deteriorate the microstructure of FGSS; precipitation of carbides and nitrides is also worried to lower the fatigue strength and corrosion toughness. Various nitriding processes were compared to explain how to lower the holding temperature in [33]. The gas nitriding with use of ammonia required for higher temperature than 973 K for industrial surface treatment. The liquid nitriding with use of cyan solutions was applied to surface modification of automotive parts under the temperature of 700–900 K.

Among those nitriding processes, the plasma nitriding has capacity to harden the surface layer in the thickness of 0.1–1 mm without significant loss of corrosion toughness of stainless steels under lower holding temperature. This processing is categorized into two regimes as shown in **Figure 18**; e.g., high-temperature nitriding above 673 K and low-temperature nitriding below 673 K as surveyed in [34] for austenitic stainless steels.

Above the master curves in each stainless steel, the nitrided surface layer is hardened by the fine CrN (chromium nitride) precipitation with large volume fraction [35]. After removal of fragile γ'-Fe_4N precipitated layer (so-called white layer) and diffusion treatment, the nitrided layer with the nitrogen solubility of 0.1–0.2 mass% is utilized as a hardened protective layer of tool-steel and stainless steel parts and dies in commercial [36]. The plasma nitrided AISI316 at "A" just above its master curve in **Figure 18** is mainly hardened by the CrN precipitation. On the other hand, the inner nitriding process at "B" and "C" below the master curve in **Figure 18** is governed by the nitrogen solid solution or the nitrogen

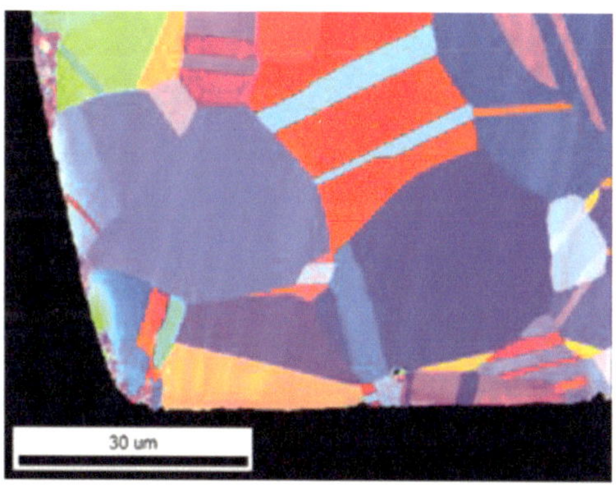

Figure 17.
Grain growth in FGSS by heat treatment for 1.8 ks at 1073 K.

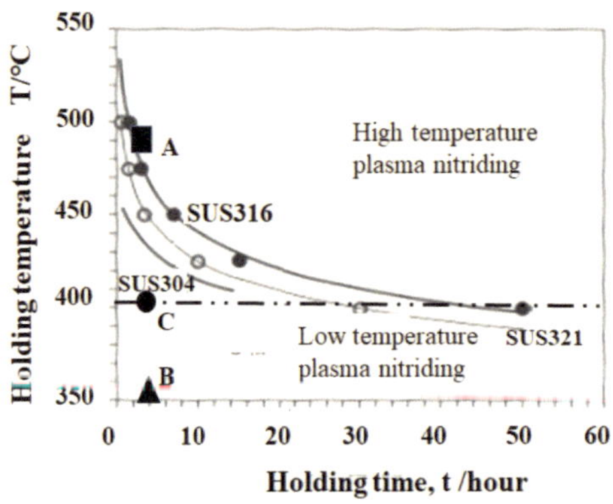

Figure 18.
Relationship of the holding temperature to the holding time in the plasma nitriding.

alloying into AISI316 without formation of nitride precipitates [37]. This low temperature plasma nitriding is proved to be suitable to the surface treatment of FGSS with high hardness and strength and without thermal distortion.

7.1 Plasma nitriding of fine-grained AISI316 at 623 K

The fine-grained AISI316 (FGSS316) specimen was prepared for plasma nitriding at 623 K [38–41]. Its chemical compositions are: [C] = 0.08 mass%, [Si] = 1.00 mass%, [Mn] = 2.00 mass%, [P] < 0.045 mass%, [S] < 0.030 mass%, [Ni] = 12.0 mass%, [Cr] = 17.0 mass%, and [Mo] = 2.5 mass% for iron in balance. The sample surface was mirror polished and cleaned by the ultrasonic cleaner before plasma nitriding. **Figure 19(A)** depicts the high density plasma nitriding system with use of the hollow cathode device to intensify the density of nitrogen ions as well as the NH-radicals. The plasma-processing conditions are also summarized in **Figure 19(B)** in correspondence to "B" in **Figure 18**.

Microstructure and nitrogen mapping on the cross section of this nitrided FGSS316 specimen, describe the average nitrogen diffusion layer from the surface. As shown in **Figure 20(A)**, the nitriding front end locates at the depth of 40 μm from the surface. Although the crystal grain size below this nitriding front end remains the same as before nitriding, these grains are significantly refined in the nitrided layer. The nitrogen mapping in **Figure 20(B)** proves that high nitrogen content uniformly distributes in the nitrided layer from the surface to the depth

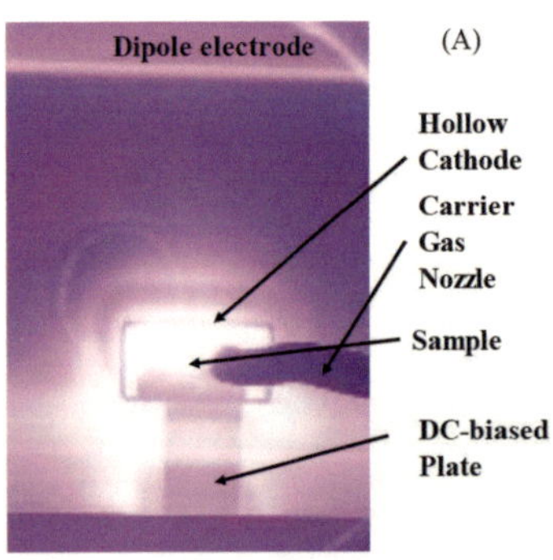

(A)

Item	Parameter
Temperature	623 K
RF-voltage	250 V
DC-bias	-500 V
Pressure	70 Pa
Carrier gas	N_2-H_2 mixture
Device	Hollow cathode

(B)

Figure 19.
High density plasma nitriding system with use of the hollow cathode for low temperature plasma nitriding of FGSS316. (A) Schematic view of nitriding system, and (B) plasma nitriding conditions in the following experiments.

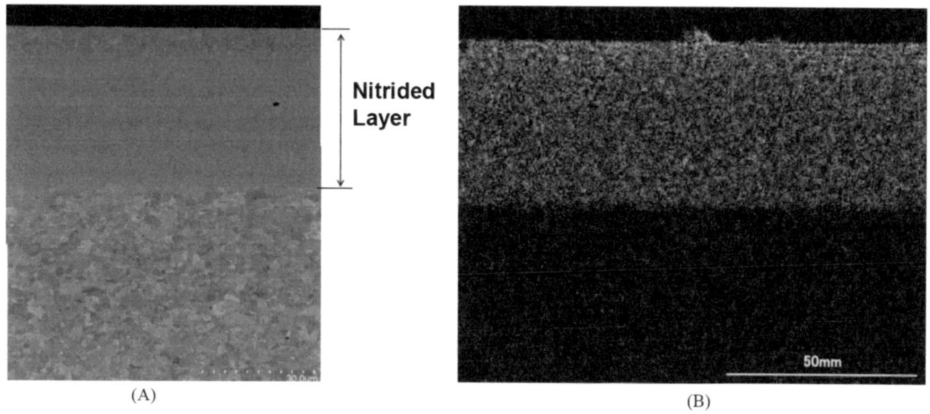

Figure 20.
Cross-sectional image of the plasma nitrided FGSS316 for 14.4 ks at 623 K. (A) SEM image and (B) nitrogen mapping analyzed by EDX.

of 40 μm. This implies that the thick nitrided layer with high nitrogen content and refined grains is formed by the present surface treatment at 623 K.

Figure 21 compares the XRD diagrams of FGSS316 specimen surfaces before and after plasma nitriding at 623 K for 14.4 ks. XRD diagram of bare FGSS316 is characterized by γ (111)-peak at 2θ = 43.5° as well as γ (200)-peak at 2θ = 50.7°, respectively. After nitriding, this γ (111)-peak shifts from 43.5° to 40.5°, and, γ (200)-peak, from 50.7° to 46.6°, respectively. A new peak detected at 2θ = 43.65° is identified as α' (110). Besides for these three peaks, no other peaks are detected even by the narrow scanned XRD in **Figure 21**; no chromium and iron nitrides are synthesized in this plasma nitriding. Nitrogen solute atoms do not react with the constituent atoms such as iron and chromium in FGSS316 but work as a constituent alloying element in the γ-lattice of FGSS316. The nitrogen solute atoms occupy the octahedral vacancy sites in the γ-lattices in FGSS316. This in situ nitrogen solute occupation with vacancy sites in the γ-lattices accompanies with the γ-lattice expansion in elasticity in **Figure 21**, and, characterizes the nitrogen supersaturation process in the low-temperature nitriding.

Difference in the holding temperature between "A" and "B/C" in **Figure 18** reflects on the inner nitriding mechanism. High-temperature nitriding is driven by the nitrogen body-diffusion process with the precipitation reaction of diffusing nitrogen atom and constituent elements in FGSS316 to nitrides. Low-temperature nitriding is controlled by the nitrogen boundary-diffusion with the nitrogen

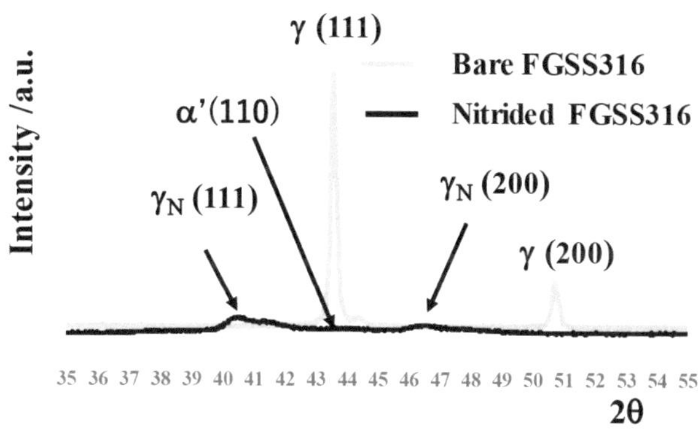

Figure 21.
XRD diagram of FGSS316 specimens before and after plasma nitriding at 623 K for 14.4 ks.

supersaturation in concurrent. This difference reflects on the microstructure evolution during the inner nitriding. Let us analyze this by using EBSD.

Figure 22(A) depicts the phase mapping on the cross section of nitrided FGSS316 at 623 K. The original microstructure below the nitriding front end has mainly austenitic phase but partially includes the martensitic phase. This phase transformation is strain-induced from the original γ-phase by intense rolling during fabrication of FGSS316 plates. Total volume fraction of α'-phase is only 6%. This implies that highly strained γ-phase grains must be massively transformed into α'-phase and that most of γ-phase remains as a matrix. On the other hand, this martensitic phase distributes finely together with γ-phase above the nitriding front end. The volume fraction of this α'-phase reaches to 70%. This implies that the nitrided layer has fine γ-α' two-phase structure. The original matrix structure, seen below the nitriding front end in **Figure 22(A)**, disappears and turns to be fine two-phase structure with nitrogen supersaturation.

As before mentioned, this nitrogen supersaturation accompanies with the plastic straining as well as the γ to α' phase transformation by elastic straining in the above. **Figure 22(B)** shows the KAM distribution on the same cross section. Besides for the upper part with remaining relatively large γ-phase regions in **Figure 22(A)**, the two-phase structured grains in the nitrided layer have high-angled misorientation; almost every grain in the nitrided layer is plastically strained. As explained by **Figure 21**, the nitrogen supersaturated zones in each grain are forced to expand themselves in elasticity; while unsaturated zones have no elastic strains. Considering that strain incompatibility might be induced between these nitrogen-saturated and unsaturated zones in every nitrided grain, the unsaturated zones are plastically strained. KAM distribution in **Figure 22(B)** represents the equivalent plastic strain distribution by this plastic straining.

Figure 22(C) shows the inverse pole figure in the nitrided layer. The original FGSS matrix, seen above the nitriding front end, turns to have much refined microstructure. The grain size is measured to be less than 100 nm or the spatial resolution limit of EBSD. This grain-size refinement is driven by intense plastic straining as seen in **Figure 22(B)**.

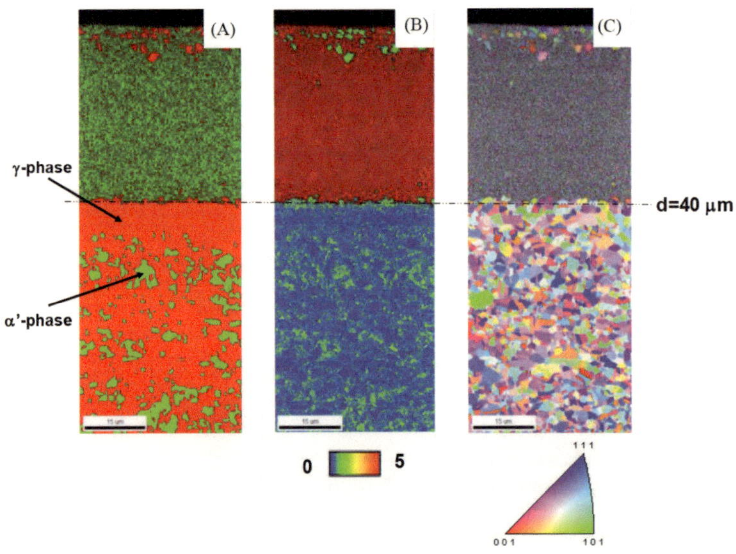

Figure 22.
Microstructure on the cross section of nitrided FGSS316 at 623 K for 14.4 ks. (A) Phase mapping, (B) KAM distribution, and (C) inverse pole figures.

7.2 In situ two-phase structuring and refinement

This in situ formation of two-phase microstructure with refined grains in **Figure 22** is suitable to industrial applications. Different from the coarse mixture of γ-phase and α'-phase grains in normal two-phase stainless steels [42], this fine γ-α' microstructure results in high hardness and toughness. In addition, high nitrogen solute concentration improves the corrosion toughness even in severe conditions. Let us evaluate on these features by hardness testing and nitrogen depth profile measurement.

Figure 23(A) depicts the hardness depth profile of nitrided FGSS316 at 623 K. Hardness in the nitrided layer with the thickness of 40 μm is uniform by 1400 HV, and, gradually decreases down to the matrix hardness across the nitriding front end. The grain-size refinement in **Figure 22** contributes to this high hardness. As reported in [31, 33, 35], the maximum surface hardness by the high temperature plasma nitriding above 750 K was often limited by 1200–1300 HV. That hardness monotonously decreased with the depth except for high chromium alloys [31, 35]. Hence, the fact that nitrided layer is uniformly hardened layer by 1400 HV in average even without nitride precipitation, is a new knowledge to surface treatment of stainless steels and tool steels. As also stated in [31, 33, 35, 43], the maximum nitrogen solubility except for the bound nitrogen atoms in nitrides is limited by 0.1–0.2 mass%. Most of chromium solutes are reacted with diffusing nitrogen atoms and bound into CrN in the nitrided stainless steels over 750 K; those nitrided steels reduce their original corrosion toughness [44]. **Figure 23(B)** shows the nitrogen solute content depth profile. The surface nitrogen content becomes higher than 6 mass%, and, this [N] is nearly constant by 4–5 mass% down to the nitriding front end. As had been studied in [43, 45], this high nitrogen solute content significantly promotes the corrosion toughness even under the dipping condition into HCl solutions. This improvement of mechanical and electro-chemical properties by the plasma nitriding is attractive to industrial applications.

7.3 Plasma nitrided FGSS316 dies at 673 K

This low temperature plasma nitriding was first applied to fabricate the FGSS316 punch for microembossing the regular-square meshing pattern with the line width of 50 μm and pitch of 250 μm into aluminum plate. In the conventional die-fabrication, this textured multi-head punch is made by drilling each regular-square microcavity with the edge of 200 μm. The number of microcavities reaches to 3200 on the punch with the size of 20 × 10 mm. When using the milling tool with the diameter of 10 μm and machining a single microcavity by the speed of 50 μm/s, including the cutting path changing time, and the threading depth of 10 μm, the

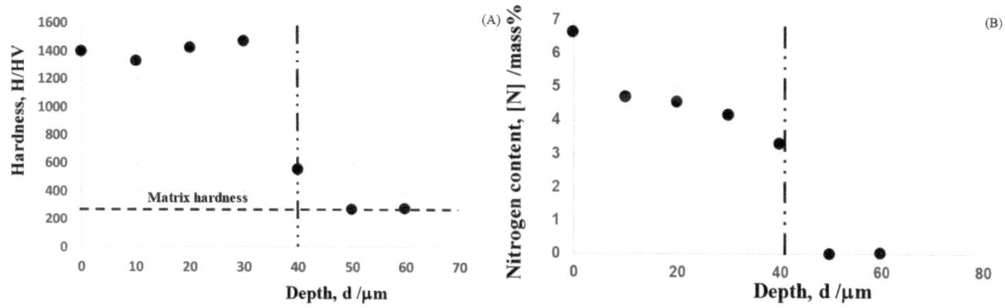

Figure 23.
Hardness and nitrogen solute content depth profiles of nitrided FGSS316 at 623 K. (A) Hardness and (B) nitrogen solute content.

total machining time is estimated to be 1, 280 ks, or 356 h. This micromachining provides no solution to fabrication of the multi-arrayed punch with regular-square meshing heads. After [46, 47], the plasma printing method with use of the low temperature plasma nitriding provides a solution to fabricate this die as shown in **Figure 24**. A meshing pattern is first printed by using the screen with a unit pattern in **Figure 24(A)**. The unprinted cross-meshing surfaces are selectively nitrided at 673 K for 14.4 ks ("C" in **Figure 18**) in **Figure 24(B)**. After sand-blasting, this unit pattern transforms to **Figure 24(C)**. The microtexture arrayed punch is fabricated as shown in **Figure 24(D)** to have regular meshing texture in **Figure 24(E)**.

The takt time of this processing is only 18 ks, 70 times shorter than micromilling even excluding the cutting tool life as well as the preparation for CAM data before actual machining.

7.4 Plasma nitrided FGSS316 wires at 623 K

FGSS316 wire is expected to be working as a tension member and component for reinforcement of prestressed concrete and for steel wires in medical equipment. These members often experience the friction and wear in severe surface chemical conditions; they have to be surface treated to improve their wear resistance and corrosion toughness. The low temperature plasma nitriding at 623 K is also effective to improve the surface hardness as well as the electro-chemical properties. FGSS316 wire with the diameter of 2.6 mm was prepared for plasma nitriding at 623 K for 14.4 ks by 70 Pa.

Figure 25 depicts the SEM image and nitrogen mapping on the cross section of FGSS316 wire. The wire surface is homogeneously processed to form the nitrided layer with high nitrogen concentration. The layer thickness approaches to 35 μm, which is a bit less than that seen in **Figures 22** and **23** for nitrided FGSS316 plates. A hard wire often experiences the bending and wiping as well as the tensile stretching in its usual operation. Owing to the surface hardening of wire by the plasma nitriding at 623 K, the bending and torsion rigidities as well as the tensile rigidity are expected to increase [48].

7.5 Summary

The plasma nitriding in the lower temperature regime than 700 K provides a means to modify the microstructure of FGSS by nitrogen alloying and to

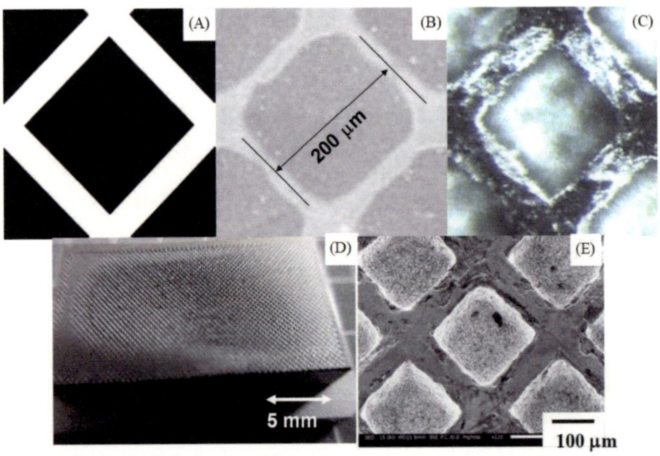

Figure 24.
Efficient production of the multi-head punch by the plasma nitriding with use of the plasma nitriding at 673 K. (A) Meshing pattern on the screen, (B) plasma nitrided pattern on the FGSS316 die, (C) three dimensionally textured FGSS316, (D) microtextured FGSS316 punch, and (E) meshing-texture arrayed punch.

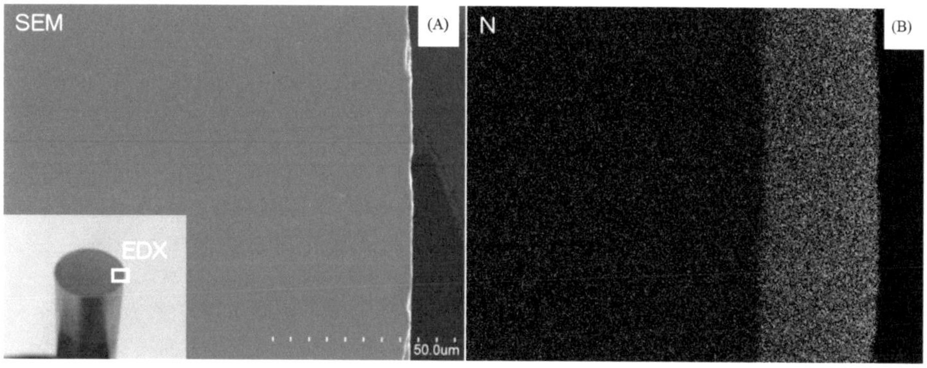

Figure 25.
Plasma nitrided FGSS316 wire at 623 K for 14.4 ks. (A) Cross-sectional SEM image, and (B) nitrogen mapping on the cross section.

significantly improve the mechanical and electro-chemical properties of FGSS. The nitrided FGSS316 layer consists of homogeneous two-phase microstructure with finer grain size than 0.1 μm. Since the original grain size of FGSS316 is 1.7 μm, the strength of nitrided FGSS316 increases by more than 10 times after the Hall-Petch relationship. This surface treatment by low temperature plasma nitriding is available not only in improvement of the wear resistance of dies for microstamping and forging but also in high strengthening of wires for medical equipment. In the former, the micropunch with multi punch head array is fabricated by blasting the un-nitrided substrate surface areas. This plasma printing is responsible to make microtextured die and punch in much less takt time for microembossing and micropiercing the metallic and polymer sheets. In the latter, the stainless steel bars, wires and fibers are surface treated to have higher nitrogen content layer. This nitrogen supersaturated surface layer works as a hardened and strengthened layer even after warm and hot drawing them to smaller diameters. Effect of supersaturated nitrogen distribution in them on their strength and ductility could be of much importance in application to industries and medicals.

8. Conclusion

Fine-grained stainless steels (FGSS) have grown up as a key material for mechanical and short-pulse laser machining, for metal forming and diffusion bonding, and for surface treatment to fabricate the miniature automotive parts and fine medical components. Their high strength without loss of ductility by grain-size refinement is not only attractive to improve the work and die material performance but also effective to enhance the workability in their integrated manufacturing. Smooth finish in mechanical machining as well as deep cutting by laser machining are intrinsic to FGSS. Precise shaping by micropiercing as well as low temperature diffusion bonding are also unique to FGSS. In their plasma nitriding, the holding temperature is significantly lowered down to 623 K for homogeneous formation of the hardened layer by 1400 HV with the thickness of 40 μm. Each advancement in this integrated manufacturing is fused and edited to construct the production line toward FGSS-based fine medical tools and precise mechanical parts. In particular, FGSS bars, wires, sheets, and plates transform to high strength, precisely shaped reliable media in mechanical units and medical equipments.

Acknowledgements

The authors would like to express their gratitude to Prof. S. Torizuka (Hyogo Prefectural University), Mr. E. Muramatsu, Mr. Y. Suzuki, and Mr. T. Yoshino (KSJ) for their help in experiments.

Conflict of interest

The authors declare no conflict of interest.

Author details

Tatsuhiko Aizawa[1*], Tomomi Shiratori[2] and Takafumi Komatsu[2]

1 Surface Engineering Design Laboratory, SIT, Tokyo, Japan

2 Komatsu-Seiki Kosakusho, Co. Ltd., Suwa, Japan

*Address all correspondence to: taizawa@sic.shibaura-it.ac.jp; shira@eng.u-toyama.ac.jp

References

[1] Hall EO. The deformation and ageing of mild steel: III discussion of results. Proceedings of the Physical Society. Section B. 1951;**64**(9):747

[2] Petch NJ. The cleavage strength of polycrystals. Journal of the Iron and Steel Institute. 1953;**174**:25-28

[3] Sakai G, Horita Z, Langdon TG. Grain refinement and superplasticity in an aluminum alloy processed by high-pressure torsion. Materials Science and Engineering: A. 2005;**393**(1-2):344-351

[4] Zhilyaev AP, Oh-Ishi K, Langdon TG, McNelley TR. Microstructural evolution in commercial purity aluminum during high-pressure torsion. Materials Science and Engineering A. 2005;**410**:277-280

[5] Torizuka S, Umezawa O, Tsuzaki K, Nagai K. Shape, size and crystallographic orientation of the ferrite grains formed at grain boundaries of deformed austenite in a low carbon steel. Tetsu to Hagane. 2000;**86**(12):807-814

[6] Tomimura K, Takaki S, Tokunaga Y. Reversion mechanism from deformation induced martensite to austenite in metastable austenitic stainless steels. ISIJ International. 1991;**31**(12):1431-1437

[7] Murata Y, Ohashi S, Uematsu Y. Recent trends in high strength stainless steel. Tetsu to Hagane. 1992;**78**(3):346-353

[8] Simoneau A, Ng E, Elbestawi MA. Chip formation during microscale cutting of a medium carbon steel. International Journal of Machine Tools and Manufacture. 2006;**46**(5):467-481

[9] Lee WB, Cheung CF, To S. Materials induced vibration in ultra-precision machining. Journal of Materials Processing Technology. 1999;**19**(89):318-325

[10] Lee WB. Prediction of microcutting force variation in ultra-precision machining. Precision Engineering. 1990;**12**(1):25-28

[11] Lee WB, Cheung CF, To S. A microplasticity analysis of micro-cutting force variation in ultra-precision diamond turning. Journal of Manufacturing Science and Engineering. 2002;**124**(2):170-177

[12] Furukawa Y, Moronuki N. Effect of material properties on ultra-precise cutting processes. CIRP Annals. 1988;**37**(1):113-116

[13] Komatsu T, Matsumura T, Torizuka S. Effect of grain size in stainless steel on cutting performance in micro-scale cutting. International Journal of Automation Technology. 2011;**5**(3):334-341

[14] Komatsu T, Yoshino T, Matsumura T, Torizuka S. Effect of crystal grain size in stainless steel on cutting process in micromilling. Procedia CIRP. 2012;**1**:150-155

[15] Komatsu T, Musha Y, Yoshino T, Matsumura T. Surface finish and affected layer in milling of fine crystal grained stainless steel. Journal of Manufacturing Processes. 2015;**1**(19):148-154

[16] Fujita M, Hashida M. Femtosecond-laser processing. Journal of Plasma and Fusion Research. 2005;**81**:195-201

[17] Siegel F, Klug U, Kling R, Ostendorf A. Extensive micro-structuring of metals using picosecond pulses—Ablation behavior and industrial relevance. Journal of Laser Micro/Nanoengineering. 2009;**4**(2):104-110

[18] Komatsu T. Effects of grain size on the groove depths in microlaser cutting of austenitic stainless steel SUS304. International Journal of Automotive Technology. 2015;**9**(6):636-645

[19] Vollertsen F, Biermann D, Hansen HN, Jawahir IS, Kuzman K. Size effects in manufacturing of metallic components. CIRP Annals—Manufacturing Technology. 2009;**58**:566-587

[20] Engel U, Eckstein R. Microforming—From basic research to its realization. Journal of Materials Processing Technology. 2002;**125-126**:35-44

[21] Kals TA, Eckstein R. Miniaturization in sheet metal working. Journal of Materials Processing Technology. 2000;**103**:95-101

[22] Shrinivas V, Varma SK, Murr LE. Deformation-induced martensitic characteristics in 304 and 316 stainless steels during room temperature rolling. Metallurgical and Materials Transactions A. 1995;**26**:661-671

[23] Tamura I. On the TRIP steel. Tetsu to Hagane. 1970;**56**:429-445

[24] Shiratori T, Suzuki Y, Nakano S, Yang M, Komatsu T. Effects of grain size on the sheared surface in micropiercing of austenitic stainless steel SUS304. Manufacturing Review. 2015;**2**:9

[25] Nomura K, Kubushiro K, Sakakibara Y, Takahashi S, Yoshizawa H. Effect of grain size on plastic strain analysis by EBSD for austenitic stainless steels with tensile strain at 650°C. Journal of the Society of Materials Science, Japan. 2012;**61**:371-376

[26] Shiratori T, Katoh M, Sato N, Yoshino T, Nakano S, Yang M. Deformation and transformation behavior in micropiercing of fine-grained SUS304. Journal of the Japan Society for Technology of Plasticity. 2017;**58**:936-942

[27] Shiratori T, Yoshino T, Suzuki Y, Katoh M, Nakano S, Yang M. Deformation and transformation behavior in micropiercing of SUS304. Procedia Manufacturing. 2018;**15**:1452-1458

[28] Ohashi O, Suga S. Effect of surface composition on diffusion welding in stainless steel. Journal of the Japan Institute of Metals. 1992;**56**:579-585

[29] Sato N, Yoshino T, Shiratori T, Nakano S, Katoh M. Accelerative effects of diffusion bonding on recrystallization with reversion of deformation-induced martensite in SUS304. ISIJ International. 2016;**56**:1825-1830

[30] Katoh M, Sato N, Shiratori T, Suzuki Y. Reduction of diffusion bonding temperature with recrystallization at austenitic stainless steel. Tetsu to Hagane. 2016;**102**:34-39

[31] Kuwahara H. Surface treatment of iron base alloys by plasma technology [PhD thesis]. Kyoto University; 1992

[32] Aizawa T, Saitoh T, Shiratori T. Micro-joining of shaped stainless steel sheets for fuel injection orifice with high misting capability. In: Proc. 22nd Metal Forming (ESAFORM2019); 8-10, May 2019

[33] Anzai M, editor. Heat and Surface Treatment of Die Materials. Tokyo, Japan: Nikkan-Kogyo-Shinbunsha; 2011

[34] Aizawa T. Characterization on the properties of nitrided layer in dies and molds: Low temperature plasma nitrided stainless steels. Bulletin of the JSTP. 2019;**2**(19):411-415

[35] Granito N, Kuwahara H, Aizawa T. Normal and anormal

microstructure of plasma nitrided Fe-Cr alloys. Journal of Materials Science. 2002;**37**(4):835-844

[36] Aizawa T, Sugita Y. High density RF-DC plasma nitriding of steels for die and mold technologies. Research Report. SIT. 2013;**57**(1):1-10

[37] Imai Y, Murata T, Sakamoto M. High Nitrogen Steels. Tokyo: Agune; 2005

[38] Aizawa T, Yoshihara S-I. Homogeneous and heterogeneous micro-structuring of austenitic stainless steels by the low temperature plasma nitriding. IOP Conference Series: Materials Science and Engineering. 2018;**372**:012049-1-012049-6

[39] Aizawa T. Chapter 3—Low temperature plasma nitriding of austenitic stainless steels. In: Stainless Steels. Rijeka: IntechOpen; 2019. pp. 31-50

[40] Aizawa T, Yoshihara S-I. Inner nitriding behavior and mechanism in stainless steels type AISI316 at 623 K. In: Proc. 13th SEATUC Conf. 2019. pp. 123-128. OS07

[41] Aizawa T, Yoshino T, Shiratori T, Yoshihara S-I. Grain size effect on the nitrogen super-saturation process into AISI316 at 623 K. ISIJ International. 2019;**69**:1-7

[42] Hiraoka Y, Inoue K. Prediction of nitrogen distribution in steels after plasma nitriding. Denki-Seiko. 2010;**86**:15-24

[43] Aizawa T. Functionalization of stainless steels via low temperature plasma nitriding. In: Proceedings of the 7th Annual Basic Science International Conference; Malang, Indonesia; 2017. pp. 1-16

[44] Aizawa T, Yoshihara S-I. Inner nitriding behavior and mechanism in stainless steels at 753 K and 623 K. SJSE. 1; 2019 (in press)

[45] Aizawa T, Yoshihara S-I. Microtexturing into AISI420 dies for fine piercing of micropatterns into metallic sheets. Journal of the Japan Society for Technology of Plasticity. 2019;**60**:53-57

[46] Shiratori T, Aizawa T, Saito Y, Wasa K. Plasma printing of AISI316 multi-punch die micro-embossing into copper plates. Metals—Open Access Metallurgy Journal. 2019;**9**(4):396/1-396/11

[47] Aizawa T, Shiratori T, Wasa K. Plasma-printed AISI316L multi-punch array for fabrication of aluminum heatsink with micro-pillar fins. In: Proc. 3rd WCMNM. 2019. pp. 220-223

[48] Aizawa T, Shiratori T, Yoshino T, Suzuki Y, Kamotsu T. Microstructure evolution of fine grained AISI316 wire plasma nitrided at 623 K through uniaxial tensile loading. ISIJ International; 2019. (in press)

Chapter 2

Isothermal Transformation Behavior and Microstructural Evolution of Micro-Alloyed Steel

Sanjeev Kumar

Abstract

In this present study, the transformation products in micro-alloyed steel have been examined as well as isothermal decomposition of austenite into various phase formation. The rapid cooling from austenitizing temperature 1200°C to 14 different isothermal temperatures between 750 and 100°C with 50°C intervals were carried out by using dilatometric strain dilatometer on thermo-mechanical simulator. The heat treatments were delayed at different times to examine the microstructure evolution at all isothermal temperatures. The transformation kinetics was recorded during isothermal treatments and designed an isothermal transformation diagram, which is verified by microstructural changes. The results show that the initial microstructure which consists of proeutectoid ferrite and pearlite transforms into a combination of proeutectoid ferrite, pearlite, widmanstätten ferrite, upper or lower bainite, or martensite phases. The austenite grain size has been found to be decreased with a decrease in the isothermal holding temperature. The nose temperature was achieved at isothermal temperature 500°C which have been taking the least time for start and end of transformation of phases. It is also worth noticing that the start and end transformation times were observed decreasing with a decrease in the isothermal holding temperatures and after the nose transformation again gradually increased.

Keywords: isothermal transformation, micro-alloyed steel, microstructure, recrystallization, bainite, ITT diagram

1. Introduction

Excellent combination of mechanical and corrosion resistance properties is an initial requirement for high strength low alloyed steel plates which have to be used for the construction of bridge, pipe lines, and ship hull applications [1]. Generally, basic steels do not have the sufficient combination of properties for a direct use for such types of applications. Desirable structural properties may become possible by changes in the chemical composition of steels or by heat treatments such as continuous/isothermal heat treatments with/without thermo-mechanical processing of steels. The compassionate evolved will support in scheming isothermal heat treatment cycles during production of such steels. During steel processing as well as fabrication through welding, the phase transformation has always been a significant research topic for steel users and metallurgists. Isothermal heat treatment is subsequently possible after casting process in industry on nominal cost. Better properties offer to users to reduce the overall cost for structures by reducing the thickness of steel plates. Presently, most of researchers focus on isothermal transformation treatments for medium and high grade carbon steel [2–5] but they lack such information in wide range of isothermal temperature on micro-alloyed steel.

In the present study, effort has been made to design isothermal transformation (ITT) phase diagram for micro-alloyed steel. Isothermal heat treatments are experimented by using physical thermo-mechanical simulator in an isothermal temperature range from 750 to 100°C. Microstructural evolution is characterized by optical microscopy.

2. Experimental details

The present micro-alloyed steel are provided by the Steel Authority of India Ltd. (SAIL), the chemical composition was predicted by using Thermo Jarrell Ash spark emission spectroscope as displayed in **Table 1**. Present steel is familiar by the name of SAILMA steel.

Samples were taken from the transverse direction of a 18 mm thick hot rolled plate. Samples were treated for isothermal physical simulation by the thermo-mechanical simulator Gleeble®3800. Resistance heating was provided by K-type (Chromel-Alumel) thermocouples to the samples by a machine working based on conductive heat transfer model. A Linear Variable Displacement Transducer, LVDT dilatometer (quartz based) was fitted above the sample (where the thermocouple was welded at mid position) to acquire diametric strain changes during thermal cycles. External water quenching was arranged on both side of the hollow region of sample to freeze the microstructure and achieve the desire cooling rates as per ASTM isothermal treatments. Desired cooling rate in programmed and actual thermal cycles was achieved in less than 1 s between the solutionizing temperature and the isothermal temperatures.

For isothermal treatment simulation, the samples were first heated by 5°C/s to the solutionizing temperature 1200°C for 300 s and followed by 200°C/s down to 14 isothermal temperatures between 750 and 100°C with a temperature gap of 50°C. The samples were kept at these isothermal temperatures until completion of phase transformation estimated through dilation strain changes. Subsequently, the sample was water quenched to room temperature by external water quenching arrangements. Details of sample design and isothermal treatment cycle for one isothermal temperature 550°C are shown in **Figure 1**.

Isothermal simulated samples were partitioned cross-sectionally from the position of thermocouples for microstructural examination. One piece of partitioned surface of sample was ground and polished according to standard technique of metallography and later etched by a 2% Nital etchant. Optical micrographs of base material and simulated IT samples were characterized by using optical microscope Leica DMI 5000 M equipped with digital imaging facility. The grain size of ferrite was measured by linear intercept method using Image J software. Average grain size of 10 readings is reported in the present work.

C	Si	Mn	P	S	Al	Nb	V	Fe
0.13	0.25	1.53	0.03	0.007	0.02	0.043	0.01	Bal.

Table 1.
The chemical composition of micro-alloyed steel under investigation, in wt%.

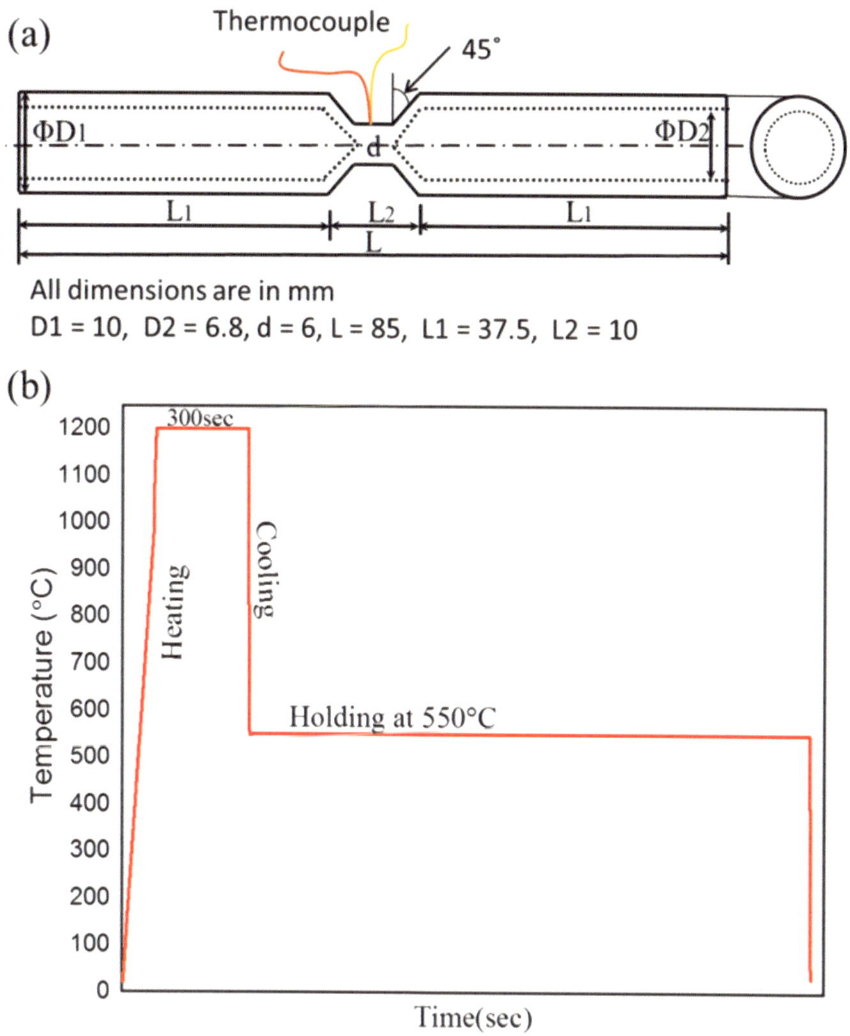

Figure 1.
Schematic diagrams: (a) sample design and (b) isothermal treatment cycle for isothermal temperature 550°C.

3. Results and discussion

3.1 Microstructural evolution

The optical micrographs for base metal and isothermally simulated steel samples are shown in **Figures 2** and **3**. The base metal micrograph consists of pro-eutectoid ferrite and pearlite as shown by arrow in **Figure 2a**. The pearlite structure is found to be in banded form in the direction of rolling [6]. The average grain size of pro-eutectoid ferrite is found to be in range of 10 ± 3 μm. The average grain size of ferrite is found to be decrease with decreasing isothermal temperature as shown in **Figure 4**. At 750°C, the banded structure of pearlite is completely dissolved and new structure is formed which consists of widmanstätten ferrite (WF), pearlite and very small amount of martensite (**Figure 2b**). Rest bright phase fraction reveals as the pro-eutectoid ferrite. It is worthy noticed that the material has not undergone any transformation due to fast cooling rate more than 175°C/s to 750°C. The austenite grain boundaries are completely illuminated through ferrite

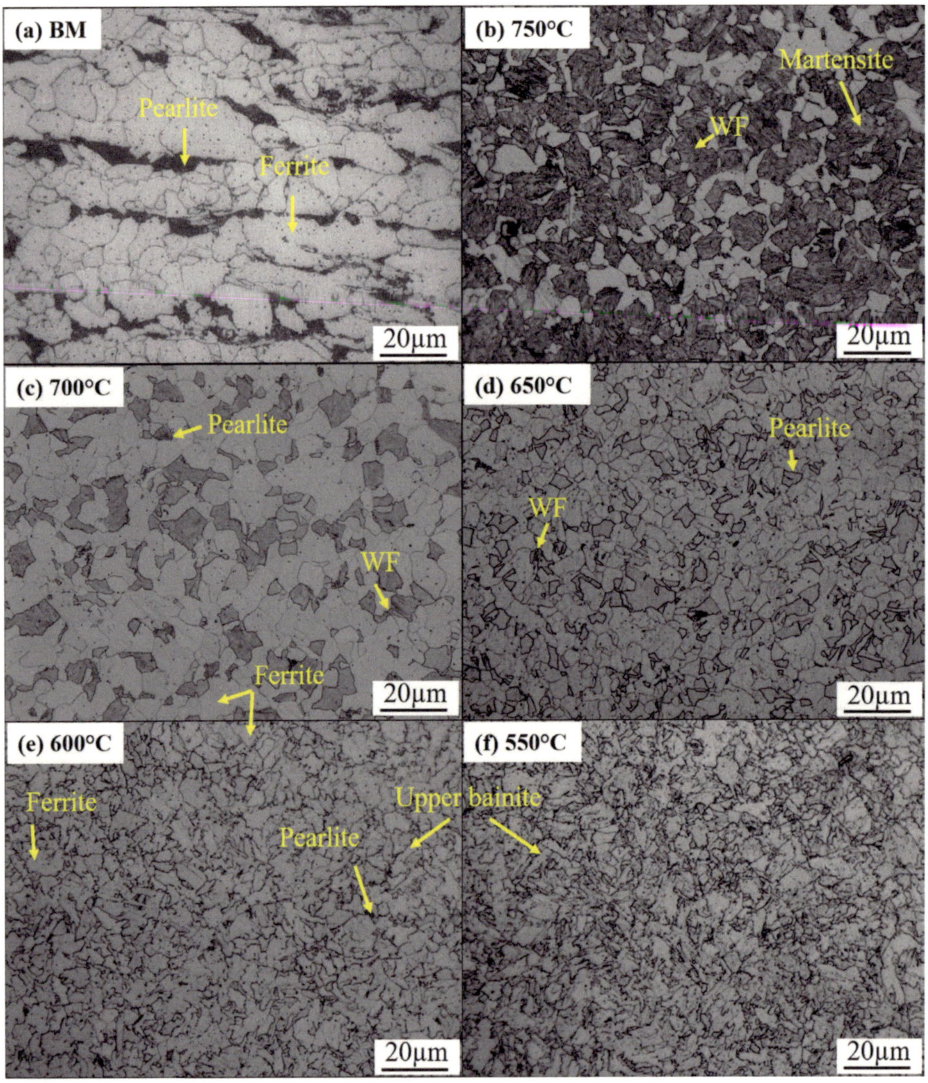

Figure 2.
Optical micrographs of micro-alloyed steel: (a) base metal steel and (b–f) isothermally treated samples subjected to temperatures between 750 and 550°C.

allotriomorphs and retained austenite has reconstructed into pearlite as can see in **Figure 2b**. This pearlite reaction fundamentally occurs at high temperature between 720 and 550°C, which is governed by the diffusion process [7]. In addition, the conventional growth occurs of widmanstätten ferrite under para-equilibrium must correspond either to interface friction control or to carbon diffusion control [8]. This widmanstätten ferrite nucleates at austenite grain boundaries and grows as plates into austenite interior in reconstructive mode [8]. The diffusion of carbon in austenite ahead of plate tip gives maximum lengthening rate due to the capillarity effects. This structure is in agreement with other micro-alloyed steels [9, 10]. The average grain size of ferrite also reformed due to solutionizing treatment and uniform grain size is found to be 8 μm at 750°C. **Figure 2c** shows the amount of widmanstätten ferrite is found to be increasing due to instability of the pearlite phase at 700°C. As the isothermal temperature decreases, the rest amount of enriched carbon islands is fully dissolved in soft pearlite phase region as shown in **Figure 2d**. The ferrite grain size is found to be decreasing gradually

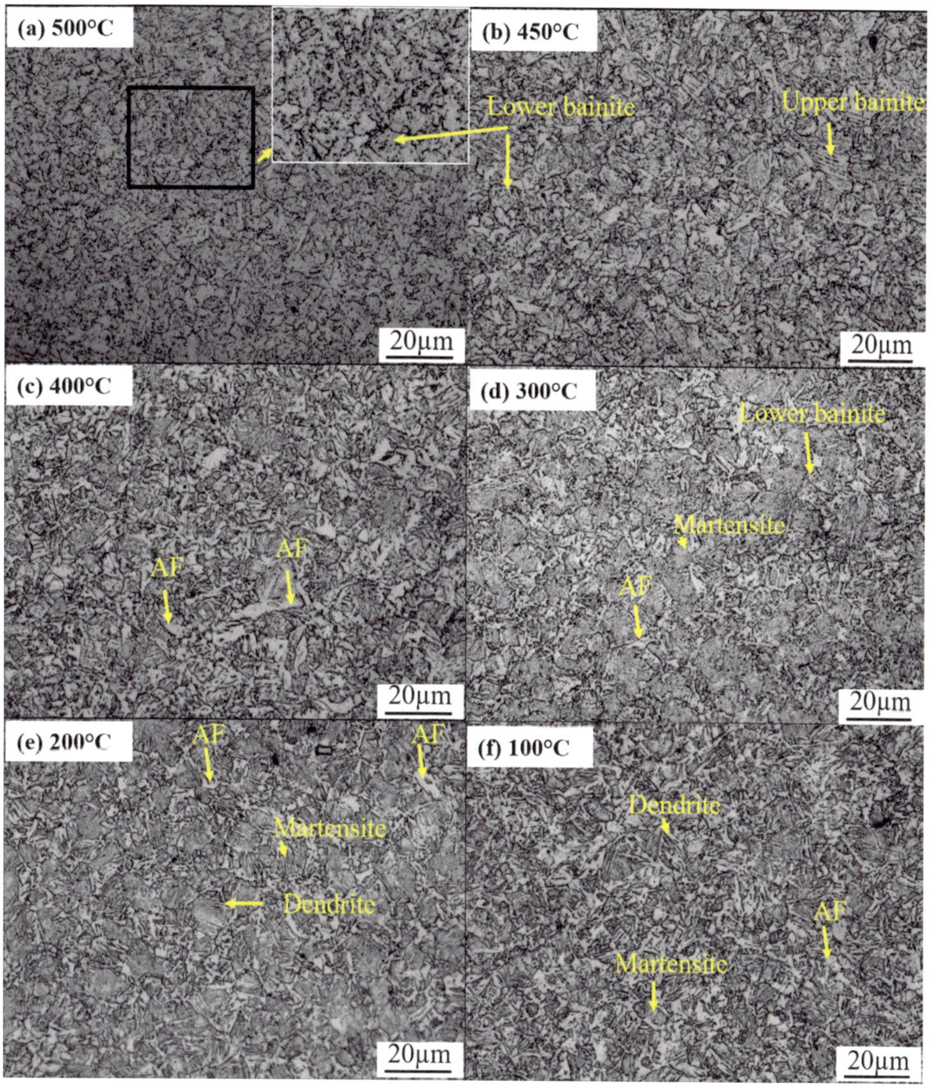

Figure 3.
Optical micrographs of isothermally treated samples subjected to temperatures between 500 and 100°C.

with a decreasing isothermal temperature (**Figure 4**). Carbon is extracted from pearlite and it transforms into residual austenite or cementite or carbides along the austenite grain boundary. This formation of ferrite laths and separation by residual austenite or cementite or carbides is known as upper bainite [5, 11, 12]. The fraction of residual austenite, etc., phases observed are decrease with a decrease in isothermal temperature (600°C) as can see in **Figure 2e**. The amount of widmanstätten ferrite is decreased with decreasing in isothermal temperature. At 550°C, the microstructure is fully transformed into upper bainite however; a small amount of proeutectoid ferrite is also observed (**Figure 2f**). Upper and lower bainite structures have formed and grown by displacive mechanism [13]. Bainite formed as an aggregate of small plates of ferrite (sub-units). These ferrite plates or laths grow in the form of clusters, which are known as sheaves. These plates are parallel and there is an identical crystallographic orientation in every sheaf. The carbon-enriched residual austenite or martensite is found between plates of ferrite in upper bainite and it is free of precipitation of carbides. However, carbide precipitates precipitate inside ferrite plates in lower bainite [11].

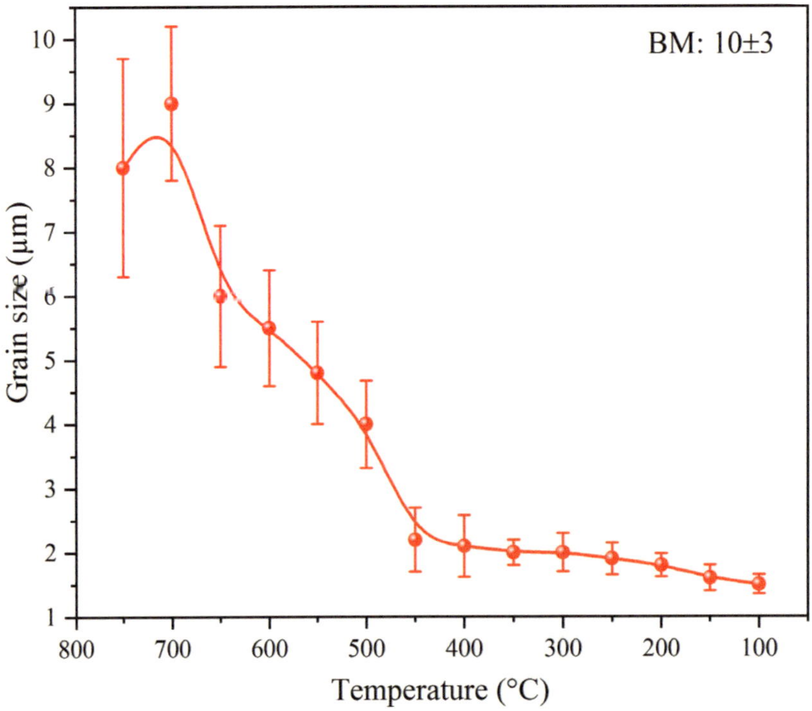

Figure 4.
The variation of ferrite grain size of micro-alloyed steel with isothermal temperature.

As there is a decrease in temperature (500°C), the amount of upper bainite is found to be decreasing and more refined due to the lower transformation temperature, and subsequently lower bainite reveals (**Figure 3a**). At 450°C, the fraction is fully transformed into lower bainite with small amount of upper bainite as can see in **Figure 3b**. It is also noticed that lower austenite grain size leads to the rate of higher amounts of bainite transformed with refinement in ferrite plates under the same isothermal temperature. At temperature 400°C, the microstructure consists of lower bainite and martensite with small amount of allotriomorphic ferrite (AF) around the austenite grain boundaries (**Figure 3c**). This microstructure is similar to allotriomorphic transformation process as reported for high temperature (at around 720°C) in the work of Ashrafi et al. [14]. The several nucleation sites form from the austenite grain boundary and simultaneously hard impingement also occurs between sheaves of bainite inside austenite grains [15]. As there is a decrease in the isothermal temperature (from 300 to 100°C), the fully lath martensite structure is observed with allotriomorphic ferrite (AF) at austenite grain boundaries. This allotriomorphic ferrite (AF) has a very similar morphology to a dendrite structure. It is acknowledged that the nucleation of ferrites form preferable locations at primary austenite grain boundaries due to the effect of boundary energy (**Figure 3d–f**). However, the martensite formed due to rapid cooling rate of austenite form of iron so that carbon atoms do not have sufficient time to diffuse out the crystal structure. Since, the nucleation of martensite is diffusionless. Laterally, the fraction of allotriomorphic ferrite (AF) observed is coarser with decreasing in martensitic transformation temperature region.

3.2 Isothermal transformation (ITT) diagram

Figure 5 illustrates the approach to calculate time for start transformation (T_s) and end transformation (T_f), and between of them such as 25, 50, and 75% phase transformation, which results are used to construct an ITT diagram (**Figure 6**). This ITT diagram is created on the basis of dilation strain data test which is in reference of phase transformation as shown in **Figure 5a** for 650°C. The phase transformation 100% is confirmed by no change in dilation w.r.t. time, it means that the amount of phase is not increasing with time at that particular isothermal temperature. The change in dilation represents to nucleation and growth phase as shown in **Figure 5b**. The fluctuation in dilation appears after start transformation of phases when temperature was lower than 550°C. It is attributed to undercooling. Further, its increase/decrease in dilation completely depends upon type of phase formation such as upper bainite/lower bainite/martensite. The incubation time is found to be decreased with decreasing in isothermal temperature. Similar trends in dilation studies are also reported in the work of Chen et al. [15]. The dilation after an incubation period is almost constant without any contraction or expansion after the start of phase transformation for isothermal temperature 400°C as can see in **Figure 5b**. Since, the most of austenite fraction is completely transformed into lower and upper bainite with small amount of martensite as confirmed from the microstructure (**Figure 3c**). Nevertheless, the dilation was found to decrease in dilation with a decrease in isothermal temperature less than 400°C. Here, both lower bainite and martensite transformation mechanism work simultaneously.

In martensitic transformation, the atom movements occur by two shears formation mechanism, i.e., long-wave as primary and shuffle type shear as secondary which take place at the same time [16]. The primary and secondary part of shears is corresponding to the Bain distortion, and the rest of the secondary shear is in agreement with the lattice invariant shear of the phenomenological theories [17]. However, widmanstätten ferrite, upper bainite, and lower bainite are formed by high-velocity shear. Since, transformation products are developed by displacive

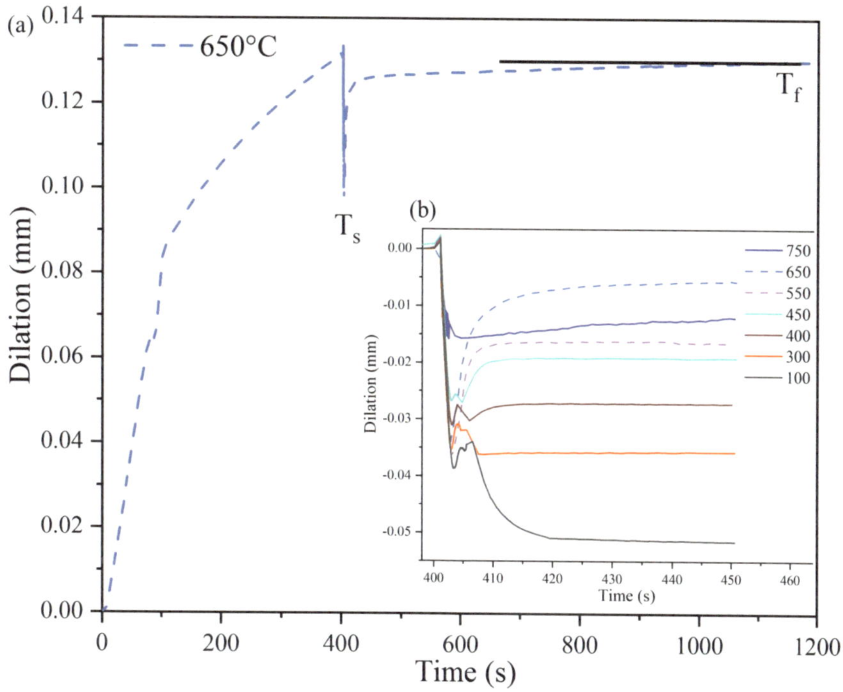

Figure 5.
Dilation vs. log time plots of micro-alloyed steel subjected to isothermal temperatures: (a) full dilation plot curve for temperature 650°C; and (b) dilation plots for start and end transformation regions of phases for selective transformation isothermal temperatures.

mechanisms, however martensite formed by shear process. The contribution of shear process increases as the temperature decreases [11].

The lower (Ac_1) and upper (Ac_3) critical temperatures during heating at 5°C/s were obtained 734 and 912°C respectively from the present study. The ITT diagram's purpose is not only to collects the statistics about the nature of phases however, it also finds out the kinetics of phase transformation for the holding period at constant (isothermal) temperature. This is a very useful set of heat treatment processes to obtain the desired properties by moderation of microstructure. This ITT diagram is also known as time temperature transformation (TTT) diagram. An ITT diagram plotted based on isothermal temperature heat treatment is displayed in **Figure 6**. The initial microstructure was consisting of proeutectoid ferrite and pearlite which transforms into proeutectoid ferrite, pearlite, and other new phases such as widmanstätten ferrite, upper/lower bainite and transformed martensite (**Figures 2** and **3**). It is worth noticing from **Figure 6** that the time of start transformation decreases with decreasing isothermal temperature until 250°C, and further starts increasing. The start transformation time is delayed by around 12 s and takes longer time to finish (end) transformation when the temperature is close to the lower critical temperature as can see in **Figure 6**. As decreasing in isothermal temperature, both T_s and T_f phase transformation times are observed decreasing. The nose of ITT is found to be at 500°C, where there was the shorter delay (less than 1 s) between the start and end transformation of phases. An analysis of this ITT diagram is divided into five regions which show a combination of phase's microstructure (**Figure 6**) as confirmed by the microstructural evolution (**Figures 2** and **3**).

IT750°C (>Ac1), the start transformation occurred after 12 s (incubation period) and the end transformation (100% transformation of phases) after 1080 s, which is predicted in the form of diametric changes in mm during experimental studies. The microstructure consists of proeutectoid ferrite, widmanstätten ferrite, pearlite and martensite at high temperature isothermal holding. Small volume fraction of

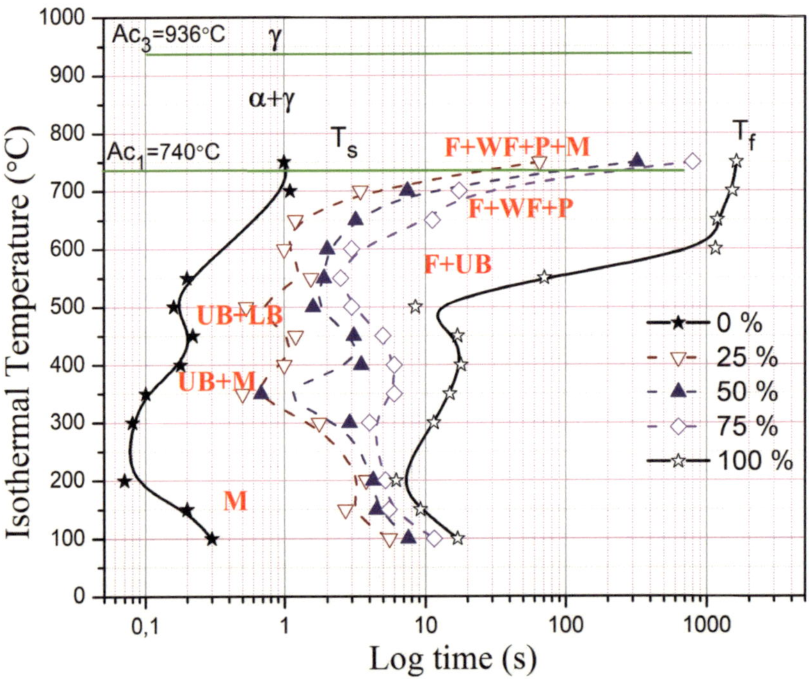

Figure 6.
Isothermal transformation diagram of micro-alloyed steel.

martensite was established at this isothermal temperature. These transformation products are developed by displacive and reconstructive mechanisms, however martensite formed by shear process which is also part of displacive mechanism.

Ac1–IT650°C denotes the formation of phases with both reconstructive and displacive transformations. The microstructural evolution is the combination of pro-eutectoid ferrite and pearlite, widmanstätten ferrite (WF), no more martensite. Start to end transformation time is found to be decreased gradually due to the formation of phases without any delayed.

IT650–500°C, the microstructural evolution was observed through the formation of upper bainite, formed by displacive or diffusionless mechanism. The ferrite plates in upper bainite are separated by the cementite as cleared from the microstructure. Nevertheless, a small fraction of lower bainite is observed. The nose temperature was achieved at isothermal temperature 500°C that has been taking the least time between the start and end transformation of phases. Later it starts increasing again gradually in incubation time with a decrease in IT temperature (**Figure 6**).

IT450–300°C, the lower bainite structure is the combination of ferrite plates with carbide and these plates are separated by retained austenite or cementite or martensite. The needle type morphology of ferrites is observed along the grain boundary in dendrite form due to higher shear strains.

IT300–100°C is once more a displacive and diffusional transformation though with tiny or no carbide precipitation which is attributed to low IT. These microstructures are transformed into fully martensite. Lower bainite amount has been found to be decreasing with isothermal temperature. These transformation products are developed by shear process mechanism.

4. Conclusions

The following outputs are constructed from the present work on micro-alloyed steel.

- The initial microstructure of base steel consists of proeutectoid ferrite and pearlite. During isothermal heat treatments, the microstructure transforms into a combination of possibly proeutectoid ferrite, pearlite, widmanstätten ferrite (WF), upper or lower bainite, or martensite phases. The austenite grain size has been found to be decreasing with a decrease in the isothermal holding temperature.

- At IT750°C, the microstructure consists of proeutectoid ferrite, widmanstätten ferrite, pearlite and martensite due to high temperature isothermal holding. Small volume fraction of martensite was found due to isothermal temperature above the critical temperature. These transformation products are developed by displacive and reproductive mechanisms, however martensite formed by shear process.

- IT700–600°C shows a significant fraction of allotriomorphic ferrite, thus the pearlite phase has apparently to be softer. However, the volume fraction of pearlite is decreased with temperature. A fraction of carbon content was observed along the grain boundary, which was extracted from the ferrite during nucleation and growth of phases.

- Most of the austenite fraction transforms into upper bainite in medium range of isothermal temperature (600–450°C). In IT450–300°C, the austenite transforms into predominantly lower bainite with increment in martensite. At less than 300°C, the microstructure was observed to be a martensite structure and acicular martensitic dendrite structure at austenite grain boundary.

Acknowledgements

The author would like to thank the Department of MME, Indian Institute of Technology Roorkee, India, for the physical experimental simulation facility Thermo-mechanical simulation Gleeble ®3800 and other characterization instruments.

Author details

Sanjeev Kumar
Christian Doppler Laboratory for Interfaces and Precipitation Engineering
CDL-IPE, TU Wien, Vienna, Austria

*Address all correspondence to: rajputskmt@gmail.com

References

[1] Ghosh A, Sahoo S, Ghosh M, Ghosh RN, Chakrabarti D. Effect of microstructural parameters, microtexture and matrix strain on the charpy impact properties of low carbon HSLA steel containing MnS inclusions. Materials Science and Engineering A. 2014;**613**:37-47. DOI: 10.1016/j.msea.2014.06.091

[2] Madariaga I, Gutierrez I, Bhadeshia HKDH. Acicular ferrite morphologies in a medium-carbon microalloyed steel. Metallurgical and Materials Transactions A. 2001;**32**:2187-2197. DOI: 10.1007/s11661-001-0194-7

[3] Francisca GC, Maria JS, Carlos GM, Carlos GA. Tme-temperature-transformation diagram within the bainitic temperature range in a medium carbon steel. Materials Transactions. 2004;**45**(12):3272-3281

[4] Wang XL, Wu KM, Hu F, Yu L, Wan XL. Multi-step isothermal bainitic transformation in medium-carbon steel. Scripta Materialia. 2014;**74**:56-59. DOI: 10.1016/j.scriptamat.2013.10.019

[5] Sajjadi SA, Zebarjad SM. Isothermal transformation of austenite to bainite in high carbon steels. Journal of Material Processing Technology. 2007;**189**:107-113. DOI: 10.1016/j.jmatprotec.2007.01.011

[6] Kumar S, Nath SK, Kumar V. Continuous cooling transformation behavior in the weld coarse grained heat affected zone and mechanical properties of Nb-microalloyed and HY85 steels. Materials and Design. 2016;**90**:177-184. DOI: 10.1016/j.matdes.2015.10.071

[7] Liang X. The complex phase transformation of austenite in high strength pipeline steels and its influence on the mechanical properties [PhD thesis]. Department of Mechanical Engineering and Materials Science, University of Pittsburgh USA; 2012. pp. 1-237

[8] Bhadeshia HKDH. Nucleation of Widmastatten ferrite, bhadeshia. Materials Science and Technology. 1990;**6**:781-784

[9] Thompson SW, Colvin DJ, Krauss G. Continuous cooling transformations and microstructures in a low-carbon, high-strength low-alloy plate steel. Metallurgical and Materials Transactions A. 1990;**21**:1493-1507. DOI: 10.1007/BF02672564

[10] Bhadeshia HKDH, Christian JW. Bainite in steels. Metallurgical Transactions A. 1990;**21**:767-797. DOI: 10.1007/BF02656561

[11] Spanos G, Fang HS, Aaronson HI. A mechanism for the formation of lower bainite. Metallurgical Transactions A. 1990;**21**:1381-1390. DOI: 10.1007/BF00653646

[12] Kumar S, Nath SK, Kumar V. Isothermal transformation behavior of a low-carbon HY 85 steel. Metallography Microstructure and Analysis. 2016;**5**:264-271. DOI: 10.1007/s13632-016-0283-z

[13] Lan L, Qiu CL, Zhao DW, Gao XH, Du L-X. Effect of austenite grain size on isothermal bainite transformation in low carbon microalloyed steel. Materials Science and Technology. 2011;**27**:1657-1663. DOI: 10.1179/1743284710Y.0000000026

[14] Ashrafi H, Shamanian M, Emadi R, Saeidi N. Examination of phase transformation kinetics during step quenching of dual phase steels. Materials Chemistry and Physics. 2017;**187**:203-217. DOI: 10.1016/j.matchemphys.2016.12.002

[15] Chen S, Wang G, Liu C, Wang C, Zhao X, Xu W. Correlation of

isothermal bainite transformation and austenite stability in quenching and partitioning steels. Journal of Iron and Steel Research, International. 2017;**24**(11):1095-1103. DOI: 10.1016/S1006-706X(17)30159-0

[16] Ahlers M. The martensitic transformation: Mechanisms and crystallography. Philosophical Magazine A. 2002;**82**:1093-1114. DOI: 10.1080/01418610208240019

[17] Bhadeshia HKDH. The bainite transformation: Unresolved issues. Materials Science and Engineering A. 1999;**273-275**:58-66. DOI: 10.1016/S0921-5093(99)00289-0

Chapter 3

Phase Transformation in Micro-Alloyed Steels

Fahim Khan and Hossain M. M. A. Rashed

Abstract

Phase transformation in crystalline solid is an important factor that designs the microstructure and plays a great role in alloy development. Iron has an allotropic form, and this unique metallurgical property leads to phase transformation. Addition of micro-alloying elements enhances the phase transformation scenarios in steels. Phase transformation due to the addition of micro-alloying elements, together with exceptional precipitation hardening capabilities, substantially improves mechanical properties of steels of different grades. Ferrite transforming to other phases reduces the hardenability of steels. Micro-addition of elements forms precipitation in ferrite and austenite, which controls the microstructure and hence the mechanical properties of steels. Besides, interactions between different deformation sequences used in the production of steel and addition of elements as solute or precipitates regulate the microstructure. Ferrite grain refinement depends on the refinement of austenite grain size in one case, and austenite grain size growth can be varied by addition of various elements. Thus, a variety of elements influences phase transformation that leads to significantly modified properties.

Keywords: austenite grain size, mechanical properties, ferrite, microstructure, precipitation

1. Introduction

Micro-alloyed steel is a type of steel containing a minute amount of alloying elements (0.05–0.15%) including niobium, vanadium, titanium, molybdenum, zirconium, boron, and rare-earth metals [1]. They are mainly used to refine the grain microstructure or facilitate precipitation hardening. Micro-alloyed steel developed in the second half of twentieth century. A lot of advancements have taken place due to the development of micro-alloyed steels. Certainly, it improves the hardening, tensile strength, and other mechanical properties of steels. Micro-alloyed substituents like niobium, vanadium, titanium, etc. cause significant grain refinement by retarding recrystallization and forms precipitation of carbon-nitride by these micronutrients. Precipitation of Nb(C,N) that forms improve the microstructure and overall properties of final product. This precipitation further inhibits the recrystallization of austenite, as a result of which reduction of ferritic grains takes place after transformation of noncrystalline austenite. On the basis of this principle later, high-strength low-alloy (HSLA) was developed.

2. Effect of alloying element on ferrite

FCC has 8 and 4 octahedral and tetrahedral voids per unit cell, respectively, whereas BCC has 12 and 6, respectively. The octahedral void in FCC is bigger than the tetrahedral. Carbon occupies the octahedral void with less distortion. The octahedral void in BCC is smaller than the tetrahedral even in the case when carbon occupies octahedral void due to lesser distortion (only top atom and bottom need to be distorted) [2].

Alloying element	Group 1 Dissolved in ferrite	Group 2 Combined in carbide
Nickel	Ni	
Silicon	Si	
Aluminum	Al	
Copper	Cu	
Manganese	Mn	Mn
Chromium	Cr	Cr
Tungsten	W	W
Molybdenum	Mo	Mo
Vanadium	V	V
Titanium	Ti	

Adapted from Ref. [3].

Table 1.
Behaviors of the individual elements in annealed steels.

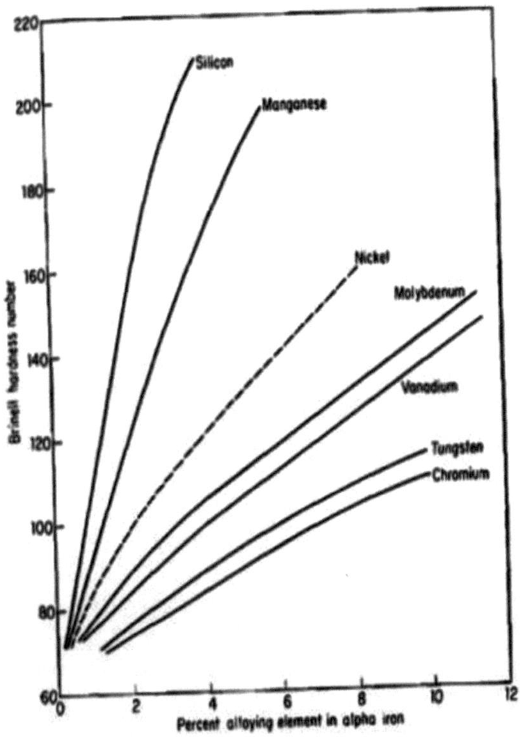

Figure 1.
Probable hardening effect of the various elements as dissolved in alpha iron (reprinted from Ref. [3]).

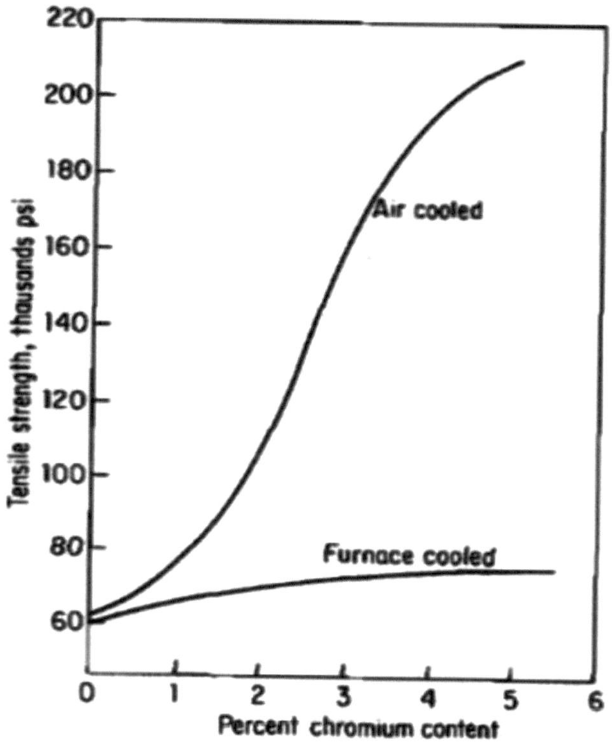

Figure 2.
The minor effect of chromium in annealed steels compared with the powerful effect as a strengthener through its influence on structure in air-cooled steels. Probable hardening effect of the various elements as dissolved in alpha iron (reprinted from Ref. [3]).

The number of voids in BCC is greater than that in FCC, whereas size of voids in BCC is significantly smaller than that in FCC. For this reason austenite have higher solubility of C than ferrite. But all the alloying elements have some sort of solubility in ferrite. It mainly depends on the amount of carbon present in the structure. Nickel, aluminum, silicon, copper, cobalt, etc. dissolve in ferrite in a large extent and play a significant role in increasing hardness and strength by solid solution hardening. Group 2 elements in **Table 1** dissolve in ferrite in the absence of carbon, otherwise it forms carbide.

In **Figure 1**, the probable hardening effect of various elements dissolved in alpha (α) iron is shown. When silicon dissolved in alpha (α) iron, then the hardening value lies in maximum among the addition of Mn, Ni, Mo, V, W, and Cr. Among the alloyed elements, the addition of chromium causes the least hardening effect.

The dissolved element has a little hardening effect in the contribution of ferrite to the overall strength of the steel. In the case of the low-carbon chromium steels, when the structural change occurs by any process, then hardening of such steels occur. The change of structure can be done in the case of annealed steels by cooling it from higher temperature by air. In **Figure 2** it is clearly understood. In **Figure 2** when the low-carbon chromium-annealed steels are air-cooled, then its tensile strength rises to a high value. On the other hand, when cooled in furnace, a little change in tensile property is seen.

3. Influence of alloying elements on iron-iron carbide diagram

The presence of alloying elements plays an important role in the change of critical range, position of eutectoid point, and location of the alpha and gamma fields indicated by the binary iron-iron carbon diagram. Besides, the presence of nickel and manganese lowers the critical temperature on heating, which stabilizes austenite. If the critical temperature lowers than that of the standard region, then austenite becomes stable at room temperature. Thus nickel and molybdenum stabilize austenite at room temperature, and the addition of such alloying element is used in the preparation of austenitic stainless steel.

On the other hand, Mo, Cr, Si, and Ti raise the critical temperature range which contracts the austenite zone and enlarges the alpha and gamma regions as well.

Austenitic stainless steel plays a great role in industrial applications by giving corrosion resistance of steels and providing well mechanical strength. It is mainly used in pressure vessels, reactors, storage tanks which are used underground, and especially in aqueous environments containing chlorides. Austenitic stainless steels are used in manufacturing pump in the oil industry that injects saltwater to expel gas and oil. For the preparation of austenitic stainless steels, Ni and Mo play an important role.

From **Figure 3**, it is clearly shown that the addition of Ti, Mo, Si, W, and Cr raises the eutectoid temperature of steels. The rise of the eutectoid temperature of steels lowers the stability region of austenite, which ultimately stabilizes austenite at the elevated temperature. At this scenario, austenite becomes unstable at lower or room temperature. Besides, the addition of Mn and Ni lowers the eutectoid temperature which passively indicates that the stability of austenite at room temperature. As a result, Mn and Ni are the helpful alloying elements in the case of the austenitic stainless steels.

Chromium in particular lowers the eutectoid temperature. Thus with the addition of chromium, the austenite zone contracts as austenite is stable above the critical temperature. If the critical temperature rises up, then austenite gets stable at elevated temperature and unstable at room temperature. Thus with the addition of chromium

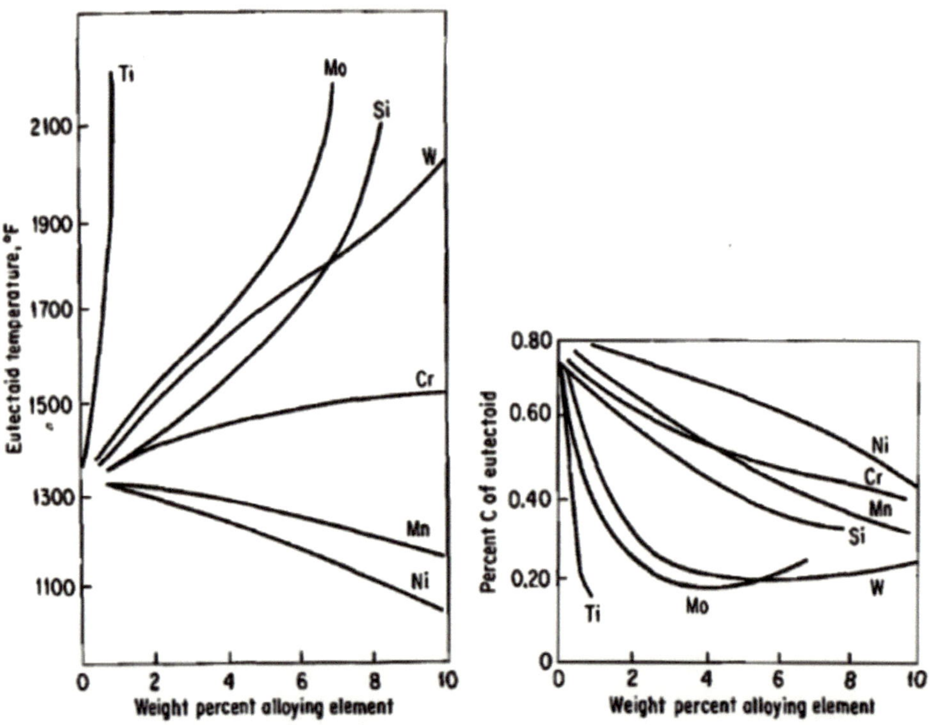

Figure 3.
Eutectoid composition and eutectoid temperature as influenced by several alloying elements. Probable hardening effect of the various elements as dissolved in alpha iron (reprinted from Ref. [3]).

the austenite zone contracts more and prone to the formation of austenitic stainless steel reduces. And austenitic stainless steels become unstable at room temperature.

4. Effect of alloying element on tempering

In the case of the tempering of the plain carbon steels, when the temperature is increased, then the hardness value is decreased; thus hardened steels are softened. At the same time, the hardness drops continuously. Some alloying elements play an important role in retarding the softening effect of the hardened steel at elevated temperature. When tempering is done at elevated temperature, then the steel may soften. Usually the elements that remain dissolved in ferrite, such as Ni, Si, and Mn, have very little effect in the retardation of the softening of steels at elevated temperature.

The complex carbide-forming elements such as Cr, W, Mo, and V retard the softening at elevated temperature while tempering. Besides, they do not only retard the softening effect but also improve the hardness of the plain carbon steels to some aspects.

5. Niobium-micro-alloyed steels

Niobium is a soft gray ductile and transition element. The main commercial source of niobium is mineral pyrochlore. Around 80% of the niobium produced is used in automotive industry, for oil and gas pipelines, and in construction. Adding niobium to steels causes the formation of niobium carbide and niobium nitride which improve grain refinement and retardation of recrystallization. Besides, it enhances precipitation hardening which increases toughness, strength, formability, and weldability of micro-alloyed steel [4].

Large-diameter pipes are used in transportation of oil and gas. It is manufactured by thermomechanical controlled processing (TMCP) [5]. Its performance can be enhanced by inducing its strength and toughness through grain refinements. Grain refinement can be done by controlling austenite parameters by the addition of niobium.

In austenitic-ferritic stainless steel, usually solidification starts at 1450°C with the formation of ferrite (α) which acts as an origin to start the formation of austenite near 1300°C. σ forms at the interphase of austenite and ferrite at 600–950°C, and the toughness of the steel gets reduced.

Figures 1–6 show the microstructural characteristics of the austenitic-ferritic stainless steel with or without niobium, after heat treatment.

Figure 5(a) shows the heat-treated steels without niobium with elongated austenitic grains in ferrite matrix. When the annealed sample is aged at 850°C/15 min then it is observed that the beginning of sigma phase forms.

When steel is modified with 0.2% niobium, then a little amount of sigma phase is observed than steel without niobium after being annealed and aged at 850°C/15 min. Besides no Laves phase is seen (**Figure 7**).

When steel is modified with 0.5% niobium after being annealed and aged at 850°C/min, then the Laves phase appears as needles associated with sigma phase.

In the aggressive environments, the preferential attack prone to the reduction of Cr and Mo near and alongside of the sigma phases. That is the reason for the reduction on pitting corrosion resistance in the steels.

The addition of niobium in supermartensitic stainless steel after tempering at 600°C for 2 h improves the mechanical resistance properties with lower degree of sensitization. Besides, given such properties, it never compromises its elongation and pitting corrosion resistance compared to the reference steels.

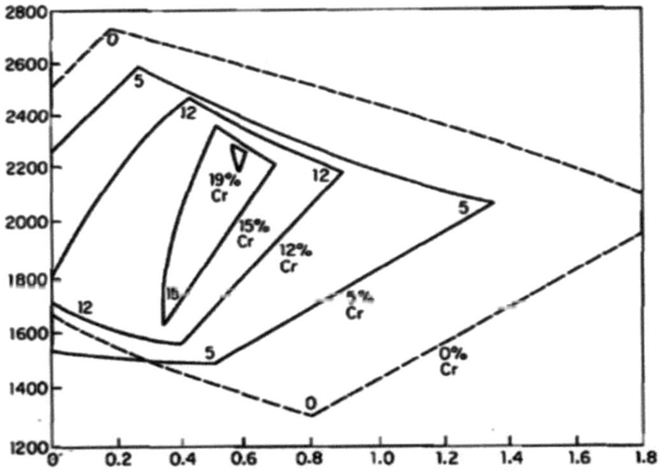

Figure 4.
Range of austenite in chromium steels (reprinted from Ref. [6] with permission).

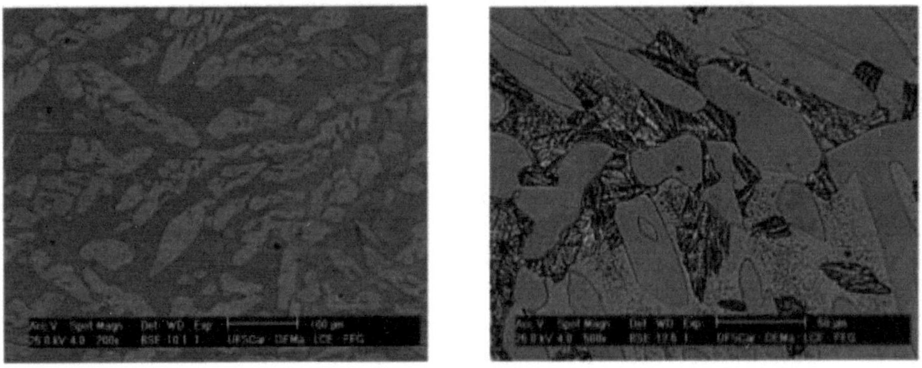

(a) (b)

Figure 5.
Annealed steel without niobium (BSE after Behara etching) with (a) Austenite grains in ferrite matrix and (b) beginning of the sigma phase (adapted from Ref. [7], p. 802).

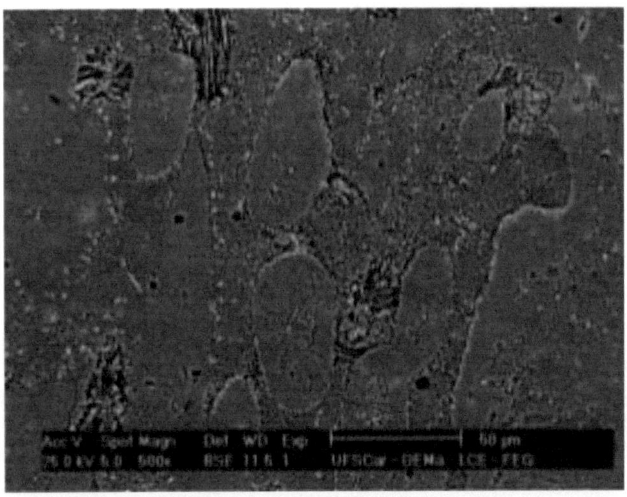

Figure 6.
Steels modified with 0.2% niobium (adapted from Ref. [7], p. 803).

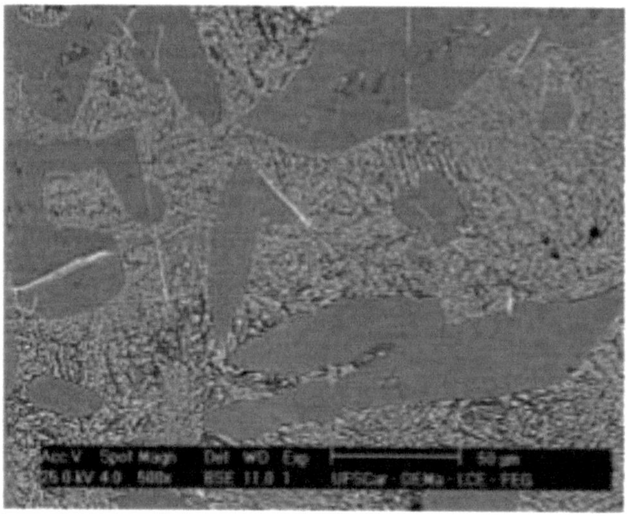

Figure 7.
Steel modified with 0.5% niobium (adapted from Ref. [7], p. 803).

6. Titanium steels

Titanium alloys are mainly used in the structural materials in the aerospace and chemical industries due to lower density, high strength, and corrosion resistance. Tensile strength/density ratios of titanium alloys are considerably greater than that of steels and Al alloys at ordinary temperature. For these reasons Ti alloys are mostly used in the aerospace industries. And for this Ti is added to the steels to improve its properties to some extent (**Figure 8**).

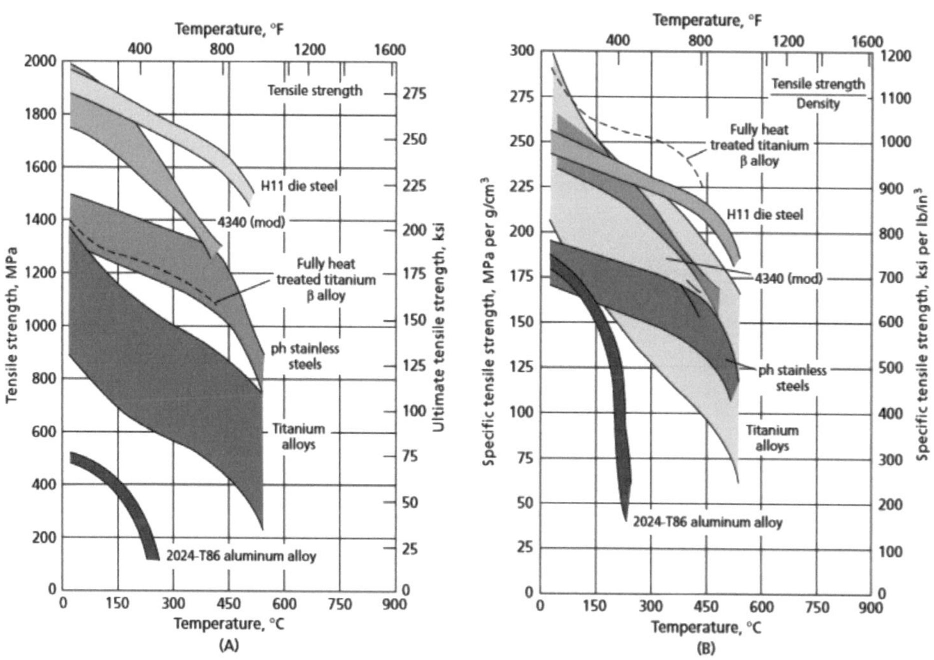

Figure 8.
Comparison of (a) short-time tensile strength and (b) tensile strength/density ratio for titanium alloys, three classes of steel, and 2024-T86 aluminum alloys included for annealed alloys with less than 10% elongation or heat-treated alloys with less than 5% elongation (reprinted from Ref. [8]).

When Ti is added to the steels, then it improves its high temperature properties as refractory metals [9]. The chemical behavior of Ti always limits its application at moderate to high temperature. At low temperature Ti passives to acids and minerals, but at elevated temperature, Ti oxidizes very fast. Besides, dissolution of hydrogen and nitrogen causes surface hardening.

Pure Ti undergoes allotropic transformations at about 1158 K. Thus if Ti remains in steels at around 1158 K, the properties may vary, because pure Ti at 1158 K transforms from a closely packed hexagonal structure to a body-centered cubic structure. The high temperature BCC of Ti is called alpha phase. On the other hand,

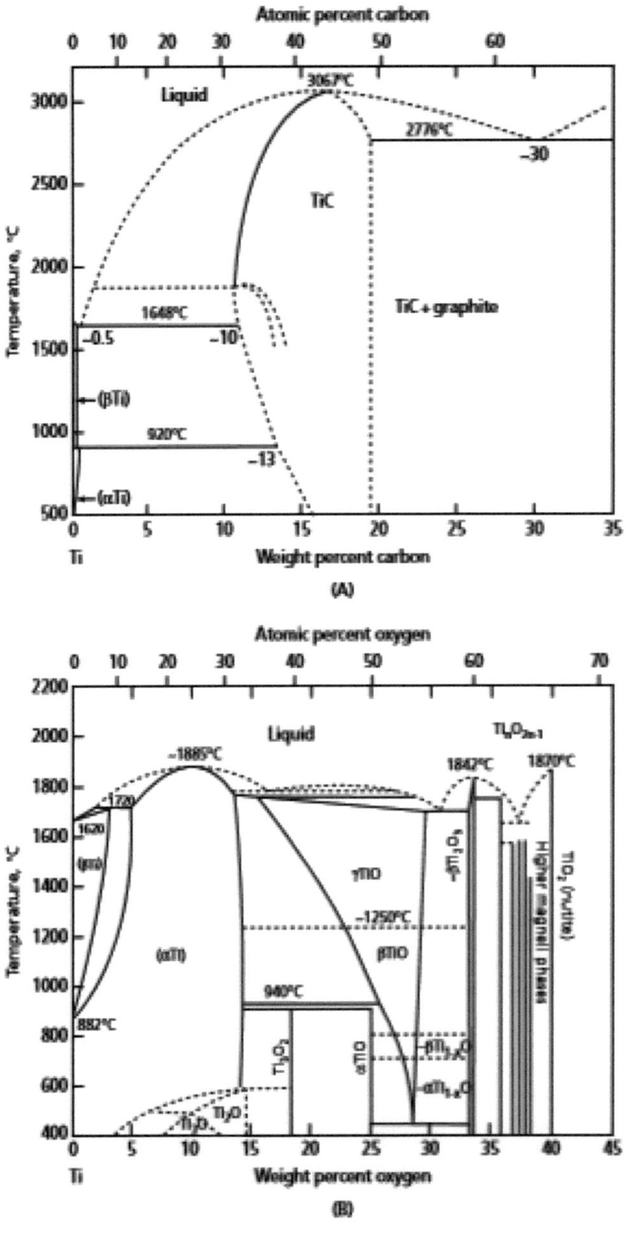

Figure 9.
Phase transformation of titanium with the weight percentage of (a) Cr and (b) Ni stabilizers (reprinted from Ref. [10]).

HCP at low temperature forms beta phase. Beta to alpha transformation happens by diffusionless martensitic transformation.

Allotropic transformation temperature depends on few alloying elements. Some alloying elements raise the transformation temperature called alpha stabilizers, whereas few lower the transformation temperatures called beta stabilizers.

Figure 9 shows that carbon, oxygen, and nitrogen are rapidly absorbed by Ti when the metal is hot. All these elements hardened and the solution hardened the alpha Ti. Al has significant solubility over alpha and beta phases. The reason to explain it is that Ti has many fold advantages on the steel acting as an alloying element.

Elements that lower the transformation temperature are of two classes. One is elements that undergo eutectoid transformations, for example, iron, Cu, N, Co,

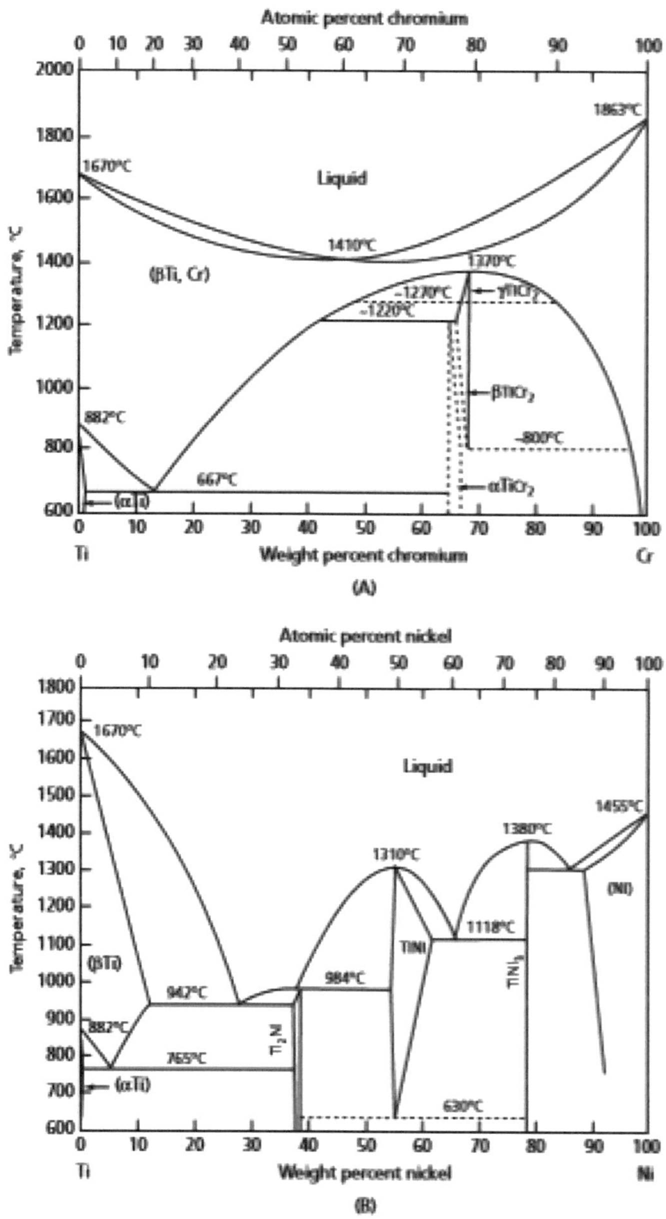

Figure 10.
Phase diagrams of titanium with selected stabilizers (reprinted from Ref. [10]).

Mn, etc. And the other one is those that are isomorphous with beta phase at high temperature and from alpha + beta equilibrium phase at ordinary temperature.

Figure 10 expresses that Mo, Ta, V, etc. have limited solubility in alpha phases.

The main role of Ti in steel is grain refinement strengthening and precipitation strengthening. The smelting of Ti-micro-alloyed steel should satisfy that most of the Ti dissolves in the molten steel and precipitates in the form of carbide or carbonitride after the subsequent solidification.

The affinity of Ti toward oxygen is less than that of the aluminum toward the oxygen. Besides, Ti has greater affinity than manganese toward oxygen. Thus if the molten steel during smelting is not deoxidized properly, then there is a large amount of titanium oxide.

The high content of nitrogen forms titanium nitride that forms inclusions in molten steels. On the other hand, titanium oxides and nitrides will obstruct the process of continuous casting. During the refining process by pyrometallurgy of Ti-micro-alloyed steel, it is required to remove sulfur, oxygen, and nitrogen. But emphasis should be given on the relationship between Ti, Al, and Ti, which are refractory elements. Compared with Nb or V, in the case of Ti, it is more difficult to control Ti-micro-alloyed steels which attributes to the more type of the secondary phase and wider temperature range of the precipitate.

During smelting Ti_2O_3 and TiN particles will precipitate in the liquid steel that improves as cast microstructure. During slab cooling process TiN and $Ti_4S_2C_2$ tens to hundreds of nanometers precipitate in the solid solution plays an important role in controlling the grain growth of austenite during soaking and recrystallization process (**Figure 11**).

During rolling TiC precipitation of TiC with the size below 10mm could result in the significantly precipitation hardening.

As a kind of micro-alloying element, Ti significantly improves the comprehensive properties of steel. However, when compared with niobium and vanadium micro-alloyed technology, Ti has not been used extensively in industry for a long time. Ti-micro-alloyed steel fluctuates largely and production process is not stable. Ti is very reactive and forms TiO and TiS that are very harmful. Formation of these phases consumes a portion of Ti that reduces the volume fraction of TiC precipitation at low temperature but also significantly changes the chemical free

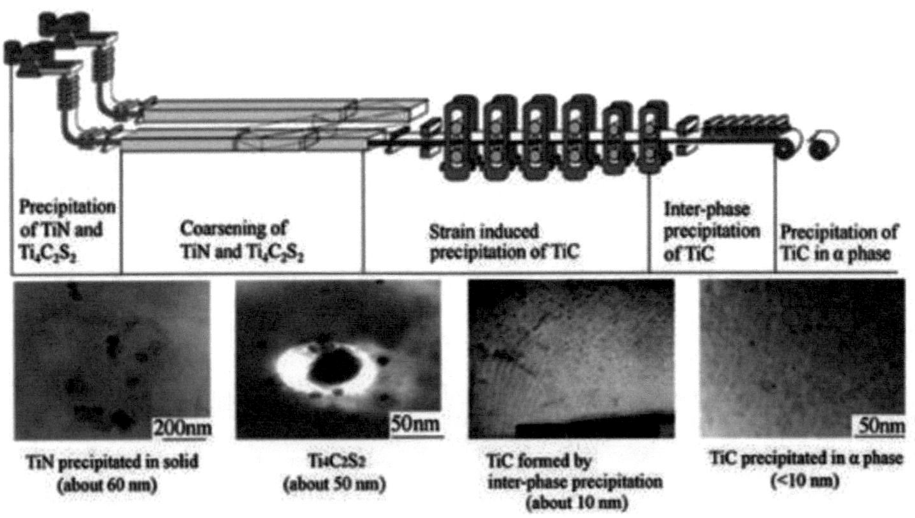

Figure 11.
Ti-bearing precipitation (adapted from Ref. [11]).

energy of TiC. Precipitation behavior of TiC changes and strengthening effect is greatly affected. Besides, TiC is sensitive to temperature and affects the properties of steel.

Besides, high-strength Ti-micro-alloyed steels are the precipitation-hardened ferritic steels. The ferrite grain refinement and TiC precipitation have a good combination in strengthening that plays an important role in obtaining both high strength and high toughness simultaneously for those steels. The ferrite grain refinement depends on the refinement of austenite grain size and on the control of transformation temperature. The refinement of austenite grain size mainly depends on the control of the austenite grain growth before hot rolling and recrystallized austenite grains during hot rolling.

7. Nickel steels

Nickel is the oldest and one of the fundamental alloying elements. It has unlimited solubility in gamma iron and is highly soluble in ferrite. As a result it gives high strength and toughness. Ni lowers the critical temperature of steels and retards the decomposition of austenite. As a result at low temperature or room temperature, austenite gets stable.

Ni does not form carbide. Besides, it reduces the carbon content of the eutectoid. As a result of which there is high percentage of pearlite forms compared to the equal composition plain carbon steels. Pearlite forms at the lower temperature thus become finer and tougher than the pearlite in unalloyed steels.

8. Chromium steels

Chromium is less expensive than Ni. Chromium is a carbide former and forms (Cr_7C_3, Cr_4C) or complex carbide [$(FeCr)_3C$]. This carbide has high hardness and wear resistance. It has 13% solubility in austenite and unlimited solubility in ferrite. In alloying steels, chromium containing more than 5% improves the corrosion resistance and high temperature properties.

9. Manganese steels

Manganese is less expensive and mostly acts as deoxidizers. The presence of manganese in alloy steel reduces the prone to the hot shortness. As a result of which, the alloy steel containing manganese can perform the hot work.

Besides, the absence of manganese in the steel may form FeS. FeS has low melting temperature. Ehen the steel sample is how rolled then due to the low melting temperature of the FeS it melted first. Thus the few places in steels containing FeS become slippery, and thus the hot-rolled samples may slip during rolling.

Mn and Ni both reduce the critical temperature and lowers the amount of carbon in eutectoid. Alloying steels containing more than 10% Mn become austenitic after slow cooling.

Hadfield Mn steel is a special type of steel (12% Mn) and has great abrasion resistance. If it is slow-cooled from 1750F, then a large brittle carbide forms surrounding the austenite grain. Ultimately forms the structure with low strength and ductility.

Nb micro-alloyed steels	Ti micro-alloyed steels	Ni micro-alloyed steels	Cr micro-alloyed steels	Mn micro-alloyed steels	Mo micro-alloyed steels	W micro-alloyed steels
Addition of Nb in steels causes formation of niobium carbide and niobium nitride which improves grain refinement, and retardation of recrystallization ultimately increases toughness, strength, formability, and weldability	Addition of Ti in steels passives to acids and minerals at low temperature and improves high temperature properties	Ni lowers the critical temperature of steels and retards the decomposition of austenite. As a result at low temperature or room temperature, austenite gets stable	In the alloying steels, chromium containing more than 5% improves the corrosion resistance and high temperature properties	The presence of manganese in the alloy steel reduces the prone to the hot shortness	Molybdenum is used in combination with Ni or Cr or both. Plain molybdenum steel is carburized to improve wear resistance	Tungsten is mainly popular for providing high temperature properties and hardenability of steels

Table 2.
Basic comparison of different types of micro-alloyed steels.

10. Molybdenum steels

Molybdenum is a little expensive alloying element. It has limited solubility in austenite and ferrite. As a result of which, it is a strong carbide former. Molybdenum is used in combination with Ni or Cr or both. Plain molybdenum steel is carburized to improve wear resistance.

A lot of research has been done in the case of interphase precipitation. In matter molybdenum plays significant roles. Four steels were manufactured with identical composition, and Ti, V, Mo, and N content is added to investigate the effect of composition on interphase precipitation. Alloys were rapidly cooled from the single austenite phase field and isothermally transformed at 630°C and 650°C for 90 min. When Mo is added, then there is a significant reduction in the austenite to ferrite transformation kinetics, particularly in the case of V steels. Interphase precipitation was observed in all alloys at both transformation temperatures. In the case of the Ti-bearing steel, two types of precipitate were observed, namely, TiC (finer) and Ti_2C (coarser), while for the V-bearing steels, VC (finer) and V_4C_3 (coarser) were observed. Where Mo was present in the alloy, it was found dissolved in all carbide types. The (Ti,Mo)C and (V,Mo)C were formed by classical planer interphase precipitation (PIP), while the $(Ti,Mo)_2C$ and $(V,Mo)_4C_3$, which had a much wider row spacing, were formed through curved interphase precipitation (CIP). Each adopted one variant of the Baker-Nutting orientation relationship. The Ti-micro-alloyed steels undergo the smallest precipitates of all the steels, which were approximately the same size irrespective of whether Mo was present in the alloy and irrespective of the transformation temperature. However, the addition of Mo to the V-bearing steels causes significant increase in precipitate volume fraction and a reduction in precipitate size.

11. Tungsten steels

Tungsten is mainly popular for providing high temperature properties and hardenability. It is mainly a carbide former. Approximately 2–3% W is equivalent to 1% Mo. Tungsten is mainly used in the tools industry (**Table 2**).

For achieving high strength and toughness, fine grain structure is essential in steels. To produce such microstructure, a carefully controlled high temperature processing of steels must be done. Hot working alone cannot refine the coarse or nonuniform grain. For example, grain coarsening behavior of laboratory heats of C-Si-Mn base steels varies with the concentration of Al, V, Ti, or Nb micro-alloy addition. Thus, steels containing the very insoluble TiN coarsen at much higher temperatures than steels containing the more soluble VCN.

The main strengthening mechanisms of micro-alloyed steels are grain refinement and precipitation [12]. It can be done by high temperature-controlled process and by adding proper alloying elements. Nowadays an economical alternative of the traditional quenched and tempered steels is micro-alloyed steels.

12. Conclusion

Strengthening mechanism can be done by precipitation forming and grain refining. Micro-alloy element hinders grain growth that causes grain refinement [13]. Precipitates forming on ferrite or austenite cause improvement of hardening or strengthening of steel. Phase transformation in some cases also causes strengthening of steels. In phase transformation, different micro-alloying elements appear to contribute considerably. Strengthening of steels can be done by different heat treatment techniques as well in addition to alloying. Besides, the most economical alternative way of improving the mechanical properties of steel is adding smaller amounts of some special elements.

Author details

Fahim Khan[1] and Hossain M. M. A. Rashed[2*]

1 Department of Materials Science and Engineering, Khulna University of Engineering and Technology, Khulna, Bangladesh

2 Department of Materials and Metallurgical Engineering, Bangladesh University of Engineering and Technology, Dhaka, Bangladesh

*Address all correspondence to: hrashed@mme.buet.ac.bd

References

[1] Abbaschian R, Abbaschian L, Reed-Hill RE. Physical Metallurgy Principles. 4th ed. Stamford: Cengage Learning; 1994

[2] Avner SH. Introduction to Physical Metallurgy. New York: MAcGraw-Hill Book Company; 1974

[3] Bain EC, Paxoton HW. Alloying Element in Steel. Metals Park, Ohio: American Society for Metals; 1961

[4] Oliveira M, Calderon–Harnandez J, Magnabosco R, Hincapie D, Alonso–Falleiros N. Effect of niobium on phase transformations, mechanical properties and corrosion of supermartensitic stainless steel. Journal of Materials Engineering and Performance. 2016;**26**. DOI: 10.1007/s11665 017-2610-1

[5] Javaheri V. Effect of Niobium and Phase Transformation Temperature on the Microstructure and Texture of a Novel 0.40% C Thermomechanically Processed Steel. Finland: Elsevier; 2018. pp. 295-308

[6] Davis Joseph R. Metals Handbook Desk. 2nd ed. ASM International. 1998. ISBN: 978-0-87170-654-6

[7] Filho AI, Silva RV, da Silva Cardoso W, Casteletti LC, et al. Effect of niobium in the phase transformation and corrosion resistance of one austenitic-ferritic stainless steel. Materials Research. 2013;**17**(4):801-806. DOI: 10.1590/1516-1439.190113

[8] ASM Handbook. Properties and Selection: Nonferrous Alloys and Special-Purpose Materials. Vol. 2. ASM International. 1990

[9] Mao X. Titanium Microalloyed Steel: Fundamentals, Technology, and Products. Beijing, China: Metallurgical Industry Press; 2019

[10] Massalski TB. Binary Alloy Phase Diagrams. 2nd ed. Vol. 3. Ohio, USA: ASM International Materials Park; December 1990

[11] Yong Q. Physical Metallurgical Principles of Titanium Microalloyed Steel—Dissolution and Precipitation of Titanium-Bearing Secondary Phases. China: Springer; pp. 71-139

[12] Leslie WC, Rickett RL, Dotson CL, Walton CS. Influence of Al content on the corrosion resistance of micro-alloyed hot rolled steel as a function of grain size. Transactions of the American Society of Metals. 1954;**46**:1470

[13] Joseph RD. Metals Handbook Desk Edition. 2nd ed. ASM International. 1998. ISBN: 978-0-87170-654-6

Section 2

Advanced Lightweight Steels

Chapter 4

Technologies of High-Temperature Insulating Coatings on Stainless Steels

Zoia Duriagina, Taras Kovbasyuk, Volodymyr Kulyk, Andriy Trostianchyn and Tetiana Tepla

Abstract

For the use of chromium steels in instrumentation, microelectronics, and electrical engineering, their surfaces are additionally protected by coatings based on glass ceramics and other insulating materials. Such materials can operate at high temperatures for a long time under the influence of the electric current or magnetic field. This chapter describes the research results on synthesized coatings based on oxide glass ceramics and oxide materials obtained by plasma chemical vapor deposition (CVD) on the surfaces of stainless steels.

Keywords: insulating coatings, glass-ceramic coatings, oxide materials, stainless steel substrate

1. Introduction

It is known that the corrosion-mechanical and functional properties of steels can be improved by alloying elements with a more negative electrode potential (chromium, silicon, aluminum) or elements with a higher potential than iron (e.g., copper, molybdenum, nickel) or elements that promote the formation of carbides, nitrides, and other phases.

The second way (which can occur simultaneously with the first) is to use various methods of surface engineering, in particular, the formation of a dense oxide film, which protects the surface of the alloy from oxidation and dissolution or acts as a functional coating, e.g., a dielectric one. To do this, you need to have the appropriate concentration of the alloying element. To ensure the proper level of functional properties, such coatings must be without defects, with high adhesion to the surface and minimal difference in thermal expansion coefficients as compared to the stainless steel substrate. The condition of formation of such coatings is a certain ratio of the volumes of oxide (V_{MeO}) and the oxidizing metal (V_{Me}): $V_{MeO}/V_{Me} > 1$. Under such a condition, as a rule, coatings are formed that improve the functional properties of the metal and slow down both the degradation process of its microstructure and deterioration of the properties.

As is known, stainless steels are obtained by introducing into the composition of low- and medium-carbon steels at least 12.5% chromium as well as nickel and other alloying elements (titanium, aluminum, molybdenum, niobium, copper, manganese). Depending on the structure, austenitic (AISI 302, 347, 316L, 316, 316Ti, and others), ferritic (AISI 430, 430Ti, 439, and others), martensitic (AISI 420, 431, 420F), austenitic-martensitic (AISI 631, AM350), and austenitic-ferritic (AISI 301) stainless steels are distinguished. Chromium ferritic and chromium-nickel austenitic steels are the most common in engineering practice [1].

Chromium ferritic stainless steels are the cheapest, but they are inferior to chromium-nickel steels in corrosion resistance. So, AISI 430 steel is stable in acidic environments, but not suitable for welding, because during welding when heated above 900–950°C and rapidly cooled, the grain boundaries become depleted in chromium. With content less than 12% Cr, the electrochemical potential of the steel becomes negative, and it loses its ability to resist corrosion. Under these conditions, AISI 430 steel has a dangerous type of corrosion—intercrystalline corrosion. Stabilization of this steel with titanium or niobium is used to prevent this. It should be noted that σ phase (Fe, Cr) can be formed in high-chromium steels as a result of delamination of the alloyed solid solution into a mixture of $\alpha + \alpha'$ phases (where α' is σ phase). Due to this, the corrosion resistance of such steels is reduced. In ferritic steels with 20% Cr, the minimum formation time of the σ phase at 600°C is 150 h, and in the steels with 25% Cr at 650°C, it is 15 h. Therefore, when using these steels, it is desirable to bypass this temperature range.

2. Features of synthesis of glass-ceramic coatings on stainless steels

Ferritic and ferritic-martensitic stainless steels are widely used in electrical engineering as heat carriers and metal substrates for the manufacturing of film heating elements (FHE) (**Figure 1**). To insulate the metal substrate from the action of electric current, a dielectric insulation coating is applied to its surface. There are a large number of materials that can be used as electrical insulators at high temperatures. But for the correct choice of technology of a quality coating on the surface of stainless steel, it is necessary to take into account the coefficient of linear thermal expansion (CLTE) and the absence of phase transformations in the structure of the coating material during operation [2].

Due to the compatibility of thermal, physical, and micromechanical properties of stainless steels to the protective coatings, a material based on the fusible glass ceramics of the ZnO-PbO-B_2O_3 system was proposed as a coating material [3–15].

For the preparation of coatings based on the PbO-ZnO-B_2O_3 glass-crystal system, mixtures of powders, the compositions of which are shown in **Table 1**, were used. Each mixture was poured into an electrocorundum crucible and melted

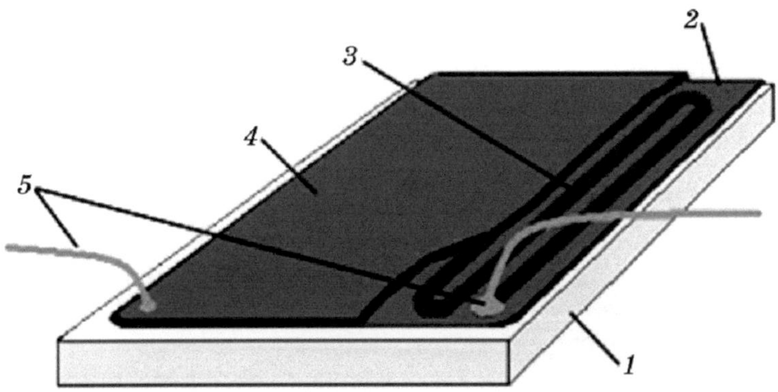

Figure 1.
Schematic diagram of a film heating element [2]: (1) stainless steel substrate, (2) dielectric layer, (3) resistive element, (4) protective shell, and (5) contact electrodes with current collectors.

Marking	Chemical composition, wt%					
	PbO	ZnO	B_2O_3	SiO_2	Al_2O_3	BaO
SC 100-1	75.5	12.0	8.4	2.1	2.0	–
SC 90-1	75.3	11.6	8.5	2.1	0.8	1.7
SC 88	75.1	11.2	9.4	1.9	–	1.9
SC 90	75.3	11.6	8.5	2.1	–	2.5

Table 1.
Chemical compositions of powders for the synthesis of the PbO-ZnO-B_2O_3 glass-ceramic system-based coatings [2].

at 1180°C. After holding at this temperature for 60 min, the melt was rapidly cooled to form an amorphous structure and prevent crystallization. The dried granules were ground and sieved to obtain powder fractions with a granule average size of not more than 56 μm. On the basis of the powder obtained, a dielectric paste was made, which was applied to a prepared surface of AISI 420 steel samples and dried at 70°C. For all specimens, standard heat treatment [16–20] was performed with two-step annealing at 380 and 440°C and holding at these temperatures for 45 min. The thermal treatment of the coatings was carried out without a protective atmosphere to activate the formation of oxides.

The coatings were applied to the surface of AISI 420 stainless steel. To determine the adhesion properties, the steel surfaces were treated with three different methods, which allowed obtaining different indices of surface roughness: S1, machine grinding; S2, electrolytic etching; S3, manual grinding [3].

Using the method of interference profilometry, surface profilograms of AISI 420 stainless steel substrates were obtained after various surface treatment methods. From the obtained profilograms using the software "Micron-Gamma," the basic parameters of the roughness of the substrates were calculated, the values of which are given in **Table 2**.

The surface of AISI 420 stainless steel with roughness S1 and S3 is characterized by a nonuniform row structure with gaps and projections of the height of 4.6–5.2 μm (**Figure 2**). The surface of AISI 420 stainless steel with roughness S3 after electrolytic etching has a more uniform structure. The height of the needle projections is uniform and averages 5.3 μm.

Traditional methods do not allow qualitative estimation of quantitative indices of adhesion strength of coatings obtained by the method of thick films, the thickness of which exceeds 100 μm. Because of this, indirect methods of investigation can be used to predict the adhesion strength of such coatings, in particular, the determination of the free surface energy of the substrates on which the coatings are synthesized (**Figure 3**). It is known that the level of surface energy will specify the adhesion properties. This technique allows determining the level of the adhesion

Marking	Arithmetical mean deviation of the assessed profile Ra, μm	Average distance based on the ten highest peaks and lowest valleys over the entire sampling length R_Z, μm	Maximum height of the profile R_{max}, μm	Average step of peaks along the mean line S_m, μm
AISI 420—S1	1.11	5.11	4.68	1.02
AISI 420—S2	1.31	5.15	5.36	1.03
AISI 420—S3	1.11	4.00	5.21	1.00

Table 2.
Surface roughness parameters of the AISI 420 steel substrates after various types of grinding operations [3].

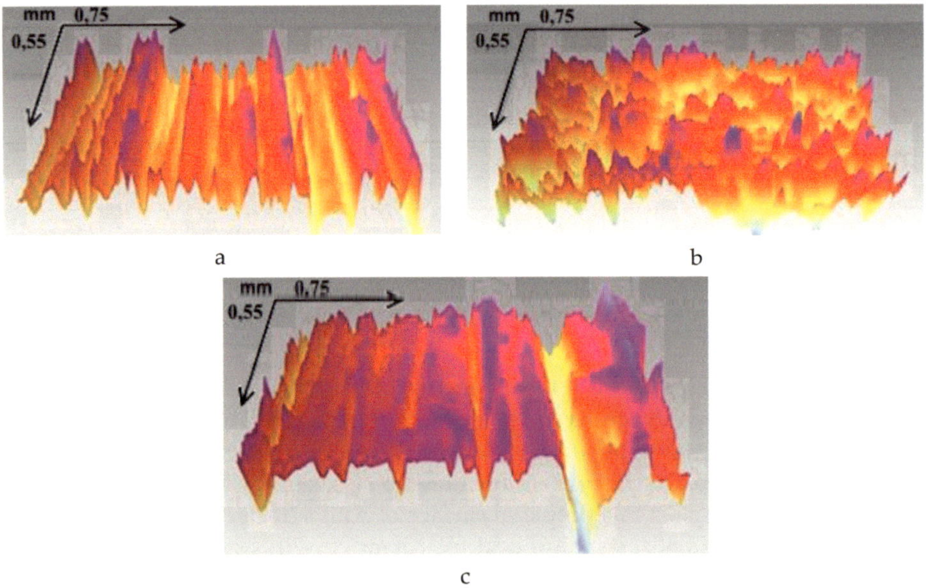

Figure 2.
Three-dimensional surface microtopography of AISI 420 steel substrates after various types of grinding operations: (a) machine grinding S1; (b) electrolytic etching S2; (c) manual grinding S3 [3].

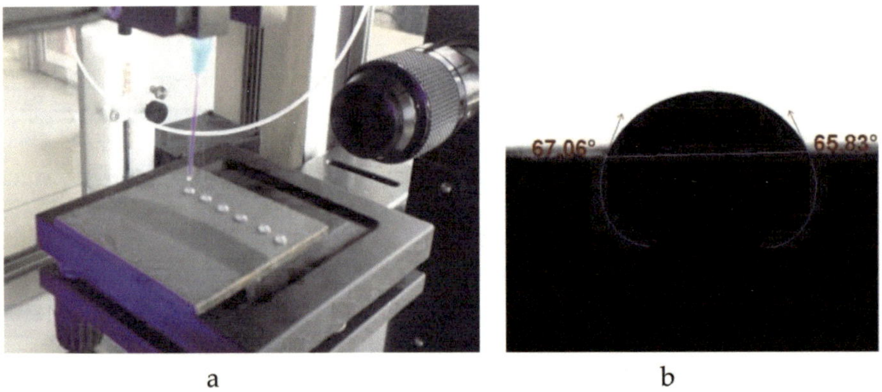

Figure 3.
General view of the desktop of the KSV Attension Theta optical tensiometer (a) and the droplet along with the calculated wetting angles (b) [3].

strength in the coating-substrate system if the actual contact area between them is known [21–24].

To predict the adhesion strength of the dielectric coating to the substrate, the surface topography of the prepared substrates was investigated, and the values of the free surface energy were calculated for them. Optical tensiometry was used to calculate the free surface energy with the measurement of wetting angles between the test surface and droplets of substances: water, glycerol, formamide, ethylene glycol, diiodomethane, and bromonaphthalene [3]. **Figure 4** shows the change in the wetting angles for the substrates after various surface treatments, depending on the type of drip fraction of the substance applied to the surface.

The substrates S1 and S2 have the highest average values of the wetting angles. This testifies to the dependence of the wetting angle not only on the physicochemical properties of the droplet but also on the roughness and microtopography of the surface.

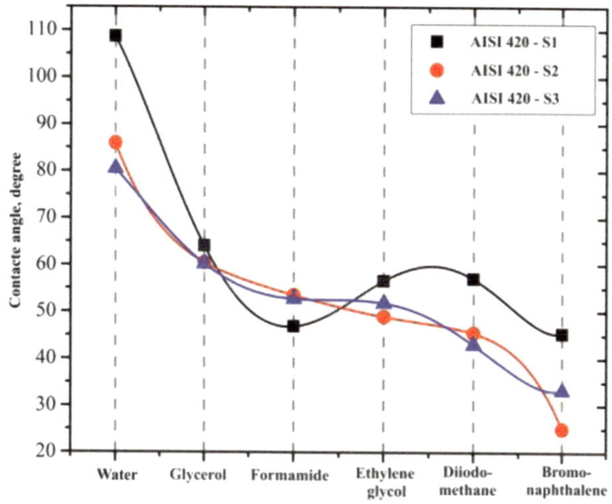

Figure 4.
Values of wetting angles for AISI 420 stainless steel substrates after machine grinding S1, electrolytic etching S2, and manual grinding S3 depending on the substance of the droplet.

By changing the values of the wetting angle using two calculation methods, the values of the free surface energy components for the substrates S1, S2, and S3 were determined (**Table 3**). The first calculation method [25] gave a large variation of the experimental data, which makes it difficult to predict the physical and mechanical properties of the surface being studied. According to the second calculation method [26], the substrates S1 and S2 have the highest values of the free surface energy. This is confirmed by the fact that the actual contact surface area increases with increasing roughness. This, in turn, will increase the free surface energy indices, which will be reflected in the adsorption component of adhesion.

Thus, the formation of the developed surface structure of AISI 420 stainless steel substrates obtained by the surface etching method will provide the maximum level of the free surface energy by forming the largest contact area between the applied dielectric layer and the substrate, which will provide the best level of adhesion uniformity [3].

To obtain a high-quality insulation coating on the surface of stainless steels, it is necessary to ensure the maximum homogeneity of its structure in both the thickness and surface area. The surface roughness of such coatings must be within *Rz* = 1–2 μm to provide the required adhesion strength between dielectric and resistive

Constituents of surface energy		Lifshitz-van der Waals energy γ^{LW} (mJ/m^2)	Energy of the electron acceptors γ^+ (mJ/m^2)	Energy of the electron donors γ^- (mJ/m^2)	Total surface energy γ^{tot} (mJ/m^2)
S1	Method 1	32.3	0.2	2.7	34.2
	Method 2	34.8	1.4	4.8	39.9
S2	Method 1	39.2	1.2	0.7	41.0
	Method 2	36.9	2.5	1.1	40.2
S2	Method 1	38.2	2.4	0.6	40.6
	Method 2	34.9	7.4	0.4	38.2

Table 3.
Values of the free surface energy constituents for the substrates S1, S2, and S3 and coatings based on the PbO-ZnO-B_2O_3 glass crystalline system [3].

layers. It should be noted that the glass-ceramic coating should have a minimum porosity to obtain perfect electrophysical characteristics, in particular, the breakdown voltage and electrical strength.

After synthesis and heat treatment, the obtained glass-ceramic coatings have a dark gray color with shades of green and are smooth to the touch. Small projections and gaps are observed along the entire area of the formed coating. No defects in the form of pores, gas bubbles, or residues of non-smelted sealant (SC, sital cement) powder have been detected. The thickness of the coatings ranges from 90 to 105 μm.

To evaluate the microgeometry parameters of the surface structure for each synthesized coating, the surface topography was investigated in five sections of the surface with a section area of 0.55 × 0.75 mm^2. Surface profilograms of the coatings were obtained using the interference profilometer (**Figure 5**).

Structural and geometric parameters of the surface roughness of the coatings, the values of which are given in **Table 4**, were calculated from the obtained profilograms using the Micron-Gamma software. According to **Table 4**, the values of *Rz* parameter of the surface roughness are 2.11, 1.87, 1.63, and 2.14 μm for the investigated coatings SC 100-1, SC 90-1, SC 88, and SC 90, respectively.

The study of three-dimensional topography of the surface of the formed coatings by the method of interference profilometry testifies to the presence of identical fragments of their structure, regardless of the microtopography of the substrate surface (**Figure 6**).

Coatings based on powders SC 100-1, SC 90-1, SC 88, and SC 90 have a homogeneous structure with shallow round gaps and needle projections of height 0.8–1.1 μm.

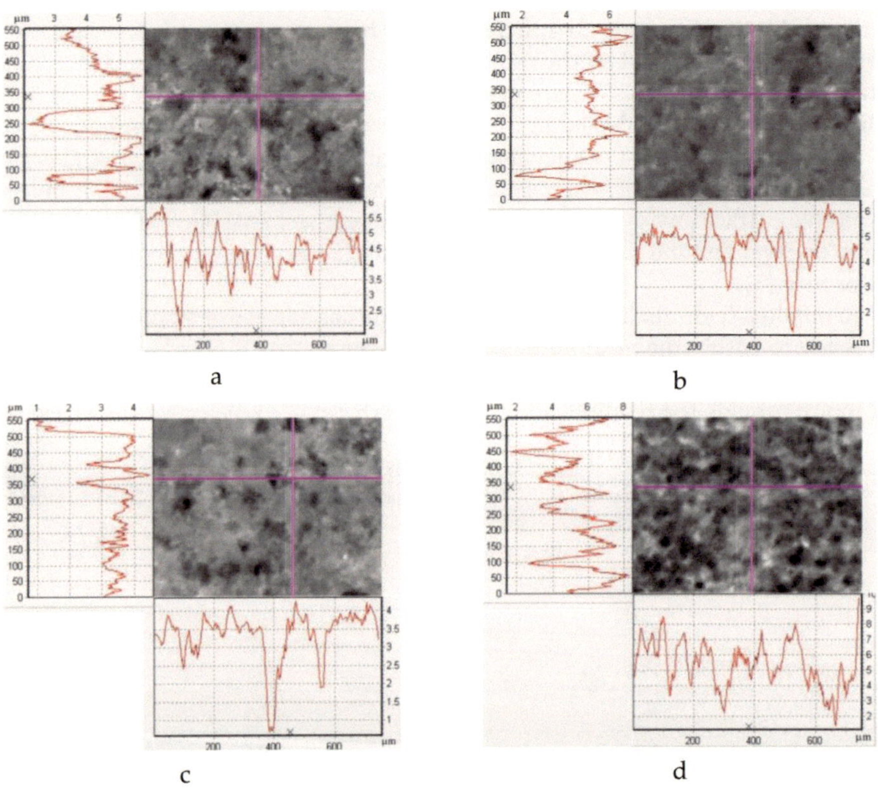

Figure 5.
Surface profilograms of synthesized functional coatings based on glass crystalline materials of grades (a) SC 100-1, (b) SC 90-1, (c) SC 88, and (d) SC 90 on AISI 420 stainless steel substrates [3].

Marking	Arithmetical mean deviation of the assessed profile *Ra*, μm	Average distance based on the 10 highest peaks and lowest valleys over the entire sampling length R_Z, μm	Maximum height of the profile R_{max}, μm	Average step of peaks along the mean line S_m, μm
SC 100-1	0.55	2.11	2.29	1.01
SC 90-1	0.49	1.87	2.21	1.01
SC 88	0.41	1.63	1.82	1.00
SC 90	0.57	2.14	2.29	1.01

Table 4.
Parameters of surface roughness of coatings SC 100-1, SC 90-1, SC 88, and SC 90 [3].

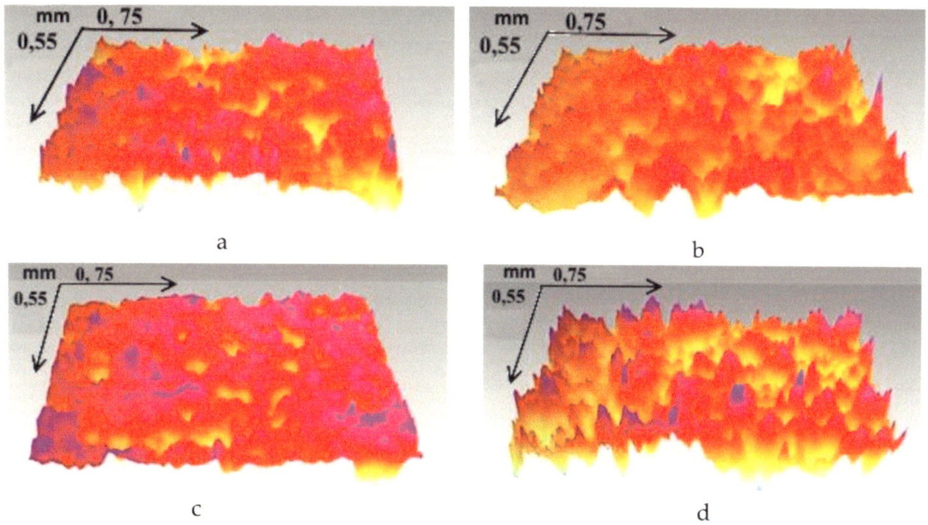

Figure 6.
Three-dimensional microtopography of the surfaces of functional coatings based on glass crystalline materials of grades (a) SC 100-1, (b) SC 90-1, (c) SC 88, and (d) SC 90 on AISI 420 stainless steel substrates [3].

Such surface microtopography of the dielectric coating will guarantee correspondingly high adhesion strength of the applied resistive layer [1–3].

2.1 Microstructure of functional coatings based on glass crystalline materials

The dendritic component has been revealed in the surface layers of the dielectric coatings using a scanning electron microscope (**Figure 7**). This is a positive feature because it is the crystalline component of the microstructure of the $PbO\text{-}ZnO\text{-}B_2O_3$ system coating that is responsible for the proper functional properties of the insulation layer [4]. The largest dendritic crystals are observed in the structure of SC 100-1 and SC 90 coatings. As the coatings of SC 100-1 and SC 90 grades have the highest surface roughness compared to the other two coating grades, it can be assumed that the growth rate of dendritic crystals will affect the final surface roughness parameters of the coatings.

It should be noted that the presence of an amorphous phase in the amount of more than 15% will cause a significant deterioration of the micromechanical and thermophysical characteristics of coatings while increasing the dielectric strength [4].

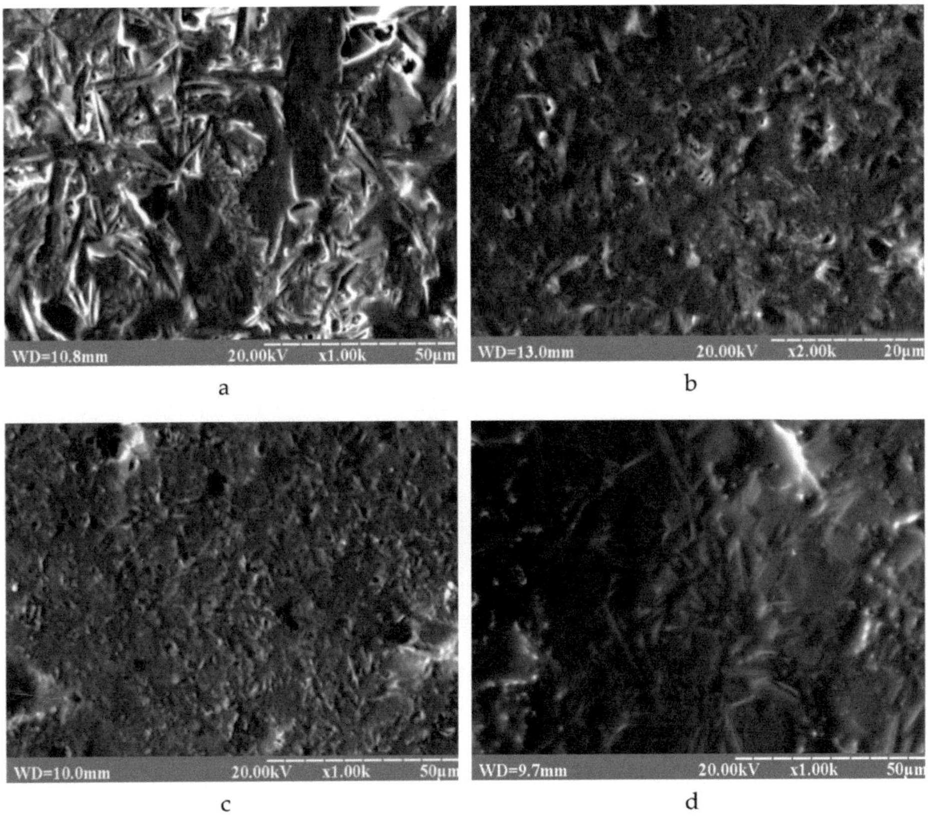

Figure 7.
Surface microstructure of dielectric coatings based on glass crystalline materials of grades (a) SC 100-1, (b) SC 90-1, (c) SC 88, and (d) SC 90 [4].

To evaluate the porosity of coatings based on glass crystalline materials of SC 100-1 and SC 90-1 grades, their microstructure after polishing was investigated (**Figure 8**). They are found to contain a large number of pores of different sizes, unevenly distributed in the bulk of the coatings. The formation of pores can be explained by the uncontrolled process of sublimation of organic solvents (butyl acetate, amyl acetate) during synthesis at temperatures of 250–350°C.

Quantitative analysis of the porosity of synthesized coatings based on SC 90-1 and SC 100-1 powders was carried out (**Table 5**). It was found out that the volume

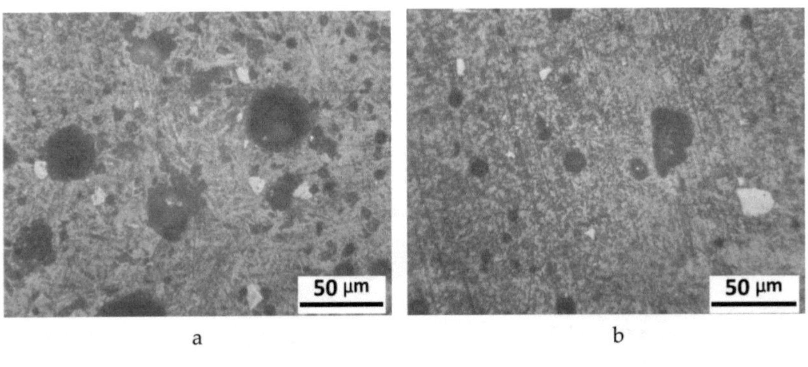

Figure 8.
The microstructure of coatings based on glass crystalline materials of grades (a) SC 100-1 and (b) SC 90-1 after polishing [4].

Parameter	Coating based on SC 90-1 powder	Coating based on SC 100-1 powder
Number of pores per field of view	157	88
Volume fraction of pores, %	4.1	19.7
Specific surface of pores	0.031	0.047
Mean chord length of pores, μm	5.3	16.9
Average distance between pores, μm	126.5	69.1
Fractal dimension	1.07	1.56
Form factor	0.867	0.825
Compact factor	0.989	0.935
Stretching factor	0.815	0.734
Cutting factor of the contour	0.867	0.859

Table 5.
Quantitative analysis of the porosity of coatings SC 100-1 and SC 90-1.

fraction of pores in the structure of the SC 90-1-based glass-ceramic material coating is 4.1%, in contrast to the coating SC 100-1, the average porosity of which is 19.7%.

The difference in the structure of the coatings can be explained by the alloying of the $PbO-ZnO-B_2O_3$ system with BaO oxide in quantities up to 1.7%, which reduces the temperature of formation of the amorphous glass material in the synthesis stage, which in turn causes acceleration of the sublimation processes of organic solvents and accelerates the crystallization process [4].

The microstructural analysis of the polished transverse sections of the synthesized coatings on AISI 420 steel substrates was carried out. Using energy dispersive X-ray spectroscopy (EDS), it was found that no diffusion or transition zones are formed between the applied coatings and the substrates. This is evidenced by a sharp drop in the content of lead in the place where the coating is bonded to the substrate (**Figure 9**). This indicates a significant influence of electrochemical processes (double electric layer) during the formation of a strong adhesion of the synthesized glass crystalline coatings of the $PbO-ZnO-B_2O_3$ system with the substrate surfaces.

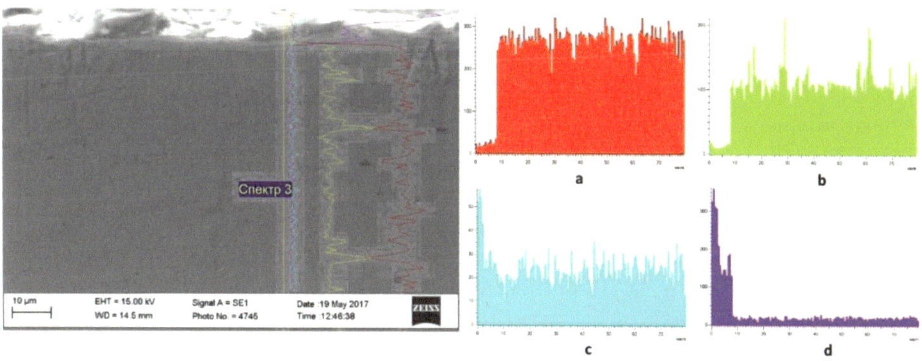

Figure 9.
The local EDS microanalysis of the coating based on the glass crystalline material of SC 90-1 grade by the depth of the layer showing the distribution of elements (a) Fe, (b) Cr, (c) Si, and (d) Pb.

2.2 X-ray structural studies of functional coatings based on glass crystalline materials

The diffraction pattern of the coating based on the SC 100-1 powder (**Figure 10**) confirms the fact of the formation of a crystalline structure during synthesis. It shows pronounced peaks corresponding to lead and zinc as pure chemical elements. The shape and height of the peaks indicate the presence of crystalline phases in the volume of the coating.

To determine the chemical homogeneity of the coatings, a local EDS micro-analysis was performed. It was detected lead, zinc, and boron oxides in all types of synthesized coatings in various proportions depending on the location of the analysis.

In the coatings based on SC 90-1, SC 88, and SC 90 powders, BaO and SiO_2 oxides were found. Due to the low accuracy of the EDS method when detecting light elements, including boron oxide, the volume fraction of each of the oxides in the coating structure could not be estimated correctly.

X-ray photoelectron spectroscopy (XPS) was used to determine the qualitative and quantitative chemical composition of coatings, as it allows doing this more accurately and detecting all chemical elements except hydrogen and helium. This technique is based on obtaining XPS spectra by irradiating a material with a beam of X-rays (Al Kα) while simultaneously measuring the kinetic energy and number of electrons that escape from the top 0 to 10 nm of the coating being analyzed. To improve the reliability of the results obtained, the evaluation of the chemical composition was performed in three different areas on the sample surfaces. According to the investigation results, experimental diffraction patterns were constructed (**Figure 11**).

It was revealed that the elemental compositions of the synthesized coatings correspond to the compositions of the original powders within the error of the device (2%). As an example, the chemical composition of the synthesized coating based on sital cement of SC 90-1 grade is as follows: 75.3 wt% PbO, 11.6 wt% ZnO, 8.5 wt% B_2O_3, 2.1 wt% SiO_2, 0.8 wt% Al_2O_3, and 1.7 wt% BaO. Besides, unwanted phases of 0.15% $BaSO_4$, 1.02 % (ZnS + ZnF_2), and 0.28% B_4C, which could be formed during prolonged isothermal holding and interaction of powder particles with residues of nitrocellulose lacquer, were detected. To avoid the formation of these phases, it is necessary to increase the annealing temperature to accelerate the crystallization process and simultaneously reduce the holding time. However, it

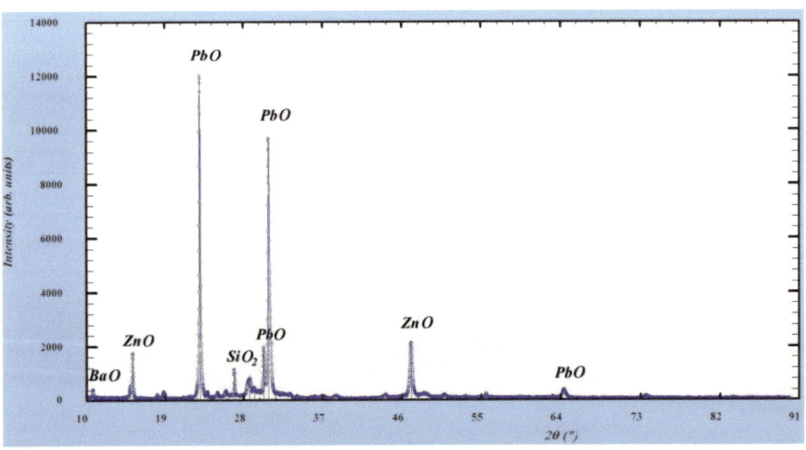

Figure 10.
X-ray pattern of the surface of the coating based on the glass crystalline material of SC 100-1 grade.

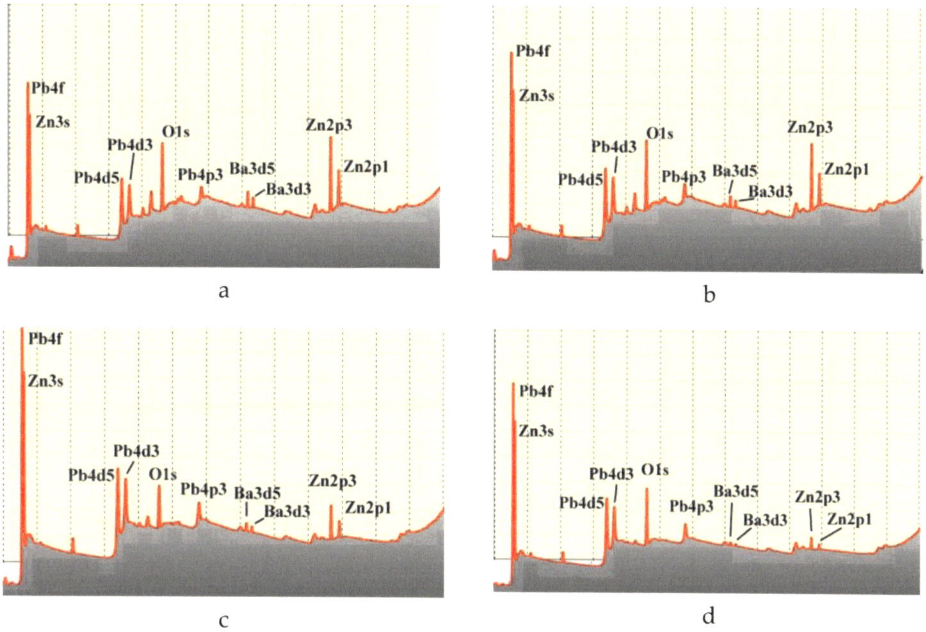

Figure 11.
Experimental X-ray patterns of the surfaces of coatings based on glass crystalline materials of grades (a) SC 100-1, (b) SC 90-1, (c) SC 88, and (d) SC 90 obtained by X-ray photoelectron spectroscopy.

should be remembered that the final crystallization temperature should not exceed the glass spill temperature more than two times (in our case up to 520°C).

3. Formation of insulating coatings by plasma chemical vapor deposition (CVD)

The ion-plasma sputtering as one of the CVD technique methods was chosen as an alternative method of synthesizing protective insulating coatings on stainless steels. As materials for the sputtering, it is advisable to use pure compounds based on nitrides and oxides of metals, which have a wide range of functional properties [27–31].

The most optimal materials for application to the surfaces of ferritic and ferritic-martensitic stainless steels are coatings of magnesia (MgO), alumina (Al_2O_3), titanium dioxide (TiO_2), and aluminum nitride (AlN) (**Table 6**). It should be noted that the thermal conductivity of beryllium oxide (BeO) is nine times higher than of the listed materials. Thus, it can significantly increase the power of FHE on stainless steel substrates. However, the use of BeO is hampered by its high cost and toxicity during processing. Traditionally, this or that type of ceramics is obtained by sintering oxide or nitride powders at various temperatures. However, high-energy methods of surface engineering are promising when forming dielectric layers on the surfaces of various structural materials. The properties of the layers synthesized in this way, which are given in various literary sources, differ significantly and depend on a method used to form these layers.

Magnesium oxide is the only oxygen compound of magnesium. It exists in only one modification and crystallizes in a cubic system. The crystalline form of MgO is called periclase. It has a potassium salt lattice with a parameter of 0.42 nm. By its chemical properties, MgO is a basic oxide, and, as a consequence, it combines with

Parameter	MgO	AlN	Al_2O_3	TiO_2
Density, g/cm^3	3.58	3.2–3.24	3.6	3.9–4.3
Dielectric constant	8–9	8.5	12.0	31–173 (114 for rutile)
Relative density	0.95–0.96	0.96–0.98	0.94–0.98	-
Melting point, °C	2800	2400	2050	1855
Operating temperature range for an atmosphere, °C	Up to 2000 in air	1300–1400	1600–1700	-
Sintering temperature, °C	1600	-	1700–1750	
Coefficient of linear thermal expansion α, $\times 10^6$ K^{-1}	11.7–14.2	4.8	5-6.7	9.8–10.8
Thermal conductivity, W m^{-1} K^{-1}	28.0	66	24–29	
Modification	Cubic a = 0.42 nm	Hexagonal a = 0.31–0.313 nm c = 0.493–0.498 nm	α-Trigonal β-Hexagonal γ-Cubic	Anatase Brookite Rutile

Table 6.
Physical properties of materials of the dielectric layer.

all acidic oxides and dissolves in inorganic acids (partially in water). Hydration of MgO limits the possibility of its fine fragmentation in water, as it increases greatly with increasing both fragmentation and temperature. MgO sintering is facilitated by TiO_2, ZrO_2, Al_2O_3, and Fe_2O_3 impurities. Magnesium oxide is a good insulator. Its crystals have ionic conductivity. The work [32] presents the study of the electrical properties of magnesium oxide layers formed by magnetron sputtering of magnesium followed by annealing in air. It is found out that, depending on the annealing temperature, the layer resistivity ranges from 1.7×10^7 to 2.81×10^{12} mOhm cm. The work [33] presents the study of MgO layers formed by the sol-gel method. The obtained results indicate the weak crystallinity and the disorientation of the magnesium oxide phase. The breakdown voltage ranges from 5 to 78 MW/cm. Leakage currents when heated to 250°C increase within 10^{-9}–10^{-3} A/cm^2. In the work [34], using the method of pulsed laser deposition, MgO polycrystalline films with various crystallographic orientations were obtained. The value of the dielectric constant was 9.67 for a density near the band gap of 4.5×10^{11} eV^{-1} cm^{-2}. In some works [32–35], magnesia ceramics are characterized by a dense fine-crystalline structure with perfect physical, mechanical, and electrophysical properties.

Another promising material widely used in radio and microelectronic technology is aluminum oxide [33–35]. It is known for its nine crystallographic modifications, the most important for the industry being α-modification. The corundum structure (**Figure 12**) can be considered as a hexagonal dense packing of O^{2-} ions in which 2/3 of the octahedral gaps are occupied by Al^{3+}.

The sintering temperature of α-Al_2O_3 of technical purity (99–99.5) and the fraction grade of 1–2 μm without additives are 1700–1750°C. At this temperature, a density of 3.75–3.85 g/cm^3 is reached. The porosity of the sintered corundum is mainly closed and internal, and the pore shape is circular. The size of initial Al_2O_3 particles has a decisive influence on the sintering temperature. The maximum size of Al_2O_3 crystallites capable of active solid phase sintering should not exceed 3–5 μm. By adding to Al_2O_3 powder some substances in the form of oxides or salts called mineralizers, it is possible to reduce the sintering temperature of corundum by 150–200°C, and the character of crystallization can become directional because the delay or growth of crystals in certain directions is ensured. Among the additives

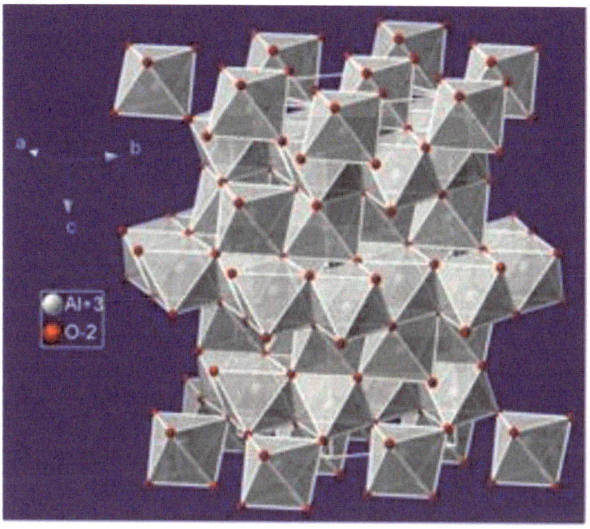

Figure 12.
Coordination polyhedra in α-Al_2O_3 structure [36].

that delay the growth of crystals, the oxides TiO_2, MnO_2, and Fe_2O_3 should be noted. The most effective is the addition of TiO_2, forming a solid solution and reducing the sintering temperature of corundum to 1500–1550°C. The introduction of the Mn^{4+} ion, whose ion radius is 0.052 nm, should also lead to the formation of an interstitial solid solution. Another group of additives affects the growth of crystals during annealing. The addition of such compounds results in crystal growth, sometimes very intense, and the sintering temperature can either decrease or remain unchanged. Thus, there are additives that simultaneously reduce the sintering temperature and affect the growth of crystals.

The formation of coatings has many technological difficulties related to the provision of the required stoichiometry, phase composition, crystalline structure, impurity level, porosity, etc. In the work [5], the influence of the parameters of the Al_2O_3 film forming process in high-frequency magnetron discharge on the structure, phase composition, porosity, and electrical strength of the films was investigated. It is shown that under appropriate conditions, multiphase polycrystalline coatings with stable α-structure are formed. Their electrical strength varies in the range of 0.4–1.6 V/m. The high chemical and thermal resistances of aluminum oxide make it possible to use it in the manufacture of heating devices.

Aluminum nitride is the only compound of nitrogen with aluminum that has high chemical resistance. It crystallizes to form a hexagonal lattice of the wurtzite type with parameters a = 0.31–0.313 nm and c = 0.493–0.498 nm. Aluminum nitride has no modification, which simplifies the technology of manufacturing products. At 1900–2000°C, AlN decomposes. AlN ceramics can be used in an inert environment up to 1800°C, in vacuum up to 1600°C, in air up to 1300–1400°C. The high thermal conductivity (140–280 W m^{-1} K^{-1}), which is a characteristic of aluminum nitride, and the heat resistance make these ceramics promising for use in conditions with a sharp change in temperature. In particular, the thermophysical properties of aluminum nitride ceramics were investigated in. The heat capacity of C_p was measured using the standard adiabatic calorimeter method. The thermal conductivity χ was determined by the absolute stationary longitudinal heat flux method. It was found that at room temperature C_p = 25 J mol^{-1} K^{-1}, χ = 123 W m^{-1} K^{-1}. The CLTE was $(2.9–3.4) \times 10^{-6}$ K^{-1}. Aluminum nitride has also been used as a protective coating for electronic components under the influence of percussion mechanical loads, which is associated with its high hardness (7–8 on the Mohs scale) [2, 5].

Ceramics, the crystalline base of which is titanium dioxide, are considered as a separate grade of technical ceramics since this compound has a high value of relative permittivity as compared to other ceramic materials (**Table 6**). Titanium oxide films, due to their physicochemical properties, are widely used as protective and optically sensitive coatings in gas sensors and during photocatalysis. They are also used in microelectronics (dynamic memory, field-effect transistors, ferromagnetic materials). Besides, TiO_2 is used in electronic engineering, in particular for the manufacture of capacitors. Titanium dioxide does not occur in nature in pure form. It is obtained by chemical processing of titanium ores $FeTiO_3$, $CaTiSiO_5$, $CaTiO_3$, and others. Titanium dioxide is available in three modifications: anatase, brookite, and rutile. Rutile is a stable high-temperature form of titanium dioxide, and the other two transforms irreversibly into this. Anatase modification of titanium dioxide is used as a catalyst and component of solar cells. Due to its high reflectance, it is used to protect spacecraft from the sun's radiation. For the ceramics industry, they produce a special brand of TiO_2 called "capacitor."

Thus, the properties of MgO, Al_2O_3, TiO_2, and AlN ceramics depend strongly on the method of their production and a set of certain parameters of the initial powders. Of the four dielectric materials presented, all to a greater or lesser extent satisfy the requirements relating to insulating layers of film heating elements based on a stainless steel substrate. Thus, titanium oxide has high relative permittivity, while aluminum nitride has high thermal conductivity. Both magnesium and aluminum oxides have satisfactory electrical strength at low CLTE. The combination of such dielectric properties with the verified technology of forming ceramic coatings on the surfaces of structural materials can serve effectively for the replacement of FHE constituents.

Among the high-energy surface engineering technologies, the ion-plasma sputtering method is the most promising for the formation of dielectric and resistive layers on the surfaces of stainless steels [37]. Therefore, the functional layers were formed by this method using a multifunctional ion-plasma discharge system. To create the dielectric layer by ion-plasma sputtering, a number of materials were analyzed that would provide the dielectric properties of the FHE and meet technological, economic, and environmental requirements. Oxides and nitrides of Mg, Ti, and Al, which corresponded to the abovementioned dielectric properties of FHEs [38], were selected for the application.

The stainless steel substrates were placed in the substrate holder of a reaction chamber at a distance of 40–50 cm from the cathode. The cathodes were made of aluminum of A1 99.7 grade (for AlN and Al_2O_3), magnesium of Mg 98 grade (for MgO), titanium of VT1-00 grade (for TiO_2), and Cr20Ni80-H nichrome alloy (for the resistive layer). They were manufactured in the form of rods with a diameter of 40 mm. The sputtering was performed under various modes in oxygen, nitrogen, or vacuum atmosphere. Each mode was characterized by certain parameters such as the gas pressure in the reaction chamber, the substrate potential, the current, and the sputtering time. To increase the adhesion strength of the layer, the substrate was heated using a furnace placed directly in the reaction chamber of the ion-plasma device. The plasma stream was directed from the cathode region toward the substrate. The beam divergence angle was about 20°. This provided optimum ion density in the plasma stream [37].

After the samples are uploaded, the reaction chamber of the hybrid ion-plasma device was evacuated. Then, using the helicon source in column mode, the final purification of the substrates in an argon plasma stream for 30 min at a pressure of 0.93 Pa, the potential on the substrate E = −100 V, and the current on the sample of

0.15 A was performed. During the purification, argon ions bombarded the substrate, removing residues of contaminants and impurities from its surface [1, 5].

The ion-plasma discharge device is designed to form nanoscale structure elements [32–37] and, due to its low operating temperatures, allows to apply functional layers to virtually any substrate (metal, plastic, ceramics, etc.). The plasma discharge chamber consists of three process modules (helicon source discharge chamber, drift chamber, treatment chamber), which are connected by flanges using vacuum seals. The scheme of the discharge plasma chamber is shown in **Figure 13**. A working chamber is connected directly to a flange of the vacuum system. A table with the substrate is located in this chamber. A drift chamber with docked vacuum arc accelerators is located above the working chamber. The drift chamber is connected to the helicon source discharge camera. Its top is covered by quartz glass, through which high-frequency energy is supplied during discharge using an antenna connected via a coordination unit to a high-frequency generator.

The magnetic coil located in the working chamber provides the plasma stream from the helicon source to the substrate. Besides, it makes it possible to direct the plasma stream from the plasma-arc accelerator toward the substrate, which significantly reduces the amount of transferred micro-droplet fraction that is formed when working with fusible materials (T_m < 2000 K).

For the formation of high-quality insulating and resistive layers, it is necessary to constantly monitor the state of the substrate surface in all stages of the technological process. Only in this case can the reproduction of the results, the stability of the process, and the high level of quality of coatings on the surfaces of stainless steels be ensured. When using standard high-vacuum equipment, the surface of the substrate always contains several adsorbed monolayers with residual components of the atmosphere. A plasma helicon source and plasma-arc accelerators ensure constant action of an argon ion flux on the substrate in all stages of the process. In this case, the density and energy of the ion flux are sufficient to effectively remove the adsorbed light residual components of the atmosphere from the substrate surface

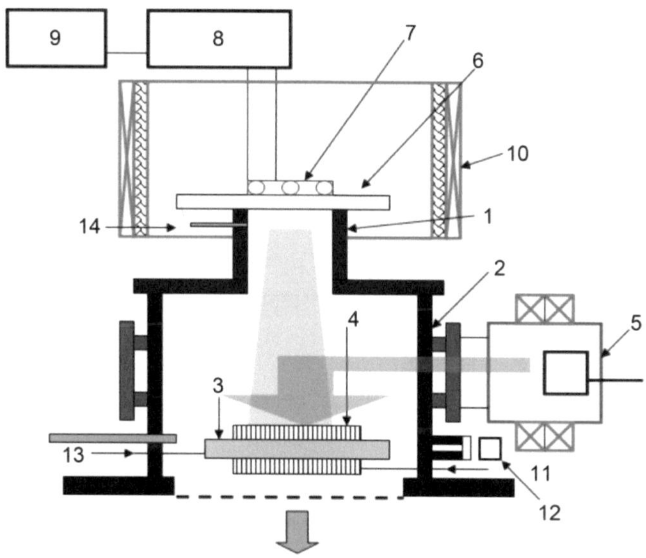

Figure 13.
The scheme of the ion-plasma discharge chamber [2, 37]: (1) discharge chamber, (2) drift chamber, (3) table, (4) substrate, (5) plasma-arc accelerator, (6) dielectric window, (7) antenna, (8) coordination unit, (9) HF generator, (10) permanent magnet or an electromagnet, (11) heating system, (12) optical spectrometer, (13) displacement probe, and (14) gas inlet.

layer by the selective sputtering mechanism and do not interfere with the basic technological operations [37].

The ion-plasma device consists of a high-frequency plasma source based on helicon discharge, for the implementation of the CVD process, and plasma-arc accelerators. The high degree of ionization of the plasma stream (more than 80 %) formed by the plasma-arc accelerator allows controlling the thickness of the layers during their application by the value of the total charge transferred to the substrate. **Figure 14** shows a general view of the discharge chamber of an ion-plasma device [2, 37].

Plasma helicon sources are non-electrode high-frequency induction magnetic field sources capable of generating dense (n = 10^{11}–10^{13} cm^{-3}) low-temperature (T_e = 2–10 eV) plasma over wide ranges of the operating gas pressure (P = 0.5–100 mTorr) and magnetic fields (B = 10–2000 Hs). These sources range in size from a few centimeters to several meters. They are excited by simple antennas over a wide frequency range (f = 7–100 MHz) and effectively generate plasma at relatively low specific HF power ($P_{rf} \geq$ 50–100 mW/cm^{-3}) [37].

In general, this method is not effective in applying dielectric and resistive layers under conditions of mass production of FHEs, because of the high cost of the device for magnetron sputtering; the small size of the camera, which does not allow sputtering on a large number of samples simultaneously; and the lack of quality cleaning of the working chamber when replacing sputtering cathodes (MgO, Al_2O_3, nichrome, etc.). The application of this method is also limited since the obtained dielectric layers have a high porosity, which increases the probability of electrical breakdown of the coating and failure of the heating device as a whole.

Figure 14.
General view of the discharge chamber of an ion-plasma device [2].

Functional dielectric coatings based on oxides of Mg, Ti, and Al for FHEs on stainless steels were obtained by ion-plasma sputtering and were investigated by various methods, namely:

- Study of peculiarities of formation of structure and phase composition of dielectric layers

- Study of adhesion and micromechanical properties

- Measurement of electrophysical characteristics of functional layers

The study of the surface morphology of the formed layers was carried out using an electron microscope JSM-6490LV (JEOL, Japan), equipped with an analytical setup for elemental analysis (spectrometer with energy dispersion INCA Energy+ Oxford). Determination of thickness and the structure and elemental composition analyses were performed using a scanning electron microscope with a PEMMA-102-02 microanalyzer. Surface topography was investigated using a Solver P47-PRO atomic force microscope.

The surface morphology of the formed dielectric layer with MgO is shown in **Figure 15**. The grain sizes varied from 20 to 70 nm. Grain growth is likely carried out by an islet mechanism when the formed grain becomes the basis for the formation of nuclei of grains and their subsequent growth (**Figure 15**a). Thus, the newly formed grains are combined into clusters of various sizes. Due to the specific structure, the grains are characterized by a large specific surface, which increases because of petals formed in the perpendicular direction (**Figure 15b**). The thickness of the identified petals is in the range of 10–20 nm. This mechanism of the layer structure formation and a significant difference between the sizes of the grain clusters cause a substantially high porosity of the dielectric layer, which can significantly increase its hygroscopicity [1].

To study the microtopography of the TiO_2 layer surface, SEM and local EDS analyses were performed (**Figure 16**). The surface of the formed layer is continuous and homogeneous, with a small porosity. In some parts of the surface layer, elliptical inclusions of size from 0.08 to 1 μm are observed (**Figure 16b**). Elemental analysis indicates that the bulk of the layer contains titanium and oxygen, the quantitative ratio of which corresponds to the TiO_2 compound of stoichiometric composition. Besides, impurity atoms of iron and aluminum were identified, the presence of which was caused by contamination [2].

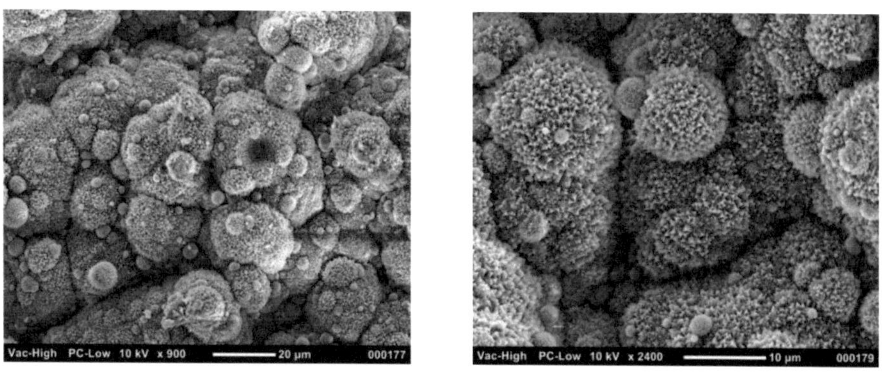

a b

Figure 15.
SEM surface microtopography of the MgO dielectric layer [2]. Magnifications: (a) 900× and (b) 2400×.

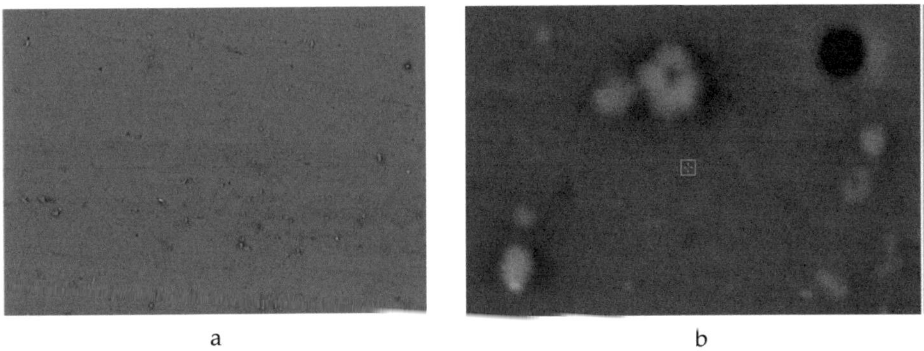

Figure 16.
SEM surface microtopography of the TiO_2 dielectric layer [2]. Magnifications: (a) 900× and (b) 2400×.

Figure 17.
SEM surface microtopography of the Al_2O_3 dielectric layer [2]. Magnifications: (a) 100× and (b) 1000×.

The Al_2O_3 layer was formed by ion-plasma deposition of aluminum onto a substrate of the AMg2 aluminum magnesium alloy in an oxygen atmosphere [5]. The result is a layer with a solid, smooth, visually nonporous surface topography (**Figure 17a**). The predominant orientation of individual elements of the surface structure in the form of projections of elongated shape was observed (**Figure 17b**) [1].

By studying SEM microstructure of polished transverse microsections by the depth of the oxide layers MgO, TiO_2, and Al_2O_3, it was revealed that all the layers except MgO have a gradient structure and consist of two sublayers (**Figure 18**). The inner Al_2O_3 sublayer, 1.5–2.5 μm thick, is the boundary section of the substrate layer system and has a highly dispersed structure with a grain size of 90–200 nm, and the outer Al_2O_3 sublayer of a thickness of 12.5–13.5 μm has a fragmented structure (**Figure 18a**). The outer TiO_2 sublayer, 12–13 μm thick, is almost defect-free, uniform, solid, and homogeneous (**Figure 18b**). It has a two-phase structure, with dispersed inclusions of a globular shape, which may belong to one of the TiO_2 modifications. The inner TiO_2 sublayer, 2–3 μm thick, adjoins the substrate surface and completely reproduces its morphology. By studying SEM microstructure of polished transverse microsections of MgO dielectric layers, it was revealed that their thickness ranges within 25–65 μm. MgO layer continuity disturbances were observed only along the boundaries of the separation of large clusters. At higher magnifications, MgO dendritic crystallites were identified, which are oriented in the direction perpendicular to the substrate surface (**Figure 18c**).

The MgO layer has a developed surface with well-defined elongated grains, which demonstrates the growth direction of individual grains. Based on the surface microtopography of the MgO layer (**Figure 19a**), it is revealed that there exist grain

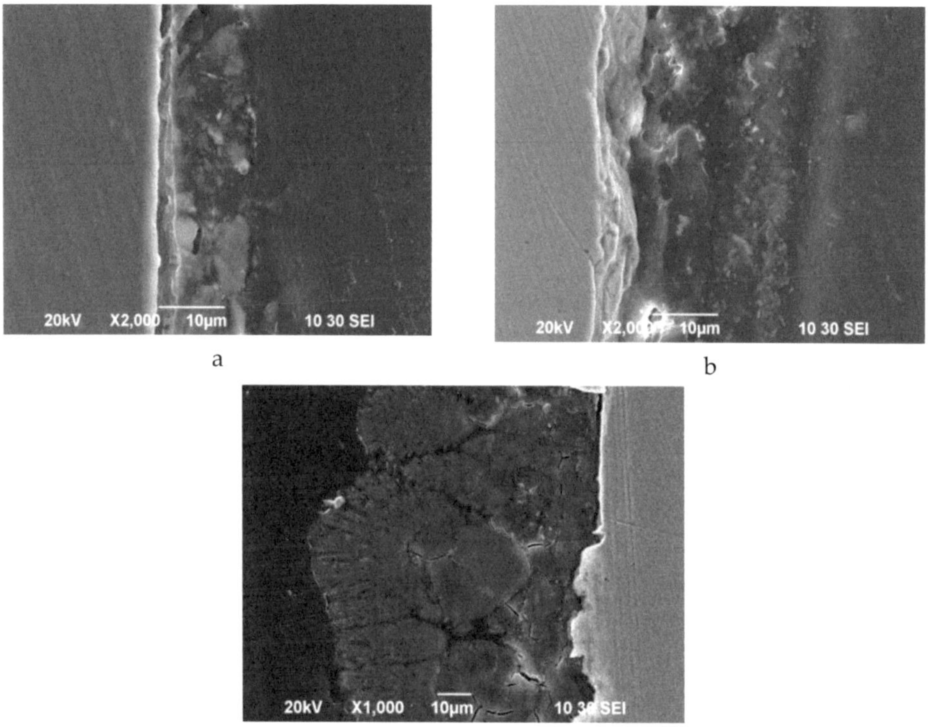

Figure 18.
SEM microstructure of polished transverse microsections by the depth of oxide layers (a) Al_2O_3, (b) TiO_2, and (c) MgO [2].

clusters along with the individual grains. Each cluster of the size of 200–400 nm is separated by smaller grains (60–150 nm). Similarly to the case of MgO, the outer TiO_2 sublayer has a developed surface with well-defined projections, gaps, and grain boundaries (**Figure 19b**). The TiO_2 surface is characterized by a spheroidal, row-like structure that could have been formed as a result of local instability of the plasma combustion. The disorientation of TiO_2 grains may be due to the high concentration of plasma on the substrate surface and the intensity of heat removal from the substrate. Grain clusters of the size of 50–100 nm in the structure of the TiO_2 sublayer are aggregated into packages up to 500 nm in size. More precise studies of the topology of the outer Al_2O_3 sublayer made it possible to clearly identify grain boundaries and quantify grain size gradation (**Figure 19c**). The microrelief of the Al_2O_3 surface has a dimple structure which is oriented perpendicularly to the surface [4].

In order to estimate the increase in the loss current and the breakdown voltage of the dielectric layers, the electrical characteristics were investigated for an alternating current case. To obtain the electrical characteristics of the coating, the upper contact made of foil was pressed against its surface, and an electrode was applied over the foil. As the lower contact, a steel substrate was used. Measurements of capacitance (C) and tangent of dielectric loss angle ($tg\ \delta$) at frequencies of 120 Hz, 1 kHz, 10 kHz, and 100 kHz were performed by bridge method using RLC meter in the mode of the parallel equivalent circuit with a measurement error of 0.05%.

The electrical conductivity values for Al_2O_3, MgO, and TiO_2 coatings at low frequencies are almost unchanged at room temperature and are approximately 3×10^{-8} $Ohm^{-1}\ m^{-1}$, 7×10^{-8} $Ohm^{-1}\ m^{-1}$, and 3×10^{-8} $Ohm^{-1}\ m^{-1}$, respectively. In dielectric layers of Al_2O_3 (at temperatures above 300°C) and MgO

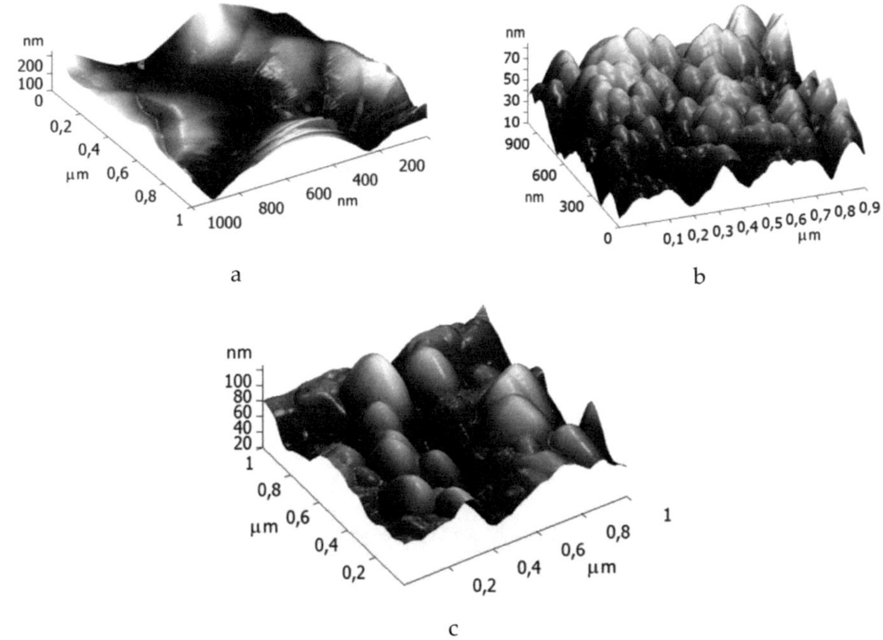

Figure 19.
AFM surface microtopography of the dielectric layers (a) MgO, (b) TiO_2, and (c) Al_2O_3 [2].

(at temperatures above 150°C), a monotonous increase in conductivity begins, caused by the thermal activation of charge carriers from energy levels of 0.63 eV and 0.35 eV, respectively. The electrical conductivity of TiO_2 coatings decreases from 4×10^{-8} Ohm^{-1} m^{-1} to 1×10^{-10} Ohm^{-1} m^{-1} with increasing temperature from room temperature to 250°C and increases to 1×10^{-7} Ohm^{-1} m^{-1} with increasing temperature up to 400°C [2].

As can be seen from the temperature dependences of the electrical conductivity of the dielectric layers for an alternating current case, in all investigated coating systems with increasing frequency, the magnitude of the electrical conductivity increases linearly. The increase in $\sigma(\omega)$ is due to the delay of slow polarization mechanisms. Moreover, the exponent n in the equation $\sigma(\omega) = \omega^n$ for the studied materials differs: for the Al_2O_3 layers $n = 0.2$, whereas for MgO $n = 0.5$. For titanium oxide coatings, it increases to 0.9.

Thus, the peculiarities of formation of the structure, adhesive, micromechanical, and electrophysical properties of the Al_2O_3, TiO_2, and MgO dielectric layers obtained by ion-plasma sputtering were established. An optimal mode of formation of Al_2O_3, TiO_2, and MgO dielectric layers was also determined: the gas pressure in the range $P = (1.5–4) \times 10^{-2}$ mmHg, the arc current in the range $I = 30–40$ A, the displacement potential on substrate $E = -60$ V, and the time $\tau = 20$ min.

When selecting the material for the substrate and dielectric layer during the design of a film heating element, it is necessary to evaluate the type and order of values of the dielectric losses that occur under operating conditions.

4. Conclusion

The existing methods for producing insulating coatings on the surface of corrosion-resistant steels are analyzed. It is shown that the CVD methods (ion-plasma deposition of metal oxide metal) of deposition on stainless steel substrates make it possible to obtain high-quality dielectric coatings for film heating elements with high micromechanical and electrophysical properties. The disadvantage of this method is its low productivity, which makes it impossible to use it in mass production. The relevance of using highly efficient and low-cost methods for the synthesis of functional coatings based on glass-ceramic glass sealants is shown. This method is simpler in execution technology and cheaper considering the use of raw materials, which opens up wide prospects for use in mass production. Given the proportionality of the sizes of coatings formed by both methods, the identity of the structure, microtopography of the surface, and the level of electrophysical properties, it can be argued that the first method can be recommended for applying dielectric coatings on the surface of stainless steels which is used for creating film heating elements.

Author details

Zoia Duriagina[1,2]*, Taras Kovbasyuk[1], Volodymyr Kulyk[1], Andriy Trostianchyn[1] and Tetiana Tepla[1]

1 Lviv Polytechnic National University, Lviv, Ukraine

2 John Paul II Catholic University of Lublin, Lublin, Poland

*Address all correspondence to: zduriagina@ukr.net

References

[1] Duryahina ZA, Borysyuk AK, Bespalov SA, Pidkova VY. Influence of the thermal cyclic treatment on the phase composition of ion-nitrided surface layers of 12KH18N10T steel. Materials Science. 2012;**48**:364-368. DOI: 10.1007/s11003-012-9514-x

[2] Duriagina ZA, Kovbasyuk TM, Bespalov SA. The analysis of competitive methods of improvement of operational properties of functional layers of flat heating elements. Uspehi Fiziki Metallov. 2016;**17**:29-51. DOI: 10.15407/ufm.17.01.029

[3] Duriagina Z, Kovbasyuk T, Bialopiotrowicz T, Bespalov S. Energy state and micromechanical properties of PbO-ZnO-B_2O_3 glass-ceramic functional coatings on AISI420 stainless steel substrate. Functional Materials. 2017;**24**: 250-255. DOI: 10.15407/fm24.02.250

[4] Duriagina Z, Kovbasyuk T, Zagula-Yavorska M, Bespalov S, Drajewicz M, Dychtoń K, et al. Comparative estimation of the structure and electrical properties of functional layers based on PbO–ZnO–B_2O_3 glass ceramic sealant. Powder Metallurgy and Metal Ceramics. 2016;**9/10**:580-584. DOI: 10.1007/s11106-017-9842-3

[5] Duryahina ZA, Kovbasyuk TM, Bespalov SA, Pidkova VY. Micromechanical and electrophysical properties of Al_2O_3 Nanostructured dielectric coatings on plane heating elements. Materials Science. 2016;**52**: 50-55. DOI: 10.1007/s11003-016-9925-1

[6] Park SS, Jung KS, Kim BW, Lee SE, Park HC. Microwave heating induced crystallisation of PbTiO3 from a PbO-B_2O_3-ZnO-TiO_2 glass joined to alumina. Glass Technology. 2002;**43**: 70-74

[7] Cheng Y, Xiao H, Guo W. Effects of GeO_2 on structure and properties of PbO-B_2O_3-ZnO glass. Journal of Inorganic Materials. 2006;**21**:533-538

[8] Cheng Y, Xiao H, Guo W. Thermal behavior of GeO_2 doped PbO-B_2O_3-ZnO-Bi_2O_3 glasses. Materials Science and Engineering: A-Structural Materials Properties Microstructure and Processing. 2006;**423**:184-188. DOI: 10.1016/j.msea.2005.09.130

[9] Cheng Y, Xiao H, Guo W. Influence of compositions on sealing temperature and properties of lead borate non-crystallizing sealing glasses. Materials Science and Engineering: A-Structural Materials Properties Microstructure and Processing. 2007;**464**:210-215. DOI: 10.1016/j.msea.2007.01.146

[10] Hasheminia S, Nemati A, Eftekhari Yekta B, Alizadeh P. Preparation and characterisation of diopside-based glass-ceramic foams. Ceramics International. 2012;**38**:2005-2010. DOI: 10.1016/j.ceramint.2011.10.035

[11] Ma R, Xiao B, Ma N, Du P. Low-loss insulating-conductive ceramic composite with giant permittivity and high permeability using glass phase as separating layer. Ceramics International. 2016;**42**:4126-4135. DOI: 10.1016/j.ceramint.2015.11.085

[12] Riley BJ, Vienna JD, Frank SM, Kroll JO, Peterson JA, Canfield NL, et al. Glass binder development for a glass-bonded sodalite ceramic waste form. Journal of Nuclear Materials. 2017;**489**: 42-63. DOI: 10.1016/j.jnucmat.2017.03.041

[13] Andreola F, Barbieri L, Lancellotti I, Leonelli C, Manfredini T. Recycling of industrial wastes in ceramic manufacturing: State of art and glass case studies. Ceramics International. 2016;**42**:13333-13338. DOI: 10.1016/j.ceramint.2016.05.205

[14] Subba Rao A, Ashok J, Suresh B, Naga Raju G, Venkatramaiah N, Ravi Kumar V, et al. Physical characteristics of $PbO-ZrO_2-SiO_2:TiO_2$ glass ceramics embedded with $Pb_2Ti_2O_6$ cubic pyrochlore crystal phase: Part-I electrical properties. Journal of Alloys and Compounds. 2017;**712**:672-686. DOI: 10.1016/j.jallcom.2017.04.100

[15] Subba Rao A, Kityk IV, Ashok J, Ravi Kumar V, Plucinski KJ, Reddy ASS, et al. Physical characteristics of $PbO-ZrO_2-SiO_2:TiO_2$ glass ceramics embedded with $Pb_2Ti_2O_6$ cubic pyrochlore crystal phase: Part-II piezo-optical acoustic and elastic properties. Journal of Alloys and Compounds. 2017; **725**:318-325. DOI: 10.1016/j.jallcom.2017.07.158

[16] Siva Sesha Reddy A, Brik MG, Suresh Kumar J, Graça MPF, Naga Raju G, Ravi Kumar V, et al. Structural and electrical properties of zinc tantalum borate glass ceramic. Ceramics International. 2016;**42**:17269-17282. DOI: 10.1016/j.ceramint.2016.08.022

[17] Marczewska A, Środa M, Nocuń M. Thermal and spectroscopic characterization of gallium-tellurite glasses doped BaF_2 and PbO. Journal of Non-Crystalline Solids. 2017;**464**: 104-114. DOI: 10.1016/j.jnoncrysol.2017.03.026

[18] Jakkula S, Deshpande V. Effect of MgO addition on the properties of $PbO–TiO_2–B_2O_3$ glass and glass–ceramics. Ceramics International. 2013;**39**:S15-S18. DOI: 10.1016/j.ceramint.2012.10.027

[19] Duan F. The role of PbO on crystallization in $PbO–SrO–TiO_2–SiO_2$ glass. Journal of Non-Crystalline Solids. 2010;**356**:2286-2288. DOI: 10.1016/j.jnoncrysol.2010.07.041

[20] Sharif A, Gan CL, Chen Z. Customized glass sealant for ceramic substrates for high temperature electronic application. Microelectronics Reliability. 2014;**54**:2905-2910. DOI: 10.1016/j.microrel.2014.07.005

[21] Białopiotrowicz T. Influence of erroneous data on the results of calculations from acid–base surface free energy theories. I. Simulations for a small input data set. Journal of Adhesion Science and Technology. 2007;**21**: 1539-1556. DOI: 10.1163/156856107782793230

[22] Białopiotrowicz T. Influence of erroneous data on the results of calculations from acid–base surface free energy theories. II. Why are negative values of square roots obtained. Journal of Adhesion Science and Technology. 2007;**21**:1557-1573. DOI: 10.1163/156856107782793221

[23] Białopiotrowicz T. Influence of erroneous data on results in acid-base surface free energy theories calculations. III. Solution of a three-equation set in the case of homoscedastic error. Journal of Adhesion Science and Technology. 2009; **23**:799-813. DOI: 10.1163/156856108X396327

[24] Białopiotrowicz T. Influence of erroneous data on results in acid-base surface free energy theories calculations. IV. Solution of an overdetermined set in the case of homoscedastic error. Journal of Adhesion Science and Technology. 2009;**23**:815-825. DOI: 10.1163/156856108X396336

[25] Della Volpe C, Siboni S. Some reflections on acid–base solid surface free energy theories. Journal of Colloid and Interface Science. 1997;**95**:121-136. DOI: 10.1006/jcis.1997.5124

[26] Jańczuk B, Białopiotrowicz T, Zdziennicka A. Some remarks on the components of the liquid surface free energy. Journal of Colloid and Interface Science. 1999;**211**:96-103. DOI: 10.1006/jcis.1998.5990

[27] Gleiter H. Nanocrystalline materials. Progress in Materials Science. 1989;**33**: 223-314

[28] Gleiter H. Nanostructured materials: Basic concepts and microstructure. Acta Materialia. 2000;**48**:1-29. DOI: 10.1016/0079-6425(89)90001-7

[29] Palumbo G, Erb U, Aust K. Triple line disclination effect on the mechanical behavior of materials. Scripta Metallurgica et Materialia. 1990; **24**:1347-1350. DOI: 10.1016/0956-716X(90)90091-T

[30] Sun NX, Lu K. Heat-capacity comparison among the nanocrystalline amorphous and coarse-grained polycrystalline in Se. Physical Review B. 1996;**54**:6058-6061. DOI: 10.1103/PhysRevB.54.6058

[31] Zhao YH, Lu K. Grain size dependence of thermal properties of nanocrystalline Se studied by means of X-ray diffraction. Physical Review B. 1997;**56**:14330-14337

[32] Sharma I, Sharma A, Barman PB. Electrical properties of magnesium oxide thin films. Optoelectronics and Advanced Materials—Rapid Communications. 2008;**2**:640-642

[33] Bondoux C, Prené P, Belleville P, et al. MgO insulating films prepared by sol–gel route for SiC substrate. Journal of the European Ceramic Society. 2005; **25**:2795-2798. DOI: 10.1016/j.jeurceramsoc.2005.03.142

[34] Kim TW, You YS. Microstructural and electrical properties of MgO thin films grown on p-InP (100) substrates at low temperature. Applied Surface Science. 2001;**180**:162-167. DOI: 10.1016/S0169-4332(01)00335-X

[35] Shin D-Y, Kim K-N. Electrical and optical properties of MgO films deposited on soda lime glass by a sol-gel process using magnesium acetate. Journal of Ceramic Processing Research. 2009;**10**:536-540

[36] Afanasov IM, Lazoryak BI, editors. High Temperature Ceramic Fibers. Moscow: Tutorial; 2010. p. 51 (in Russian)

[37] Stevens JE, Sowa MJ, Cecchi JL. Helicon plasma source excited by a flat spiral coil. Journal of Vacuum Science & Technology A. 1995;**13**:2476-2482. DOI: 10.1116/1.579491

[38] Duriagina ZA, Pidkova VY, Olshevska SO. Structure peculiarities and properties of magnesium oxide layers formed by a hybrid ion-plasma discharge system. Functional Materials. 2013;**20**:192-197. DOI: 10.15407/fm20.02.192

Chapter 5

Morphology Evolution of DF2 (AISI-O1) Surface Micromachined by Pulsed Nd:YAG Laser

Kelvii Wei Guo

Abstract

Pulsed Nd:YAG laser was taken to premicromachine DF2 (AISI-O1) cold work steel. The effect of laser-irradiated parameters on the morphology evolution of the processed surface was investigated by 3D profilometer, atomic force microscope (AFM), scanning electron microscopy (SEM), and optical microscopy (OM). Results show that when DF2 (AISI-O1) specimens were irradiated with various parameters, the morphology of DF2 (AISI-O1) cold work steel was changed correspondingly. Moreover, it demonstrates that for a given laser, various kinds of morphology of a laser-machined surface could be established successfully to satisfy with the desired finish surface for the practical applications later.

Keywords: surface morphology, pulsed Nd:YAG laser, micromachining, DF2 (AISI-O1)

1. Introduction

The quality requirements of the dies and molds are high during the glass, metal, and plastic product fabrication. Small abrasive tools or wheels that can rotate and vibrate by themselves can be used to improve the easily accessible areas of some surface. But, in practical situation, the surface of these molds and dies is mainly finished or processed by the hand. The processing is relied on the successive size decrement of particles taken for abrasion distributed on abrasive papers or cloths in order to make the finished surface meet with the practical applications.

The shortcomings of operators make a lot of limitations in these methods during the manual processing. It should be noted that operators have to be seriously trained to gain the considerable experiment and meet with the requirements of practical operations at various conditions, especially with the small radii equipment to attain the effective application. Moreover, it also requires the consistent and repeatable operation. In addition, it requires a skilled operator to take 10 days to process 10 dies with the complicated forms. Meanwhile, it also needs to make sure that all the finished dies are identical after completion. More importantly, the accuracy of the die dimension is not lost after the finished surface is achieved. Therefore, an operator with high experiences and skills is definitely expensive. It is well known that a well-trained operator is essential to the process of manufacturing.

As mentioned above, it can be seen that it is also extremely tedious and more time-consuming for this process. Research shows that the time consumed on a mold finishing can be about 37% of the total time of the entire mold production [1]. Consequently, it is crucial and urgent to improve or enhance the quality of die or mold at the final step with the less number of the trained and skilled operators or the decrement of processing time in order to dramatically cut the overall cost.

For the processes with automatic techniques, there are also some limitations to the closed die processing. Research shows that it is very slow for the precision machining by a single-point diamond tool. Moreover, it is not easily available for industrial conditions and hard for the flat surfaces [1–3]. Chemical micromachining and electrochemical micromachining are limited in their application and can be difficult to control [3–6]. Some research has been carried out into the use of robot-controlled finishing tools [1, 3–5, 7–9]. However, the use of a rotary wheel or ultrasonic chisel requires that the micromachined surface should be almost parallel to the axis of the wheel or the chisel angle. It should be noted that it is not suitable for almost flat workpieces and limits its applications.

The ultrasonic machining can be used to achieve the removal of the controllable material [1, 3]. By this technique, the hammering force between the abrasive particles (in water-based slurry) and the workpiece at high frequency can remove the material from the processed surface. As a result, the shapes with the complicated forms can be achieved by the suitably designed tools. Although it is potential for processing the closed molds or dies, there are still some disadvantages, especially for the indentation of workpiece and the drilling of the fixed location.

Currently, it is efficient and successful to apply the abrasive flow machining to process the open molds or dies. In this process, a mix of abrasive particles is taken to finish the surface by hydraulic power in the form of suspension in a pliable polymer base. However, limitation is still to restrict this process from finishing the molds or dies without an exit and an entry for the flowing mix in the hollow or tubular form [1, 2].

To date, laser has been widely used as a machine tool to modify the surface of the engineering materials, such as laser surface alloying, laser cladding, surface texturing, laser physical vapor deposition, etc. [1, 3–5]. Recently, micromachining by laser technique is gaining more attention to be an interesting method to machine lens, diamonds, etc. [7–12].

On the basis of the abovementioned, an investigation has been taken to micromachine DF2 steel with the excellent non-contact possibility and the unique characteristics of Nd:YAG laser radiation.

2. Experimental material and procedures

2.1 Experimental material

The chemical composition of DF2 cold work steel is shown in **Table 1**.

2.2 Experimental procedures

The materials were machined into 25 × 10 × 10 mm and carefully cleaned by acetone and pure ethyl alcohol to remove any contaminants on its surface. **Figure 1**

Element	C	Si	Mn	Cr	W	V	Fe
(wt.%)	0.9	0.3	1.2	0.5	0.5	0.1	Bal.

Table 1.
Chemical composition of DF2 (AISI-O1) cold work steel.

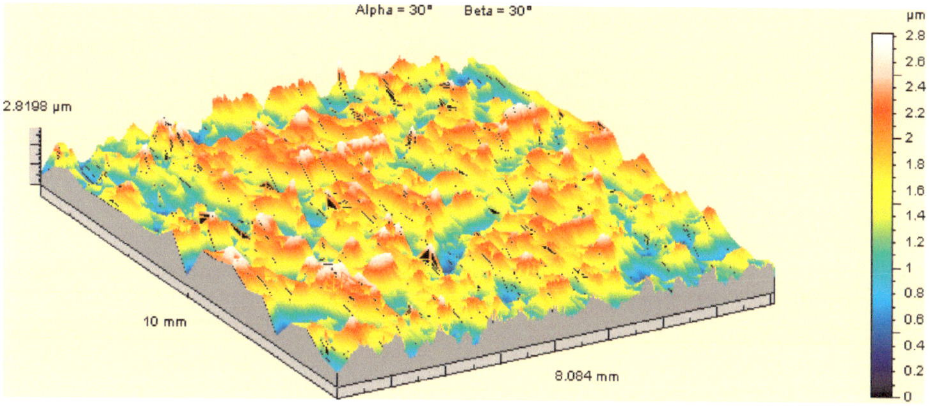

Figure 1.
Initial topographies of specimens Ra = 0.4 µm.

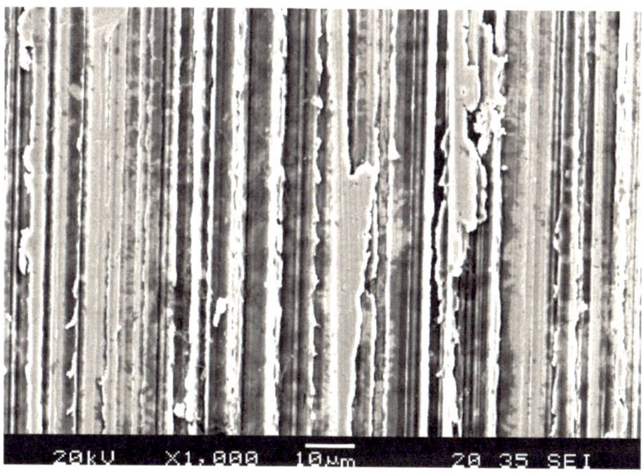

Figure 2.
SEM of the DF2 (AISI-O1) initial surfaces.

Figure 3.
Microstructure of DF2 (AISI-O1) steel.

shows the Talysurf 3D topography of the original DF2 tool steel specimens categorized with the initially machined surface roughness *R*a = 0.4 μm, while **Figure 2** shows the corresponding 2D SEM morphology. The microstructure of DF2 steel is shown in **Figure 3**.

A GSI Lumonics Model JK702H pulsed Nd:YAG TEM_{00} mode laser system, with wavelength of 1.06 μm, defocused distance of 15 mm, and focus spot diameter of ~1.26 mm on the substrate surface, was used to irradiate the DF2 steel.

After being processed, the surface morphology was observed by atomic force microscope (AFM), Taylor Hobson profilometer/Talysurf PGI, and scanning electron microscope (SEM) JEOL/JSM 5600.

3. Results and discussion

3.1 Influence of laser scanning speed (feedrate)

Comparing **Figure 1** with **Figure 4**, if the parameters are not set properly, the surface will be rather roughened than smoothened though most of the high plateaus are removed from the initial specimen surface. The initial surface with the surface roughness of 0.4 μm was irradiated with laser pulse frequency of 25 Hz, pulse energy of 1 J, and pulse duration of 3 ms. Results are shown in **Figure 4**. It suggests that when the feedrate is 300 mm/min, the achieved 3D topography of the irradiated surface is the smoothest. It also shows that with the variation of laser feedrates, the irradiated surface changed correspondingly. During the period of between 100 and 300 mm/min, the irradiated surface is becoming more and more smoother. However, when the feedrate is higher than 300 mm/min, the quality of the irradiated surface will be coarser. **Figure 4a–c** illustrates the 3D morphology of the improved smoothness of processed surface with the laser feedrate of <300 mm/min, and **Figure 4c** and **d** shows 3D morphology of unsmooth processed surface with the laser feedrate higher than 300 mm/min.

The relationship between the laser feedrate (scanning speed) and the roughness of the processed surface is shown in **Figure 5**. It indicates that when the laser scanning speed increases, the roughness of the processed surface will be decreased correspondingly. Moreover, when the laser scanning speed is up to 300 mm/min, the roughness of the processed surface is at its minimum. After that, with the laser scanning speed increasing further, the roughness of the processed surface will be increased again. Moreover, the roughness of the processed surface is generally higher than the initial condition when the relative feedrate is below 300 mm/min. However, it decreases with the increase in the relative feedrate. When the relative feedrate is at 400 mm/min, the irradiated roughness is increased to some level.

3.2 Effect of pulse energy of laser irradiating

According to the abovementioned study, the laser scanning speed is taken as 300 mm/min in the subsequent tests as the optimal feedrate. The relationship between the roughness of the processed surface and the laser irradiating energy is shown in **Figure 6**. It is well known that if the laser pulse energy raises, the energy interaction between the workpiece and laser will be raised simultaneously. When the laser pulse energy is so high, the energy input into the workpiece will be enhanced. As a result, the surface of the workpiece will be melted and the roughness of the processed surface will change seriously, especially for deteriorating the mechanical properties of the processed workpiece in the subsequent practical application afterward. **Figure 7** shows the relationship between the roughness of the

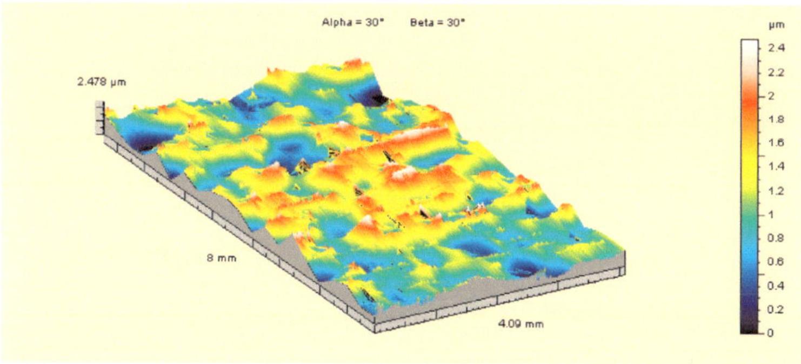

(a)

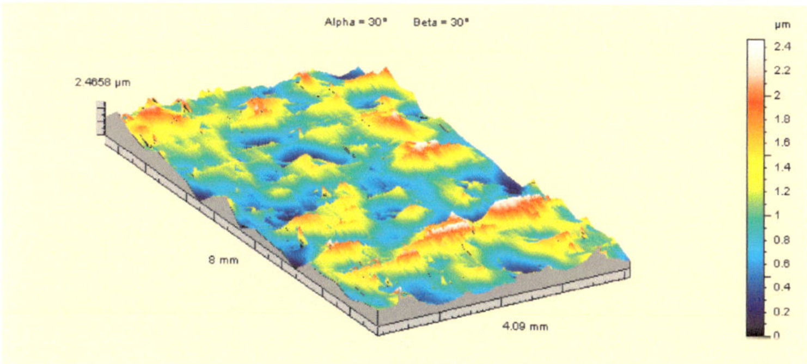

(b)

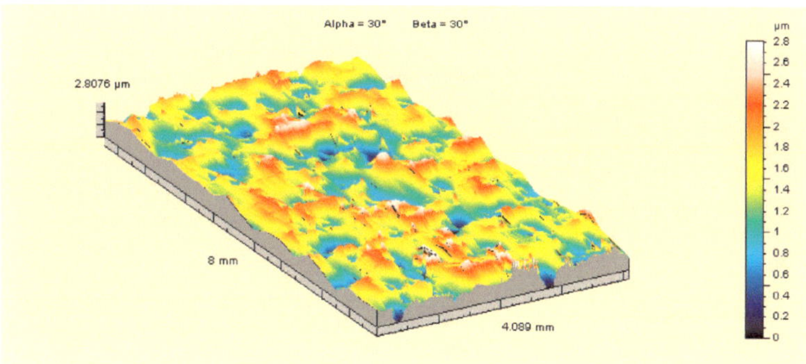

(c)

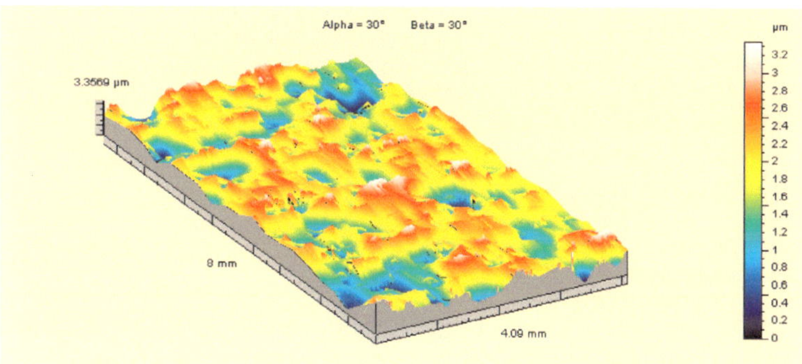

(d)

Figure 4.
Surface textures of laser micromachining at various feedrates. (a) 400 mm/min, (b) 300 mm/min, (c) 200 mm/min, and (d) 100 mm/min. Pulse energy = 1 J, f = 25 Hz, PD = 3 ms.

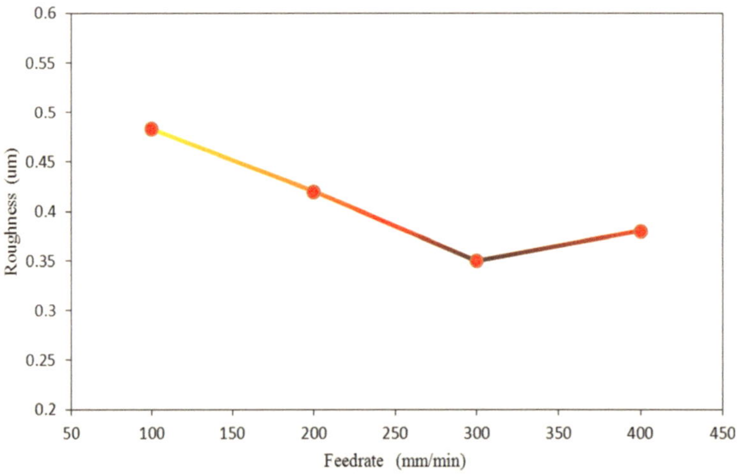

Figure 5.
Relationship between surface roughness and feedrate.

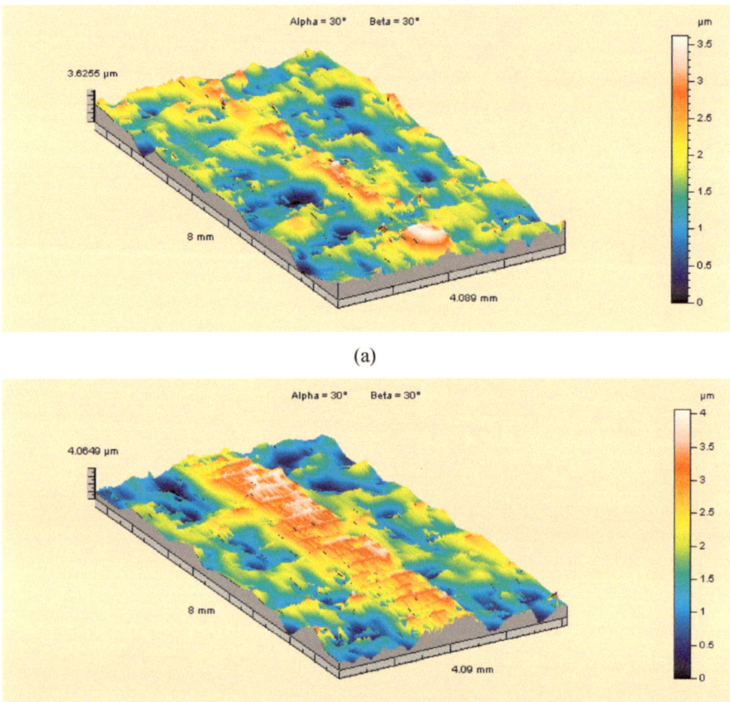

(b)

Figure 6.
Surface textures of laser micromachining at various pulse energies. (a) Pulse energy = 2 J and (b) pulse energy = 3 J. Feedrate = 300 mm/min, f = 25 Hz, PD = 3 ms.

processed surface and laser pulse energy. Results indicate that the roughness variation of the processed surface is correlated with the laser pulse energy. It expresses that when the laser pulse energy is 2 J, the roughness of the processed surface is 0.49 μm with some protrusions distributed on the surface as shown in **Figure 6a**. With the laser pulse energy increasing, more protrusions are scattered on the processed surface. Consequently, the roughness of the processed surface is up to 0.7 μm. Meanwhile, the laser-affected zone is becoming wider and larger as shown

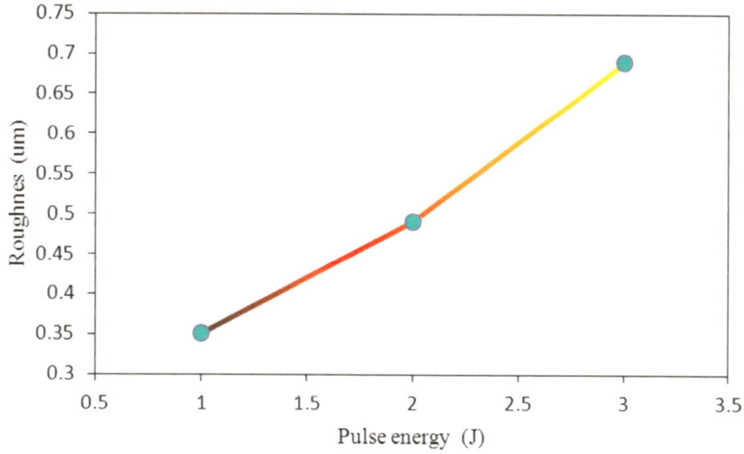

Figure 7.
Relationship between surface roughness and pulse energy.

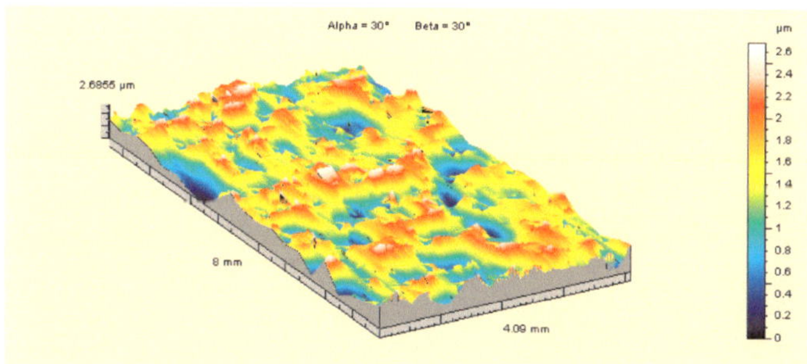

(a)

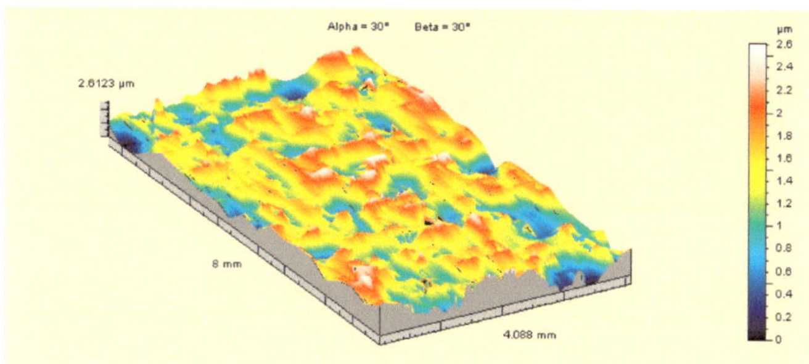

(b)

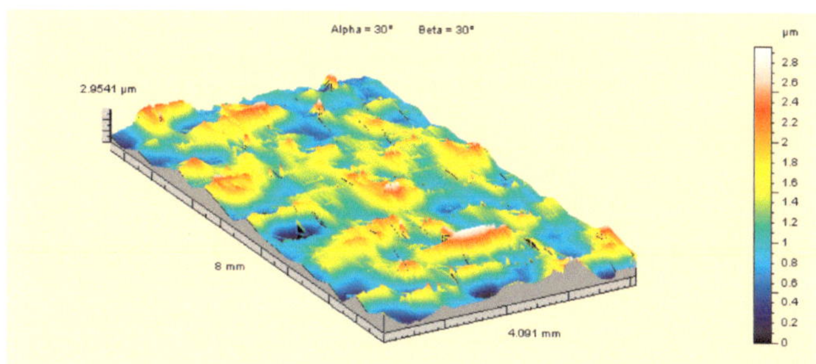

(c)

Figure 8.
Surface textures of laser micromachining at various pulse frequencies. (a) f = 15 Hz, (b) f = 20 Hz, (c) f = 30 Hz. Pulse energy = 1 J, feedrate = 300 mm/min, PD = 3 ms.

in **Figure 6b**, which indicates that if laser pulse energy is equal to or higher than 2 J, the results of the final processed surface is generally not acceptable.

3.3 Influence of laser irradiation frequency (repetition rate)

Figure 8 shows the laser-irradiated surface scanned at various pulse frequencies. It shows that the roughness of the irradiated surface decreases with the pulse frequency increasing. However, at a pulse frequency of 30 Hz, the roughness of the processed surface is distinctively higher again resulting in a coarser finished surface. Its corresponding roughness of the processed surface varied with the laser pulse energy shown in **Figure 9**. Results illustrate that when the laser pulse energy varies from 20 to 25 Hz, the roughness of the processed surface is <0.4 μm, and the processed surface is smoother at the optimal level. However, the roughness of the processed surface will be increased obviously in the case the laser pulse frequency increases, e.g., 30 Hz.

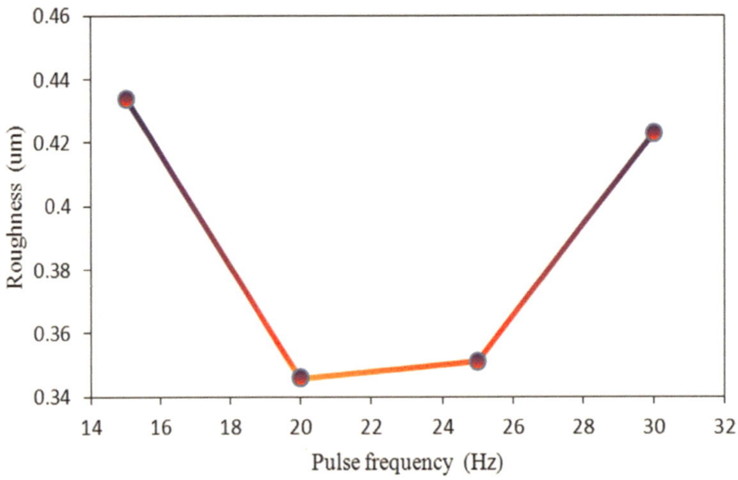

Figure 9.
Relationship between surface roughness and pulse frequencies.

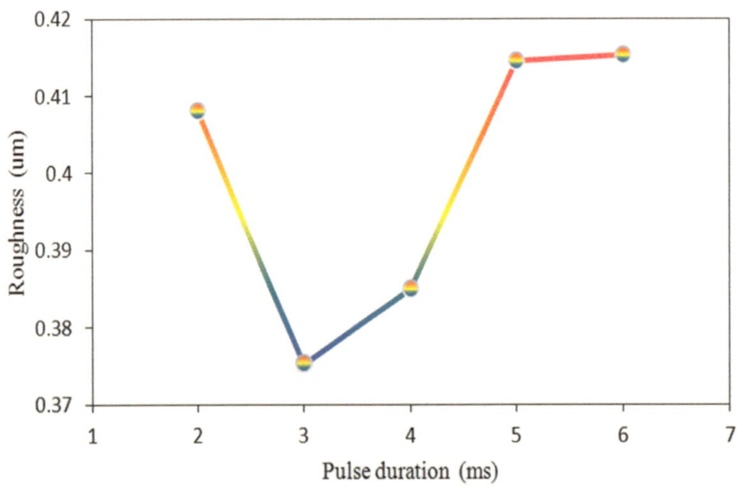

Figure 10.
Relationship between surface roughness and pulse duration.

3.4 Influence of laser micromachining pulse duration (PD)

Figure 10 shows the relationship between the laser pulse duration and the roughness of the processed surface. Results elucidate that when the laser pulse duration is decreased, the roughness of the processed surface will be decreased correspondingly. It also shows that the roughness of the processed surface is at its minimum when the laser pulse duration is 3 ms. With the decrement of the laser pulse duration further (e.g., 2 ms), the roughness of the processed surface will be increased remarkably. In addition, the roughness of the processed surface will be coarser when the laser pulse duration exceeds 4 ms.

4. Discussions

4.1 Influence of feedrate and pulse frequency

As the irradiation path of a continuous scanning mode using a pulsed laser with pulse duration PD and frequency *f* is essentially consisted of a series of overlapping spots (**Figure 11**), the basic condition to realize the micromachining by a pulsed laser is that two successive spots in a continuous scanning mode should be tangential to each other, viz.:

$$0 < \delta \leq D \tag{1}$$

where D is the laser spot diameter, δ is the distance of one pulse, δ_1 is the distance with the laser-on duration, and δ_2 is the distance with the laser-off duration, and it is defined as

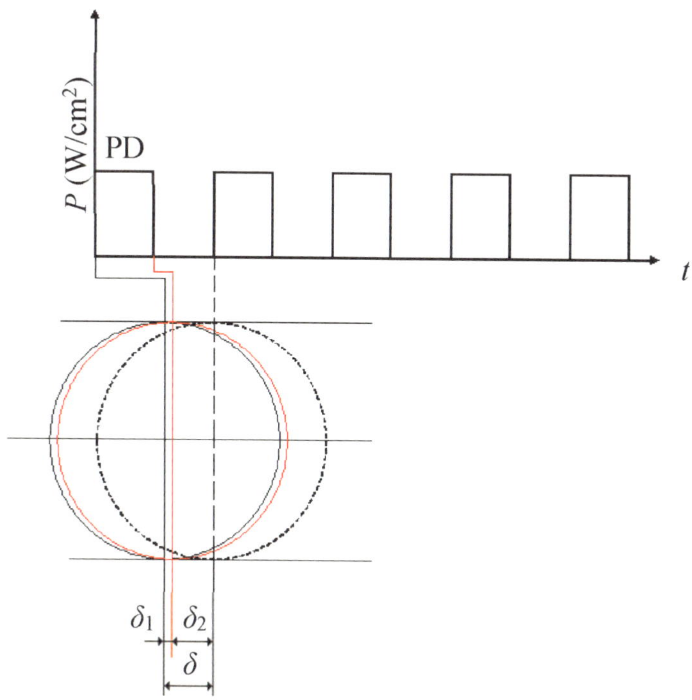

Figure 11.
Schematic image of a laser pulse irradiated on the surface.

$$\delta = \frac{Feedrate}{f} \tag{2}$$

Substituting Eq. (2) into Eq. (1) subsequently gives

$$0 < \frac{Feedrate}{f} \leq D \tag{3}$$

which can then be expressed as

$$0 < Feedrate \leq Df \tag{4}$$

Figure 12 shows the relationship between δ and pulse frequency f (see Eq. (2)) at various feedrates and its corresponding percentage of overlapping of two successive laser pulses in the continuous scanning mode. It shows that the laser frequency increased (i) with the decrease of δ (see the solid lines in **Figure 12**) and (ii) simultaneously with the increase in the overlap percentage (see the dashed lines in **Figure 12**). When the feedrate is at 300 mm/min, the pulse frequency of 20–25 Hz leads to the δ about 0.25–0.20 mm and the overlap percentage about 60–68%.

Under the conditions of the experimental feedrates and pulse frequencies in this study, the distances of a single laser pulse (δ) are thus smaller than the laser spot diameter (D), which meets the constraint of Eq. (1) for the realization of laser micromachining phenomenon. The mechanisms involved in achieving the micromachining effect at the operational parameters of pulse frequency f = 20–25 Hz and feedrate = 300 mm/min need to be explored.

In view of the overlapping geometry of the irradiation spot under continuous scanning mode, the laser micromachining process is in quasi-steady state (**Figure 13**) when the moving distance of the spot within a single pulse is equal to or greater than the spot diameter. The solid lines with different colors in **Figure 13** present the laser spot during the "on"-duration (δ_1 in **Figure 11**) of a PD, while their individual counterpart dashed lines, respectively, stand for the shifting of the laser

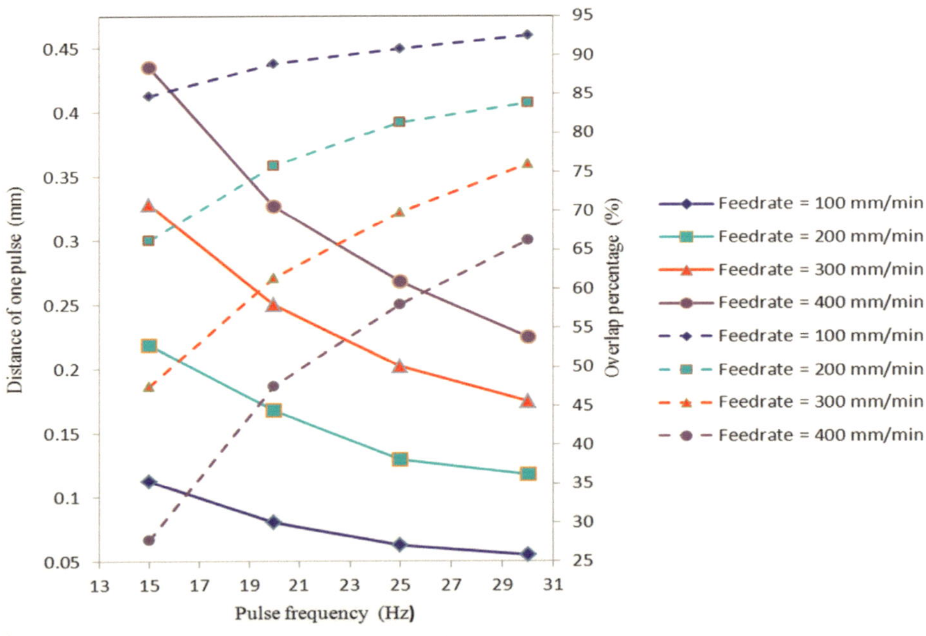

Figure 12.
Relationship between δ and pulse frequency at various feedrates.

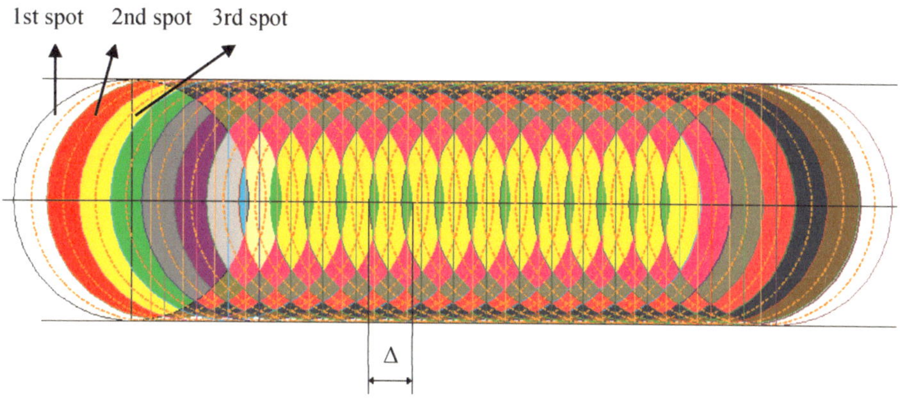

Figure 13.
Schematic diagram of laser micromachining.

spot during the "off"-duration (δ_2 in **Figure 11**) of the PD. The areas filled with various colors in **Figure 13** stand for the level of different overlapping (typically: the area in red color stands for the occurrence of only one overlap; the area in light yellow stands for the occurrence of two overlapped, and so on). Let Δ be the effective micromachining size along the central line; it can then be expressed as

$$\Delta = D - \left\| \frac{D}{\delta} \right\| \times \delta + \delta \tag{5}$$

Figure 14 shows the behaviors of effective overlap percentage versus pulse frequency at the experimental feedrates. It shows that the relative overlap percentage for the feedrate = 300 mm/min is lower than that of other feedrates when the pulse frequency is at and beyond 20 Hz. When the pulse frequency is below 20 Hz, the effective overlap percentage is increased distinctly with the decrease in pulse frequency. Experimentally, irradiated specimens with smaller overlap percentage seemed to give better metallurgical aspect. This suggests that the surface can be micromachined successfully according to various feedrates and pulse frequencies. When the effective overlap percentage inclined to null, the laser micromachining

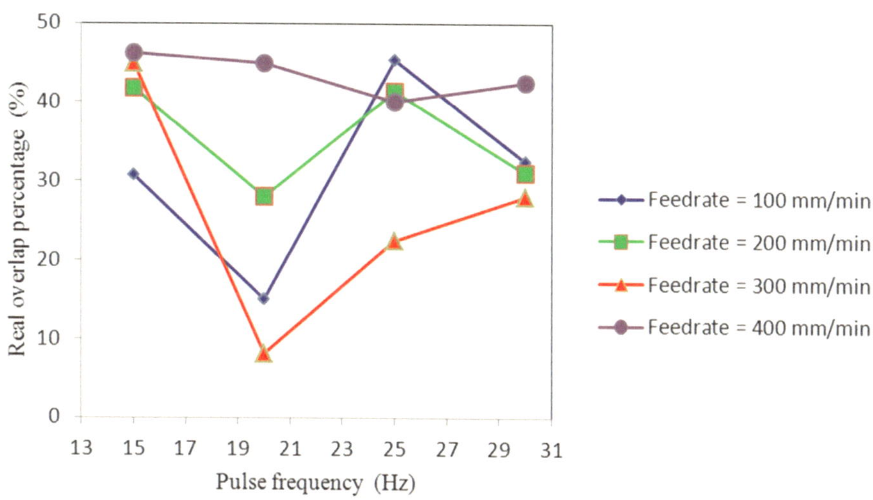

Figure 14.
Effective overlapping percentage versus pulse frequency at various feedrates.

will be of the ideal conditions as generated by a single spot irradiation. Under such conditions, it gives

$$\left\|\frac{D}{\delta}\right\| = 0 \tag{6}$$

Subsequently, substituting Eq. (2) into Eq. (6) gives

$$\left\|\frac{D \times f}{Feedrate}\right\| = 0 \tag{7}$$

Taking D = 1.26 mm, Eq. (7) is therefore transformed into

$$126 \times f = 100 \times n \times Feedrate(f, n = 1, 2, 3) \tag{8}$$

in which n stands for numbers of macroscopical overlaps. When n is equal to 1, the two continuous laser spots are tangential to each other, implying both macroscopical overlap percentage and effective overlap percentage are equal to 0. When f is 20 Hz and feedrate is 300 mm/min, it can be deduced that the value of n is therefore 5 that subsequently leads to 80% (4/5) macroscopical overlap percentage and 20% (1/5) effective overlap percentage. When f is 20 Hz, the ideal feedrate is thus estimated as $\frac{126 \times 20}{5 \times 100} \times 60 = 302.4$ mm/min that is very close to 300 mm/min. This is the reason why the texture of the micromachined surface is better when the laser is set at pulse frequency of 20 Hz and feedrate of 300 mm/min.

4.2 Influence of laser pulse energy and pulse duration (PD)

4.2.1 Physical model of laser micromachining

Figure 15 shows the physical model of Nd:YAG pulsed laser micromachining a specimen along its X-axial direction with certain specific feedrate.

4.2.2 Temperature field of laser micromachining

Figure 16 shows the model with a point source irradiating on a heavy substrate; the assumption of instantaneously releasing the laser energy on the surface of a substrate at time t = 0 leads to a temperature rise in the material (point P) as [13]

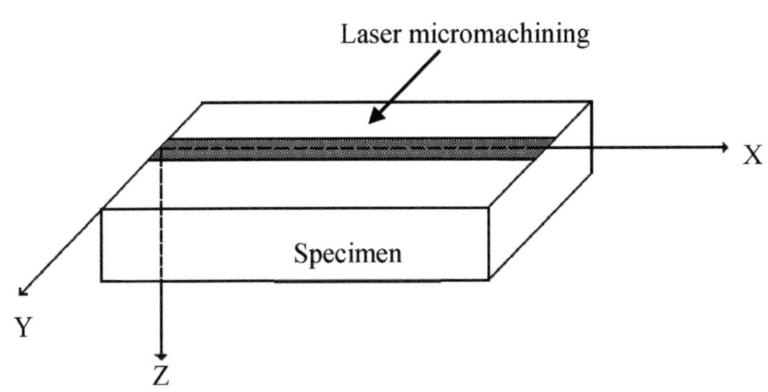

Figure 15.
Schematic diagram of laser micromachining process.

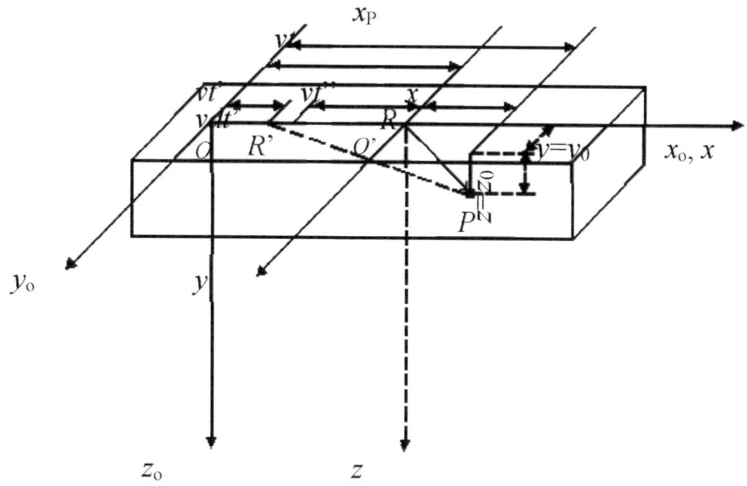

Figure 16.
Moving point source on a semi-infinite substrate.

$$T - T_0 = \frac{Q}{\rho c(4\pi\alpha t)^{\frac{3}{2}}} \exp\left(\frac{-R^2}{4\alpha t}\right) \tag{9}$$

where ρ is the material density, C is specific heat, α is thermal diffusivity, and Q is the input energy. Let the initial temperature, at time $t = 0$, of the generally thick plate be T_0, and the point source of constant power q_0 moves in the x-direction at a constant speed (feedrate) v from the position O.

Within a very short interval from time t' to $t' + dt'$, the amount of heat released at the surface is $dQ = q_o dt'$. By Eq. (9), it produces an infinitesimal temperature rise for point P at time t as

$$dT = \frac{2q_o dt'}{\rho c(4\pi\alpha(t - t'))^{\frac{3}{2}}} \exp\left(-\frac{(R')^2}{4\alpha(t - t')}\right) \tag{10}$$

By defining $t'' = t - t'$, Eq. (10) is thus transformed into

$$dT = \frac{-2q_o dt''}{\rho c(4\pi\alpha t'')^{\frac{3}{2}}} \exp\left(-\frac{(R')^2}{4\alpha t''}\right) \tag{11}$$

Geometrically, **Figure 16** allows $R' = \sqrt{(x_o - vt')^2 + y_o{}^2 + z_o{}^2}$ to be expressed. Subsequently, the transforming of the coordinate system from O to O': $y = y_o$, $z = z_o$, $x = x_o - vt$, and $x_o - vt' = x + vt - vt' = x + vt''$ allows the following expression to be obtained:

$$\begin{aligned} dT &= \frac{-2q_o dt''}{\rho c(4\pi\alpha t'')^{\frac{3}{2}}} \exp\left(-\frac{\left((x + vt'')^2 + y^2 + z^2\right)}{4\alpha t''}\right) \\ &= \frac{-2q_o dt''}{\rho c(4\pi\alpha t'')^{\frac{3}{2}}} \exp\left(-\frac{vx}{2\alpha} - \frac{R^2}{4\alpha t''} - \frac{v^2 t''}{4\alpha}\right) \end{aligned} \tag{12}$$

where $R = \sqrt{x^2 + y^2 + z^2}$.

When temperature distribution is quasi-steady state, Eq. (12) can be expressed as

$$T - T_o = \frac{q_o}{2\pi\rho c\alpha R} \exp\left(-\frac{v}{2\alpha}(R+x)\right) \tag{13}$$

where $q_0 = \eta(N(f,v,)) \times \frac{P}{\frac{\pi D^2}{4} \times PD} = \eta(N(f,v,)) \times \frac{4P}{\pi D^2 \times PD}$ in which $\eta(N(f,v,))$ is the coefficient of laser micromachining input energy and is directly proportional to either the numbers of macroscopical overlaps or the pulse frequency and feedrate. Hence, Eq. (13) can be expressed as

$$T - T_o = \frac{2 \times \eta(N(f,v,)) \times P}{\pi^2 D^2 \rho c\alpha R \times PD} \exp\left(-\frac{v}{2\alpha}(R+x)\right) \tag{14}$$

Define $\xi = \frac{1}{2\pi\rho c\alpha}$, and then Eq. (14) can be written as

$$T - T_o = \frac{\xi \times \eta(N(f,v,)) \times P}{\frac{\pi D^2}{4} \times PD \times R} \exp\left(-\frac{v}{2\alpha}(R+x)\right) \tag{15}$$

Equation (15) expresses the relationship between the laser micromachining temperature and the processing parameters like pulse energy, pulse duration, pulse frequency, and scanning speed (feedrate). The equation indicates that the increase in pulse energy and pulse frequency generally raises the micromachining temperature up. However, the increase in pulse duration and scanning speed (feedrate) accompanies with the decrease in micromachining temperature, while the increase in velocity leads to the decrease in numbers of overlaps under the constant pulse frequency condition, implying a synchronous decrease in $\eta(N)$ and micromachining temperature. Scanning speed (feedrate) is generally having more remarkable impact on the micromachining temperature than the other parameters. At a particular feedrate, there tends to have a threshold of pulse energy, pulse duration, and pulse frequency for laser micromachining successfully. Let us define

$$q = \frac{P}{\frac{\pi D^2}{4} \times PD}, \tag{16}$$

Equation (15) can then be written as

$$T - T_o = \frac{\xi \times \eta(N(f,v,)) \times q}{R} \exp\left(-\frac{v}{2\alpha}(R+x)\right) \tag{17}$$

4.3 Morphology of laser micromachining DF2 (AISI-O1) surface

On the basis of Eq. 17, it shows that the temperature of laser processing is relied on various parameters, such as laser input energy, laser feedrate (scanning speed), laser pulse duration, laser pulse frequency, the properties of the processed materials, and so on. For example, for a certain laser processing temperature, the laser scanning speed (feedrate) is correlated with the input energy of laser processing. In other words, the scanning speed (feedrate) of laser processing should be increased simultaneously with the increment of the input energy of laser processing, and vice versa. In the viewpoint of enhancing or heightening the laser processing efficiently, lowing the input energy of the laser processing or increasing the scanning speed (feedrate) must be taken to minimize the laser-affected zone on the substrate and

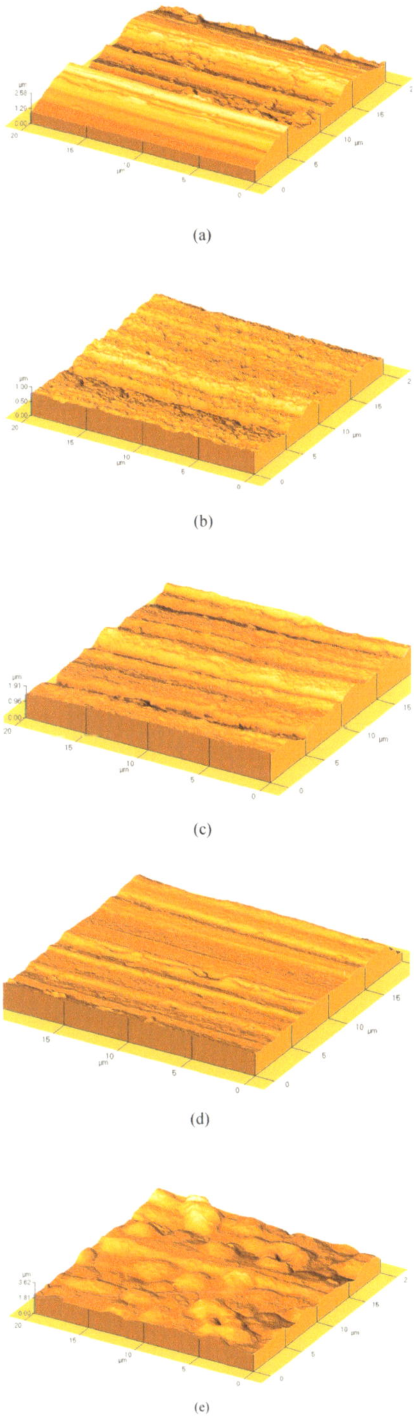

Figure 17.
Three-dimensional morphologies of laser-micromachined surface. (a) Laser-micromachined DF2 (AISI-O1) steel with lower input energy. Processing parameters: v (feedrate) = 400 mm/min, P = 1 J, PD = 2 ms, f = 15 Hz; (b) laser-micromachined DF2 (AISI-O1) steel with higher input energy than (a) processing parameters: v (feedrate) = 300 mm/min, P = 1 J, PD = 3 ms, f = 20 Hz; (c) laser-micromachined DF2 (AISI-O1) steel with medium input energy. Processing parameters: v (feedrate) = 400 mm/min, P = 2 J, PD = 3 ms, f = 20 Hz; (d) laser-micromachined DF2 (AISI-O1) steel with higher input energy than (c). Processing parameters: v (feedrate) = 200 mm/min, P = 2 J, PD = 4 ms, f = 25 Hz and (e) laser-micromachined DF2 (AISI-O1) steel with more higher input energy. Processing parameters: v (feedrate) = 100 mm/min, P = 3 J, PD = 4 ms, f = 30 Hz.

save the source of energy. The three-dimensional morphologies of the laser-micromachined surface with various input energy measured by AFM were shown in **Figure 17**.

Figure 17 shows that the morphology of laser-micromachined surface is changed with laser input energy. When the input energy is lower, the effect of laser processing parameters on the morphology of DF2 is not distinct, as shown in **Figure 17a**. With the increase of input energy, there are large amount of micro-pits distributed on the micromachined surface (**Figure 17b**) which could efficiently improve the load-carrying capacity and permit ease of release for mold made by DF2 [14]. Also, for the surface material removal, the main mechanism of the laser-processed surface relies on evaporation, melting, and its subsequent re-solidification. When the input energy of laser processing is lower, the influence of laser micromachining in this stage will mainly depend on the substrate melting. If the input energy of laser-processed surface of substrate is higher than that shown in **Figure 17b**, the 3D topography of the processed surface will vary accordingly. The relevant results are shown in **Figure 17c**. It shows that the micro-pits distributed on the processed surface as shown in **Figure 17b** are mostly melted and the morphology of the processed surface becomes relatively smooth. The effect of laser micromachining in this period is the synergetic action of the melting and little vaporizing of the substrate. With the input energy becoming higher, the morphology of laser processing will lie on the metal melting and medium vaporizing as shown in **Figure 17d**. It could be found that the surface is covered by melted metal with smoother texture and the property of substrate is changed to some extent because there are more laser energy inputs into the substrate. With the further increase of the input energy into the substrate, the surface of DF2 steel will be melted and vaporized seriously as shown in **Figure 17e**. It can be seen that there are

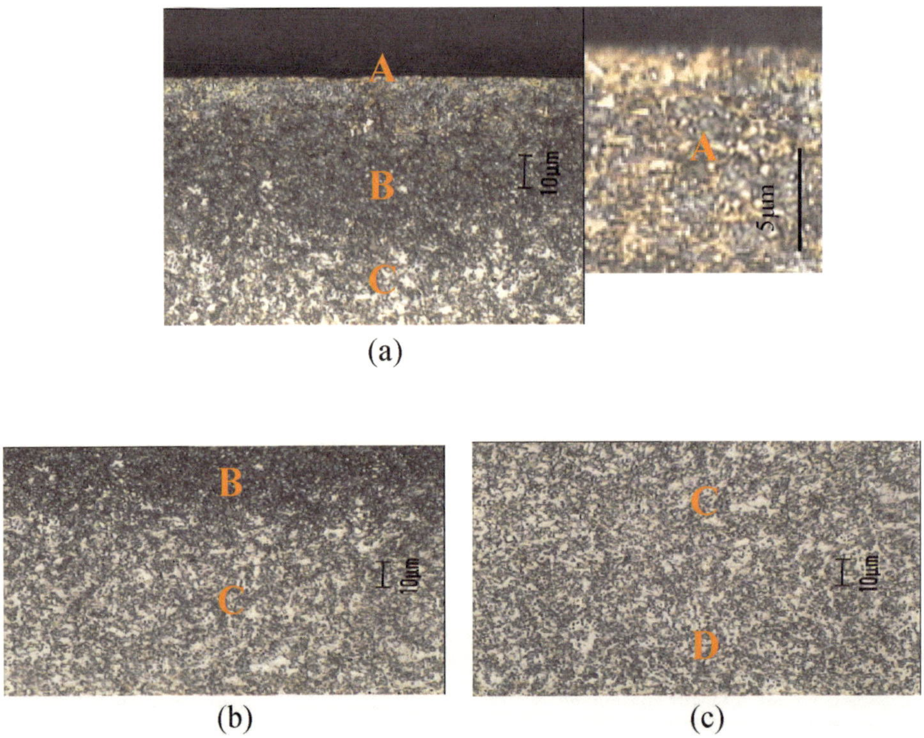

Figure 18.
Cross-section microstructures of laser-micromachined specimen (a) A-Harden layer, B-Fine grained zone, C-Heat affected zone, (b) B-Fine grained zone, C-Heat affected zone, (c) C-Heat affected zone, D-Substrate.

lots of cracks, craters, and protuberances on the processed surface resulting in coarser surface texture, which led to the poor properties of surface. It illustrates that the surface of DF2 steel could be processed successfully under the controlled input energy.

4.4 Microstructure of laser micromachining

Figure 18 shows the cross-section micrograph of laser-micromachined specimen. Obviously, the processed substrate is altered and can be detected directly. The smoother processed surface is achieved by laser micromachining with the microstructure variation as shown in **Figure 18a** labeled as zone A. Zone A, known as the hardened layer, distributes certainly enlarged and spheroidized carbide particles (high magnification in **Figure 18a**). Compared with **Figure 18a** and **b**, there are finer spheroidized carbide particles distributed in the matrix of ferrite as shown in zone B. Consequently, the property of the processed surface of the mold or die is improved to meet well with the practical applications. The heat-affected zone (HAZ), zone C, is shown in **Figure 18b** and **c**. The microstructure of the substrate, zone D, is shown in **Figure 18c**. Comparing **Figure 18c** with **Figure 3**, the microstructure does not change and keeps the same. Furthermore, the microstructure shown in zone C is almost the same as that shown in zone D, which means the energy input into the substrate by laser micromachining is too low to effectively change the mechanical property of materials. Additionally, the average thickness of zone A and zone B is ~40 μm as shown in **Figure 18a**, while the average thickness of zone A is about 10 μm. As a result, the machined surface is suitable for micro-industry application.

5. Conclusions

1. DF2 (AISI-O1) cold work steel could be successfully micromachined by pulsed Nd:YAG laser. The optimum laser micromachining parameters achieved by experiments are $P = 1$ J, feedrate = 300–400 mm/min, $PD = 3$–4 ms, and $f = 20$–25 Hz.

2. According to the onset of laser processing temperature, $T - T_o = \frac{\xi \times \eta(N(f, v)) \times q}{R} \exp\left(-\frac{v}{2a}(R + x)\right)$, morphologies of the laser-irradiated surface can be controlled with the relevant processing parameters which is attractive and beneficial to the practical applications in the micromachining world.

Author details

Kelvii Wei Guo
Department of Mechanical and Biomedical Engineering, City University of Hong Kong, Kowloon, Hong Kong

*Address all correspondence to: kelviiguo@yahoo.com

References

[1] Steen WM. Laser Material Processing. 3rd ed. London: Springer-Verlag; 2003

[2] Kurt S, Wolfgang R, Oskar P. Formation of paint surface on different surface structure of steel sheet. Iron and Steel Engineer. 1997;**74**(3):43-49

[3] Ryk G, Kligerman Y, Etson I. Experimental investigation of laser surface texturing for reciprocating automotive components. Tribology Transactions. 2002;**45**(4):444-449

[4] Temmler A, Willenborg E, Wissenbach K. Design surface by laser remelting. Physics Procedia. 2011;**12**(1): 419-430

[5] Lugomer S. Laser Technology-Laser Driven Processes. Englewood Cliffs, NJ, USA: Prentice Hall Inc; 1990. pp. 419-439

[6] Duley WW. Laser Processing and Analysis of Materials. New York and London: Plenum Press; 1983

[7] Nüsser C, Wehrmann I, Willenborg E. Influence of intensity distribution and pulse duration on laser mico polishing. Physics Procedia. 2011; **12**(1):462-471

[8] Galarneau P, Levesque M, Beaulieu R, et al. Laser micromachining of optical fibre: An instrumentation enabler. Physics Procedia. 2010;**5**(1): 477-482

[9] Murthy Peri MD, Varghese I, Cetinkaya C. Laser cleaning for removal of nano/micro-scale particles and film contamination. In: Developments in Surface Contamination and Cleaning. Methods for Removal of Particle Contaminants. Elsevier, UK: William Andrew; 2011. pp. 63-122

[10] Guo KW. Effect of polishing parameters on morphology of DF2 (AISI-O1) steel surface polished by Nd: YAG laser. Surface Engineering. 2009; **25**(3):187-195

[11] Guo W. Effect of irradiation parameters on morphology of polishing DF2 (AISI-O1) surface by Nd:YAG laser. Research Letters in Materials Science. 2007;**51316**:1-5. DOI: 10.1155/2007/51316

[12] Shen N, Matthews MJ, Fair JE, et al. Laser smoothing of sub-micron grooves in hydroxyl-rich fused silica. Applied Surface Science. 2010;**256**(12): 4031-4037

[13] Janna WS. Engineering Heat Transfer. Boston: PWS Engineering; 1986

[14] Jastrzebski ZD. The Nature and Properties of Engineering Materials. 3rd ed. New York: John Wiley & Sons; 1987

Chapter 6

Development of Lightweight Steels for Automotive Applications

Miklós Tisza

Abstract

The automotive industry plays a determinant role in the economy of developed countries. Sheet metal forming is one of the most important processes in car manufacturing. Recent trends in car production may be characterized by the application of lightweight principles. Its main priority is to fulfill both the customers' demands and the increased legal requirements. The application of high strength steels may be regarded as one of the potential possibilities. Applying high strength steels has a positive response for many of the requirements: increasing the strength may lead to the application of thinner sheets resulting in significant mass reduction. Mass reduction is leading to lower consumption with increased environment protection. However, increasing the strength can often lead to the decrease of formability, which is very unfavorable for the forming processes. In this chapter, an overview of recent material developments in the automotive industry concerning the use of new-generation high strength steels will be given. In this paper, the material developments are emphasized from the point of view sheet metal forming; therefore, our focus is on the body-in-white manufacturing in the automotive industry.

Keywords: high strength steels, automotive industry, low-cost manufacturing

1. Introduction

The ever-increasing competition in the automotive industry requires the application of low-cost production, which is strongly connected with lightweight manufacturing. The need for lightweight manufacturing in the vehicle industry is supported by several reasons: among them, the stringent and further increasing environment restrictions, the need for the reduction of harmful emissions, and the higher safety requirements should be mentioned. In the fulfillments of these requirements, the weight reduction has a determinant role. In the total weight of an automobile, the car body has a decisive role. Sheet metal forming is regarded as one of the most important manufacturing processes in the production of car body elements. Therefore, the elaboration of new, low-cost manufacturing processes is one of the main objectives in sheet metal forming: in these developments, the lightweight production principles are of utmost importance. The two main trends for producing lightweight automotive parts are the application of high strength steels or lightweight materials—especially various high strength aluminum alloys [1]. In this chapter, we will mainly focus on high strength steel materials.

2. Main requirements in the automotive industry as the driving forces of development

Considering the major requirements for the automotive industry in the recent decades, the main driving forces of material developments can be clearly defined, too.

The competition in car manufacturing is extremely strong, and the requirements are often contradictory: for example, from the customers' side, more economical, safer, and higher comfort together with better performance are the most important issues. These are further increased by legal requirements as the ever-increasing rigorous environment restrictions as the reduction of harmful emissions and higher safety requirements. Some of the legal requirements are in accordance with the customers' demands; some imposes further requirements on car manufacturing. Due to the worldwide competition in car manufacturing, the automotive industry has to find the appropriate answers for these challenges. To meet all these requirements is impossible with conventional materials and conventional manufacturing methods. This is one of the main reasons that the development needs in the automotive industry are the main driving forces in material development, too.

In the fulfillment of these manifold requirements, the weight reduction has an important role: reducing the overall weight of vehicles results in lower consumption and thus less harmful emissions together with more economical vehicles and increased environmental protection. If we analyze the potential weight reduction in various parts of a regular automobile [2], it can be seen that about 45% of the total weight is covered by the body parts, chassis, and suspension elements (**Figure 1**); thus we have to focus on these components. These parts are mainly produced by sheet metal forming: this is why the sheet metal forming as a key technology has a critical role in the weight reduction of automobiles and why lightweight design principles are in the forefront of research and development in the automotive industry.

Applying lightweight design principles in the body-in-white production necessitates the application of thinner sheets; however, both the customers' demand and the legal prescriptions require higher safety. To solve these contradictory requirements, higher strength materials are needed. However, applying higher strength materials, it leads to further contradictions: increasing the strength results in the decrease of the formability. It is well known that strength and ductility (formability) have a hyperbolic relationship. Therefore, it is important to find a good compromise between strength and formability properties. This is a great challenge in material developments that will be analyzed in the next sections.

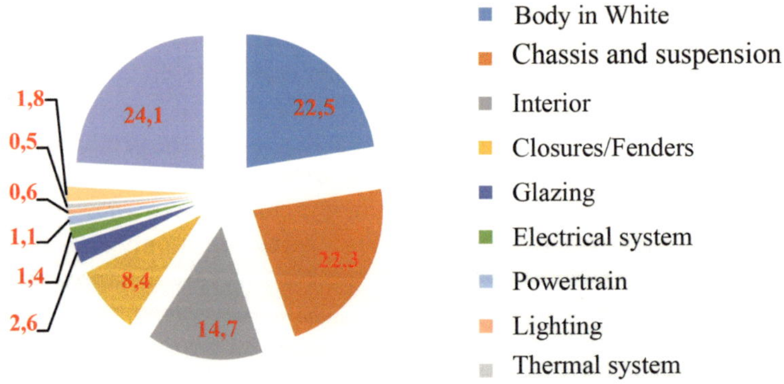

Figure 1.
Weight ratio of various vehicle components [2].

3. Material development tendencies in sheet metal forming with regard to the automotive industry

In the last 40–45 years, the reduction of fuel consumption led to the intensive development of new materials. These developments resulted in the widespread application of various grades of high strength steels. The origin of these developments can be traced back to the mid-seventieth, when the first micro-alloyed steels arrived to the industrial application. Since then, due to the continuous pressure on material development, several new high strength steel grades appeared and reached already the everyday industrial application. Systematic analysis of these developments can be found in several papers from various authors in the literature [3–7]. In the next sections, a systematic classification of these developments will be summarized.

3.1 Classification of steel developments

Steel developments—in general and particularly for the automotive industry—may be classified in several different ways. One usual way of classification is done according to the metallurgical designation. These may be grouped into low strength steels (including mild steels, interstitial free (IF) steels), conventional high strength steels like carbon-manganese (C-Mn) steels, bake-hardenable (BH) steels, high strength low alloyed (HSLA) steels, and the newer types of Advanced High Strength Steels (AHSS), e.g., Dual Phase (DP) steels, Transformation Induced Plasticity (TRIP) steels, Twinning Induced lasticity (TWIP) steels, Complex Phase (CP) steels, martensitic (MS) steels. In recent years, new AHSS grades have been developed, for example, Extra Advanced High Strength Steels (X-AHSS) and Ultra Advanced High Strength Steels (U-AHSS), and various types of the so-called third-generation AHSS steels, e.g., TRIP-aided bainitic ferrite (TBF) and Quenching & Partitioning (Q&P) or different types of NanoSteels: all these with the primary aim to provide even higher strength parameters with significantly increased formability.

Another classification introduces various mechanical properties—mainly strength and formability parameters as the Ultimate Tensile Strength (UTS) and Total Elongation (TE). This type of classification is often used together with the designation of development steel generations, as well. In **Figure 2**, the well-known relationship between strength and ductility parameters is shown applying the abovementioned classification method with a graphical representation. From **Figure 2**, it may be also seen that the product of the ultimate tensile strength and the total elongation (UTS × TE) follows a hyperbolic function. The constant (C = UTS × TE) provides further possibility to classify the steel developments by their generation.

In **Figure 2**, the material group shown by gray color includes the conventional mild steels (IF and mild steels) formerly widely applied in Body in White (BiW) production in the automotive industry. The group of conventional high strength steels (colored by light blue) includes bake hardening (BH), isotropic (IS), high strength interstitial free (HS IF), carbon-manganese (C-Mn), and high strength low alloyed (HSLA) steels. Following the conventional high strength steels, an intensive development started in the steel industry in close cooperation with the automotive industry to develop different types of Advanced High Strength Steels (AHSS) to meet the needs of lightweight automotive structures.

The development of first generation of Advanced High Strength Steels for automotive application may be regarded as the first step in this development process. DP steels, Complex Phase (CP) steels, certain types of TRIP steels, and martensitic steels (MART) belong particularly to this group. For these steels, the C constant

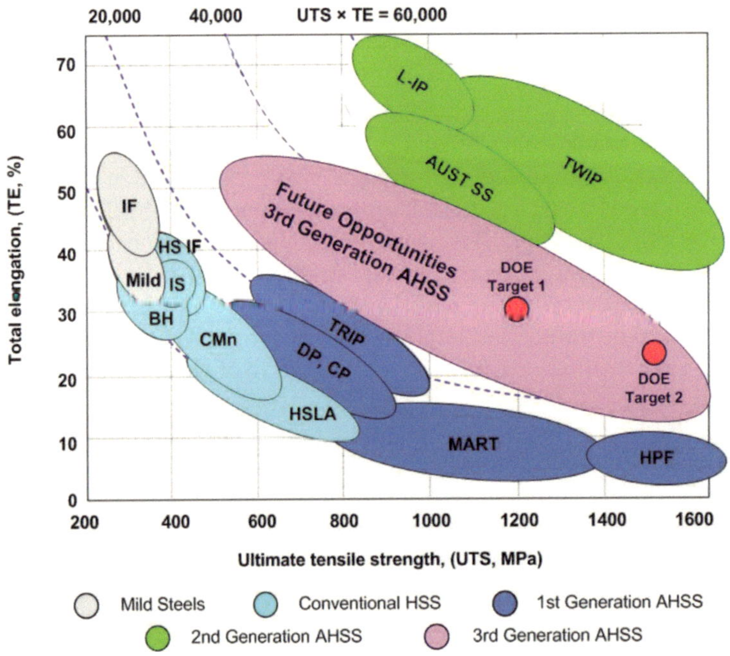

Figure 2.
Relationship between ultimate tensile strength (UTS) vs total elongation (TE) for various generations of high strength steels [4].

defined above can be found between 10,000 and 25,000. The first-generation AHSS (often referred as conventional AHSS) grades have good strength but limited ductility.

However, it is worth mentioning that for these Advanced High Strength Steels, the increase of strength parameters is much more significant than the decrease of the ductility parameters. This is particularly valid for the Dual Phase (DP) steels, Complex Phase steels (CP), Martensitic Complex Phase (MART/CP) steels, and TRIP steels. This is the reason why this group gains wide application in car body production.

The group of steels that can be found around the C = 40,000–60,000 MPa × % may be regarded as the second generation of Advanced High Strength Steels. This group includes the Twinning Induced Plasticity (TWIP) steels and some austenitic Stainless Steels (AUST SS) with high manganese content. These steel grades provide superior combination of strength and ductility. TWIP steel had successfully been trial-produced at POSCO in the early 1990s, but the trial was not extended to commercialization due to limitations in facilities and productivity [4]. Trial productions have also been made at some European steel companies. These attempts demonstrated the outstanding mechanical performance of TWIP steels; however, these trial productions turned to be commercially unsuccessful due to low productivity and high cost [5]. New approaches and further developments are done continuously to reduce these difficulties and make them suitable for automotive parts manufacturing.

The third generation of AHSS is still in the phase of development—though there are already industrial realizations, too. In this development stage, several new concepts have been already proposed. The main target in developing the third generation of AHSS is twofold, i.e., to achieve mechanical properties in the range between the first- and second-generation AHSS shown in the strength-ductility diagram (**Figure 2**) but with less alloying elements and, hence, with less expensive production than the second-generation AHSS steels [6]. The microstructure of these steels consists of a high strength phase (e.g., nano/ultrafine-grained ferrite, martensite, or bainite) combined with a further phase or constituent that provides substantial ductility and work hardening (e.g., austenite). With this development concepts, very high strength steels in the GPa range with even though remarkable formability can be produced [7].

The projected changes in the application of Advanced High Strength Steels is given by Matlock et al. [8] for North American vehicle industry (**Figure 3**), but similar trend may be estimated for other geographical regions, e.g., the European Automotive manufacturers and the Far East countries (China, Japan, and Korea). In **Figure 3**, the changes of the absolute content of AHSS applications (in kg) and percentage changes related to the total weight of vehicles are shown. Both changes show an exponential increase with a slightly higher one concerning the absolute values.

3.1.1 Development projects for Advanced High Strength Steels

In the last 30–40 years, there were several projects studying and initiating the development of new grades of Advanced High Strength Steels. Most of these projects were initiated by automotive companies, and in most cases various consortiums were established for this purpose. Each of these projects aimed to meet as much as possible the requirements analyzed in Section 2.

Among these projects, the Ultralight Steel Automotive Body (ULSAB) satisfied most of the requirements stated for a lightweight automotive structure and proved to be structurally sound, safe, executable, and affordable. Though it was a highly expensive project with the participation of 35 companies representing 18 countries, it could meet the challenges to reduce the weight of steel auto body structures at no additional costs, while maintaining or even improving the overall performance [9].

Further projects followed the ULSAB concept, among them the Ultralight Steel Auto Closures (ULSAC) [10], or the Ultralight Steel Auto Body-Advanced Vehicle Concept (ULSAB-AVC) [11], and the Future Steel Vehicle (FSV) [12]. All these projects led to the further development of Advanced High Strength Steels reaching the GigaPascal range of strength together with increased ductility.

In the previous section, we could see the main material development tendencies and their classification that included various kinds of conventional steels as well,

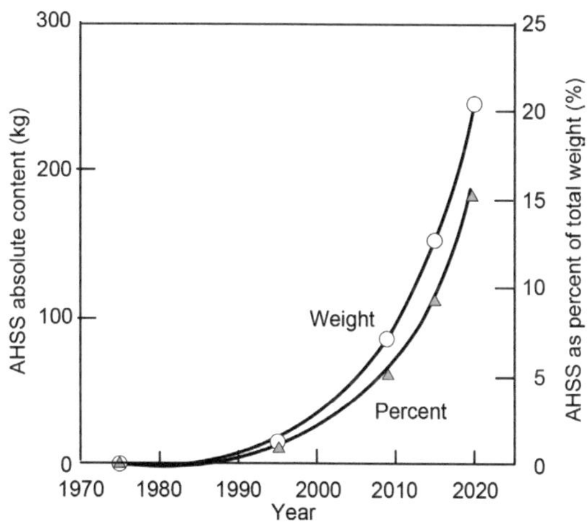

Figure 3.
Projected changes of the absolute amount (kg) and the percentage values for the total weight of vehicles.

which had a prominent role in the history of car making in the last century. In the next sections, we will mainly focus on the main types of Advanced High Strength Steels and their properties.

4. Main types and properties of Advanced High Strength Steels

AHSS are complex, sophisticated materials, with carefully selected chemical compositions and multiphase microstructures, achieved by precisely controlled heating and cooling processes. Various strengthening mechanisms are applied to get significantly increased strength, better formability, improved toughness, and fatigue properties to meet the various requirements that are defined for automotive body structures [13].

The group of AHSS materials includes Dual Phase (DP), Complex-Phase (CP), Ferritic-Bainitic (FB), Martensitic (MS), Transformation-Induced Plasticity (TRIP), Hot-Formed Press Hardened (HF, or PHS), Twinning-Induced Plasticity (TWIP) and some austenitic stainless steels (AUST SS) with high manganese content. These first- and second-generation AHSS grades due to their unique mechanical properties are well qualified to meet many of the functional and performance requirements in automobiles. From these generations of AHSS, DP and TRIP steels are excellent in the crash zones due to their high-energy absorption [14], while structural elements of the passenger compartment can be made from extremely high strength steels, such as martensitic and boron-alloyed Press Hardened Steels (PHS), and hence, resulting in improved safety performance.

Recently, there is an increased research interest for the development of the third generation of AHSS. These steels usually apply special alloying and thermomechanical processing to provide improved strength-ductility combinations compared to the present first- and second-generation AHSS grades but at lower costs. There are several good examples for these, e.g., in the USA, a program sponsored by the Department of Energy made available the development of 1200 MPa steels with threefold improvements in ductility [15]. New generation of Advanced High Strength Steel (AHSS) grades contains significant alloying and multiple phases. The multiple phases provide increased strength and ductility not attainable with single-phase steels, such as the high strength, low alloyed (HSLA) grades. In the next sections, these AHSS will be discussed.

4.1 Dual-Phase (DP) steels

As we could see from the historical analysis, Dual-Phase steels have a dominant role in the last 40 years of automotive industry; therefore, we start the overview of Advanced High Strength Steels with this group.

The development of Dual-Phase (DP) steels was right at the beginning of the new age of steel development following the conventional high strength steel era. Current commercially available and widely applied AHSS steels have evolved from significant early work on Dual-Phase steels in the late 1970s and early 1980s. Dual-Phase steels are one of the more widely applied Advanced High Strength Steels in todays' car making industry. This is mainly due to their better strength and formability parameter combination than the conventional high strength steels like HSLA steels. DP steels possess high specific strength, good initial work hardening rate, continuous yielding behavior, and superior ductility compared to conventional steel grades. These properties make them particularly suitable for body structures, closures, fuel tanks, etc. in vehicles [16].

Dual-Phase (DP) steels generally consist of ferrite matrix containing mainly hard martensite or in some cases bainite second phase as islands as shown in **Figure 4**. It is very characteristic that the ferrite phase is generally continuous providing excellent ductility. During forming, strain is concentrated in the lower strength ferrite phase surrounding the martensite islands providing unique work hardening rate that is experienced in DP steels.

4.1.1 Processing of DP steels

There are various commonly used processing routes for producing DP steels. One of the methods (Route A in **Figure 5**) involves rapid cooling from the intercritical temperature to room temperature directly. The resulting microstructure comprises ferrite and martensite [17]. Higher intercritical temperatures, for the same soaking period, result in larger amounts of martensite with increased tensile strength and decreased percentage elongation [18]. It is reported by several papers [19] that the increase in martensite fraction in DP steels promotes crack initiation and thus results in worse ductility. Therefore, martensite fraction should be kept in the range of 10–40%.

Another method for processing of DP steels (Route B in **Figure 5**) applies first slow cooling from the austenitic region to the desired ferrite transformation temperature, followed by quenching to room temperature for transforming the

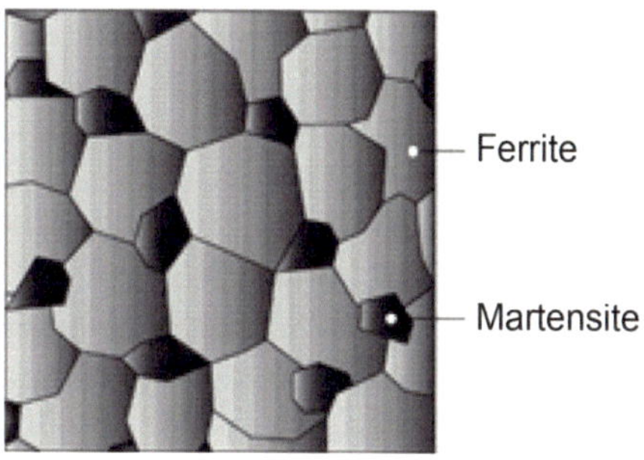

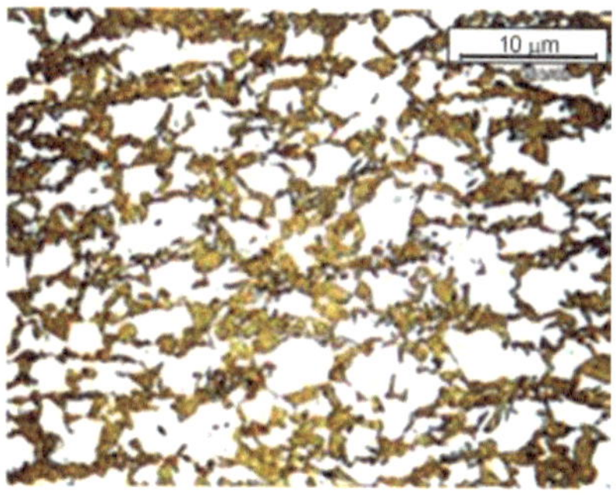

Figure 4.
Schematic view and real micrograph of a DP steel. Left: Schematic view of a microstructure of a DP steel containing martensite islands in ferrite matrix. Right: Micrograph of a DP 690 steel containing martensite islands in ferrite matrix.

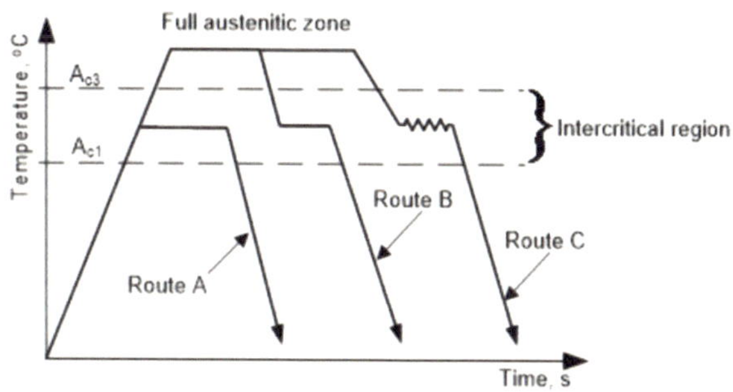

Figure 5.
Processing routes for producing DP steels.

remaining austenite to martensite [20]. The properties obtained by this method include lower tensile strength and higher ductility than those of gained by the first method (Route A).

The third method for producing DP steels (Route C in **Figure 5**) involves hot rolling of steel, followed by first slow cooling to the intermediate temperature, followed by second cooling at a very fast rate and finally slow cooling (i.e., coil cooling) to room temperature. This method of cooling is known as ultrafast cooling (UFC), and the processing route is referred to as new-generation thermomechanical controlled processing [21]. The properties obtained by Route C are better (as compared to those obtained by Route A and Route B) because higher grain refinement is achieved during rolling.

Several authors have reported that DP steels with ultrafine bainite and fine ferrite-bainite/martensite microstructure with precipitation hardening can achieve good strength without loss of ductility, making this steel category more suitable for third-generation AHSS [5].

4.2 Transformation-induced plasticity (TRIP) steels

Advanced high-strength transformation-induced plasticity (TRIP) steels are well suited for lightweighting car body construction with added advantage to reduce the safety problems. TRIP steels can be found already in the 1st+ generation AHSS as shown in **Figure 2**. One of the main features of TRIP steels that the strain or stress-induced transformation of retained austenite present in the microstructure in a sufficient amount can substantially harden the steel during deformation depending on the processing route and therefore results in a higher ductility [22].

The microstructure of TRIP steels contains retained austenite embedded in a primary matrix of ferrite. **Figure 6a** shows schematic microstructure of TRIP steel, while **Figure 6b** is a micrograph of a typical TRIP steel (TRIP 700).

In addition to a minimum of 5 vol.% of retained austenite, hard phases such as martensite and bainite are present in varying amounts. TRIP steels typically require an isothermal hold at an intermediate temperature, which produces some bainite.

TRIP steels are characterized by a relatively low content of alloying elements. For example, in TRIP 790 steel (UTS ≈ 790 MPa), the total content of alloying elements is about 3.5 wt.%. Thus, the selection of suitable alloying elements and the amount required to produce the intended properties is critical in the alloy design stage. The carbon content in TRIP steels is higher than in DP steels. Carbon

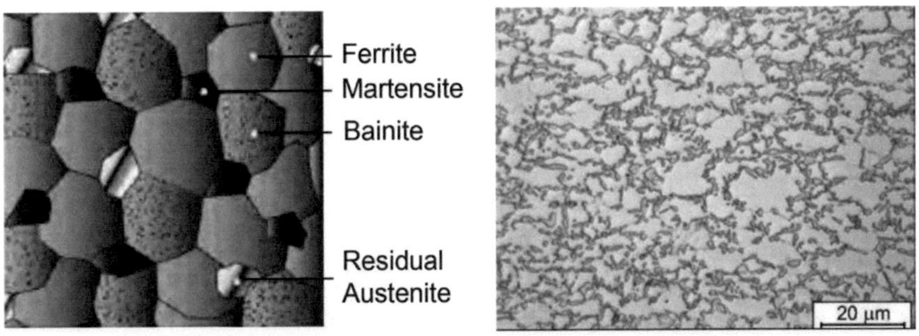

Schematic view of the microstructure of a TRIP steel

Micrograph of a typical TRIP steel (TRIP 700)

Figure 6.
Schematic view and micrograph of the microstructure of TRIP steel. (a) Schematic view of the microstructure of a TRIP steel and (b) micrograph of a typical TRIP steel (TRIP 700).

is generally kept in the range of 0.20–0.25% because of weldability reasons. The higher carbon content is necessary for stabilizing the retained austenite phase to below ambient temperature. In TRIP steels, austenite stabilizers are present, mainly C, Mn, and/or Ni. These elements assist maintaining the necessary carbon content within the retained austenite. TRIP steels mainly contain multiphase microstructures composed of about 50–55% ferrite, 30–35% bainite, 7–15% retained austenite, and 1–5% martensite.

The outstanding combination of ductility and strength in TRIP steels is a result of deformation based on transformation of retained austenite to martensite [23]. This transformation (on deformation) of phases is called the TRIP effect that provides excellent strength and elongation combination together with high impact resistance. These characteristics predestinate TRIP steels as good candidate for the third-generation AHSS, too. Dispersed hard second phases in soft ferrite provide high work hardening rate, as experienced in DP steels, too. Furthermore, in TRIP steels, the retained austenite progressively transforms to martensite with increasing strain, thereby increasing the work hardening rate at higher strain levels.

4.2.1 Processing of TRIP steels

The main processing of TRIP steels consists of heating the steel to the austenitic zone, cooling down to the intercritical region followed by deformation here, and quick transfer to the bainitic zone with subsequent soaking there, and finally quenching to room temperature (as shown in **Figure 7**).

The deformation in the intercritical region increases the rate of austenite (γ) to ferrite (α) transformation. The remaining austenite is enriched with carbon content, which stabilizes the γ phase. Furthermore, this deformation increases the nucleation rate of bainite but decreases its growth rate that results in small plates of bainite. This part of the T–t cycle also helps to enrich the γ phase with carbon and further increases the stability of γ phase. The stability of retained austenite is enhanced by the high carbon content, and the more carbon in γ phase results in more stability of γ during the TRIP effect, too, since more stable austenite needs more time to transform into martensite; these processes contribute to the increase of the ductility. The austenite to martensite transformation increases the tensile strength of the final microstructure. With this process, an improved strength–ductility combination is achieved [24]. Obviously, this processing route of TRIP steels is more time-consuming. This is because it needs special arrangements to deform the material at high temperature, to hold the specimen in the bainite region, and so on. This limits the use of TRIP steels in industrial applications. Some authors [25] using this route reported that rolling in the intercritical region improves TRIP steel

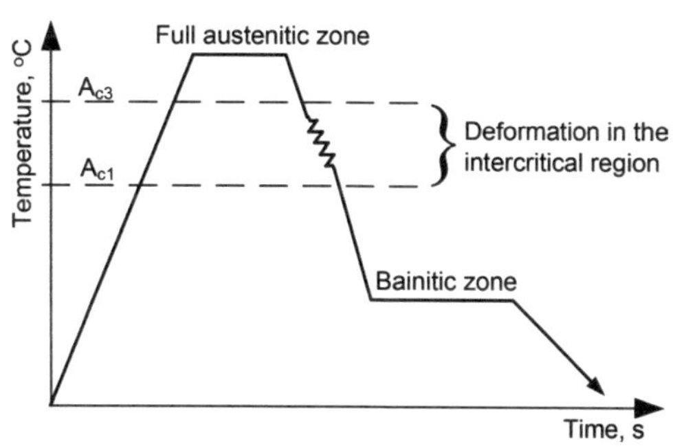

Figure 7.
Conventional processing route of TRIP steels.

properties by enhancing the carbon content and dislocation density, decreasing the grain size, and resulting in a granular type morphology.

4.3 Complex Phase (CP) steels

Complex Phase (CP) steels belong to the group of steels with usually very high ultimate tensile strength (UTS ≈ 800 MPa or even greater). CP steels generally have chemical composition and microstructure similar to TRIP steels, but it contains some quantities of other elements, e.g., Nb, Ti, and V. These additional elements enhance the precipitation strengthening effect. CP steels typically do not have retained austenite, but contain more hard phases like martensite and bainite within the ferrite/bainite matrix.

The mechanical properties of CP steels may be characterized by continuous yielding and high uniform elongation. CP steels with the bainitic matrix have excellent formability. It is primarily due to the relatively small difference between the hardness of bainite and martensite. In CP steels, the bainitic ferrite is strengthened by high density of dislocations (dislocation density is above $\rho > 10^{12}/cm^2$) together with fine dispersion of martensitic second phase and carbo-nitrides or carbides. This bainite microstructure of CP steels exhibits better strain hardening and strain capacity than that for fully bainitic microstructure. In its microstructure, the martensite and bainitic ferrite phases are separated by a third phase of intermediate strength.

4.4 Martensitic (MS) steel

Martensitic steels (MS) have mostly martensitic microstructure with some small amounts of ferrite and bainite. These steels have the highest strength but lowest formability. Martensitic steels, currently available with strengths of 900–1800 MPa, are used for body parts where deformation may be limited [26].

Producing MS steels, the austenite is transformed almost entirely to martensite during quenching on the run-out table or in the cooling section of the continuous annealing line. MS steels may be characterized by martensitic matrix containing small amounts of ferrite and/or bainite. Within the group of multiphase steels, MS steels have the highest tensile strength level. Martensitic steels show the highest ultimate strength in final products, up to 1800 MPa or even higher [27]. Their concept is based upon well-established rules with respect to chemical composition and processing technology. In order to improve ductility and provide adequate formability even at extremely high strength values, MS steels are often subjected to post-quench tempering.

Additional carbon in MS steels increases the hardenability and contributes to further strengthening the martensite. Further elements (like manganese, silicon, chromium, molybdenum, boron, vanadium, and nickel) are used in various combinations to further increase hardenability. Microstructure of martensitic steels is mainly composed of lath martensite, which is developed by the transformation of austenite during quenching after hot rolling or annealing. Martensitic steels are very hard to form, so they typically are roll formed or press hardened (hot stamped): it will be detailed in the next section where the Press Hardening Steels (PHS) will be described.

4.5 Press Hardening Steels (PHS)

Among the Advanced High Strength Steels, Press Hardening Steels (PHS) form a unique group: these are mostly different kinds of boron-alloyed manganese steels and gain wide application to produce high strength structural body elements (e.g., A- and B-pillars, etc.). Press Hardening Steels are widely used in car body manufacturing in hot forming processes. There are several grades of Press Hardening Steels; among them, the 22MnB5 alloy is regarded as the basic type of PHS steels. Here, the hot forming of Press Hardening Steels will be analyzed.

4.5.1 The hot forming process of PHS steels

Hot forming of steels is a complex forming and tempering operation: it is often termed as hot press forming or press hardening of steels, too. The full austenitization of the material is regarded as the first step in hot press forming. Forming is performed in this state when the material has good formability; then the part is cooled down rapidly in the tool applying the critical cooling rate, hence resulting in martensitic microstructure.

The usual temperature–time diagram for hot press forming is shown in **Figure 8**. Through the above-described combination of heating, holding, forming, and rapid cooling, very complex parts can be produced with excellent strength properties [28]. There are various process variants in hot press forming: among them, the so-called direct and indirect hot forming may be regarded as the basic ones. In direct or single-stage hot forming, the blank sheet is directly austenitized, then transferred to the stamping tool, and cooled down rapidly in the forming tool providing excellent strength properties [29]. In indirect or often termed as two-stage hot press forming, the initial blank is formed in cold state, and then either a hot forming is used to produce the complex parts or just a calibration process occurs in hot forming condition. The austenitization and the subsequent quenching are the inherent parts of this process chain, too, to provide the required high strength properties.

There are further process variants besides these two basic ones: the final microstructure, as well as the mechanical properties of the part, can be effectively controlled by the holding temperature and the controlled cooling process. These process variants may be derived either by altering the holding temperature or by changing the cooling rate. Depending on the holding temperature, two further process variants can be proposed: full austenitization is the basic alternative, i.e., when the holding temperature is selected in the homogeneous γ-zone. A further process variant depending on the holding temperature is derived if the holding

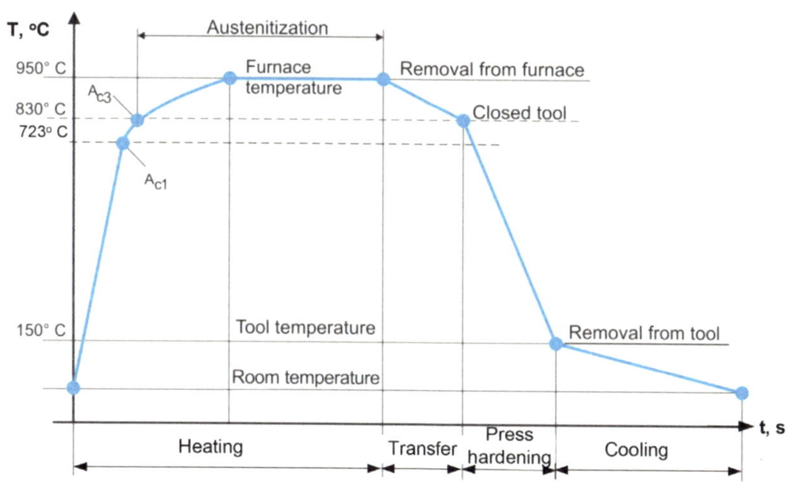

Figure 8.
Temperature vs process time for hot press forming of PHS.

temperature is in the ($\alpha + \gamma$) intercritical zone (i.e., between the A_1 and the A_3 temperature). In this case, there is no full austenitization; the starting microstructure contains, besides austenite, ferrite, too. Obviously, just the austenite content can be transformed into martensite, and the final microstructure after the hot forming and cooling process is completed has a certain amount of ferrite, too. Obviously, it results in lower strength than the full austenitization; however, it also leads to a certain amount of ductility leading to better toughness properties, as well.

Further process variants can be also derived by changing the cooling rate after the forming process. If the cooling rate is higher than the upper critical one, the final microstructure is martensite; when the cooling rate is lower than the upper critical one, besides martensite, bainite can be also found in the microstructure. However, it also results in somewhat lower strength depending on the quantity of bainite; however, it also results in the increase of toughness that may be advantageous, for example, increasing the crashworthiness of the part due to the better energy absorption properties of bainite [30].

It is essential that the forming could be finished above the M_s temperature: at this stage, these material grades still have suitable formability. After forming, the component is cooled down together with the tool: this cooling should provide the critical cooling rate to get high strength of martensitic microstructure. By this process, springback is eliminated, and very strong components can be formed to complex geometries.

Typical press hardened steels (PHS) have tensile strength of 1500–2000 MPa. In the last decades, they are already extensively used in safety and crash-resistant car body components. New-generation PHS are expected to have higher strength even above 2000 MPa. However, it should be noted that these PHS grades are used where only very small deformation is allowed. These steels have been adopted for use in many parts, including, for example, sill structures, or A- and B-pillar reinforcements. Recently, many floor panels also are made by hot forming to save weight.

4.6 Twinning-induced plasticity (TWIP) steels

TWIP steels belong to the second generation of AHSS and are based on the potential mechanism of obtaining a superior balance of tensile strength and elongation using the TWIP effect. The name of this steel is originated from this characteristic deformation mode, i.e., the twinning. The twinning causes high value of the instantaneous hardening rate (n-value) as the microstructure becomes finer and finer. The resultant twin boundaries serve as grain boundaries and strengthen the steel (**Figure 9**).

TWIP steels have high manganese content (Mn = 17–24%) that causes the steel to be fully austenitic even at room temperatures. TWIP steels are normally composed of Fe, Mn, or Ni (15–35%), Si (1–3%), and Al (1–3%) [31]. These steels exhibit outstanding tensile strength-ductility combination (e.g., a TWIP steel with tensile strengths above 1000 MPa may possess 50–60% ductility) [32]. The n-value may increase to a value of 0.4 that may result in 50–60% uniform elongation. The tensile strength may be even higher than 1500 MPa [33].

In TWIP steels, the strain hardening is strongly dependent on the stacking fault energy (SFE). This parameter controls the deformation behavior of the steel. Alloying elements generally decrease SFE leading to enhanced twinning behavior during deformation and hence lead to improved ductility. It is also known that SFE $< 20\ mJ/m^2$ causes austenite to martensite conversion and by this results in the TRIP effect. For pure twinning, SFE is desired to be greater than $20\ mJ/m^2$. Aluminum is added to steel to raise SFE, to retard the TRIP effect and to result in pure twinning.

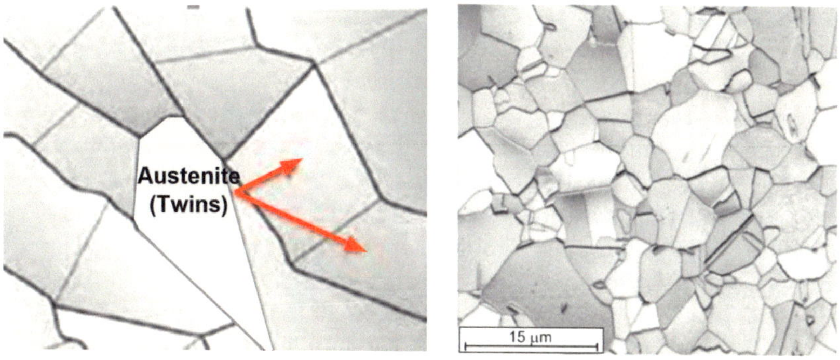

Figure 9.
Schematic view and micrograph of TWIP steel microstructure. Left: Schematic view of TWIP steel microstructure. Right: Micrograph of a TWIP steel in annealed condition.

TWIP steels show superior mechanical performance, but this category is not practically viable for industrial applications because of its limitations: poor productivity and high production costs. The main production route of TWIP steels includes homogenizing above the upper critical temperature for a long period and quenching to room temperature [34]. TWIP steels can be also produced by homogenizing, followed by deformation at a temperature above the upper critical one, with subsequent quenching to room temperature. Deformation at higher temperature provides fine grain size and high volume fraction of twins. The finer the grain structure, the more twinning occurs that improves ductility and strength.

Two types of twins are observed in the TWIP steels: (a) annealing twins caused by heat treatment and (b) deformation twins caused by deformation. The yield stresses of coarse-grained TWIP steels usually result in less than 400 MPa strength, which restricts the use of TWIP steels in the automotive sector, particularly for those parts that are supposed to be active during a crash. Many attempts are reported in the literature to increase the yield strength of TWIP steels. These attempts include, for example, grain size refinement by using V, Ti, and Nb as alloying elements to enhance precipitation of carbides, cold rolling followed by annealing treatment, and partial recrystallization [35].

5. Recent results and future trends in AHSS development

As it was already discussed at the Classification of AHSS Developments (Section 3.1), the main target in developing the third-generation AHSS is to achieve the properties in the range between the first- and second-generation AHSS with less alloying elements, hence, with less expensive processing that are suitable for early commercialization. The range of third-generation AHSS (3GAHSS) development maybe clearly identified on the diagram of tensile strength vs total elongation in between the first- and second-generation AHSS regions as shown in **Figure 2**.

However, it is also obvious that potential production requires a systematic design methodology to identify the possible combinations of microstructural constituents, which may lead to the required mechanical properties.

One of the possibilities to apply a systematic design methodology is the application of a simplified composite model [36] considering various combinations of multiphase (ferrite, austenite, bainite, and martensite) materials. With the variations of phase fractions in the hypothetical microstructure, the predicted mechanical properties can be calculated.

Another possibility to use a systematic design methodology is the application of the Integrated Computational Materials Engineering (ICME). It provides a framework for utilizing computational multiscale material development driven by multidisciplinary engineering design, analysis, and performance requirements [37]. This concept is initiated and supported by the National Research Council in the USA [38]. The ICME model can be used to guide both the material selection and the design optimization. Alternatively, it can be also used to new material development to get the best-suited macroscopic properties for a given structural application, through the determination of chemical composition and microstructural characteristics in a "reverse engineering" approach. In the automotive industry, the potential of the ICME method for vehicle lightweighting was recognized by the United States Department of Energy (DOE), too, which funded the project "Integrated Computational Materials Engineering Approach to Development of Lightweight Third Generation Advanced High Strength Steel (3GAHSS) Vehicle Assembly [39]. The abovementioned ICME approach was implemented in this project in two ways. First, the ICME principles were applied in the development of a material modeling tool set by combining material models at different length scales. Second, the Combined Constraints Crystal Plasticity (CCCP) model was implemented as a microscale constitutive model [40]. In this project, two targets were set by the United States Department of Energy: one is the 1200 MPa strength with 30% total elongation (which means C = 1200 × 30 = 36,000 MPa × %), and the other one is 1500 MPa strength with 25% total elongation (which means C = 1500 × 25 = 37,500 MPa × %) [41].

Similar projects were initiated by other steel companies and research institutes in the world. Among others, ArcelorMittal announced systematic developments of third-generation steels [42]. The microstructure of these steels consists of a high strength phase (e.g. nano/ultrafine-grained ferrite, martensite, or bainite) combined with a further phase or constituent that provides substantial ductility and work hardening (e.g. austenite). In the next sections, some results and representatives of these 3G developments will be introduced.

5.1 Medium manganese steels

The development of austenitic steel grades with high alloying contents of manganese (15–30%) was already applied during the development of second-generation AHSS. It resulted in outstanding mechanical properties (high strength with excellent elongation), which made it attractive for the automotive industry. These high strength and ductility grades were based on the austenitic single-phase concept. Their deformation mechanisms were mainly the twinning-induced plasticity (TWIP). Additionally, it was also discovered that combining specific proportions of TWIP and TRIP mechanisms allows precise control of strength and ductility [43].

In recent steel developments, it was experienced that a further deformation mechanism is provoked when different alloying concepts are used. Microband-induced plasticity (MBIP) is one of these newly discovered mechanisms, which localizes the deformation within arrays of precipitates and, thus, retarding the onset of mechanical instability and supporting homogenous yielding. Beside the outstanding mechanical properties, the steels offer processing challenges compared to low carbon steels; however, they are very expensive due to the high alloy additions required to produce austenitic microstructure.

However, these high manganese content steels initiated the development of another new group of steels belonging to the third-generation AHSS grades, namely, the medium manganese steels. The microstructure of these steels consists of a high strength phase (e.g. nano/ultrafine-grained ferrite, martensite, or bainite) combined with a further phase or constituent that provides substantial ductility and work hardening (e.g. austenite). The carbide-free bainite (CFB) or ultrafine lamellar bainite (ULB) is another possible concept. By choosing the alloying concept and the cooling condition, it is possible to suppress the carbide formation and, thus, to produce a very fine lamellar bainitic structure with austenite films between the bainite leaves. This concept provides very high strength steels right above 1 GPa and with remarkable formability.

5.2 Quenched and partitioned (Q&P) steels

Quenched and partitioned (Q&P) steels are the result of the recent developments of third-generation AHSS steels. The elaboration of Q&P steels is partly based on the knowledge of duplex stainless steels and the quenching and partitioning process [44], as well as on the properties of medium manganese steels [45]. The Q&P steels usually contain carbon, manganese, silicon, nickel, and molybdenum alloying elements. The amount of alloying elements can be around 4%, which is much lower than that of in the second-generation AHSS. During heat treatment of Q&P steel, quenching is interrupted and is reheated for partitioning. With this reheating process, a unique microstructure is created containing 5–12% stable retained austenite, 20–40% ferrite, and 50–80% martensite.

Baosteel was one of the first companies to apply Q&P steels, initially with 980 MPa and later 1180 MPa strength [46]. It was demonstrated that a B-pillar reinforcement could be cold-formed using Q&P 1180. Auto/Steel Partnership (A/SP) also has tested Q&P 980 using GM's B-pillar die, proving that this steel has better formability and is less prone to edge cracking than DP 980.

Recently, Q&P steels were developed up to 2100 MPa tensile strength with 9% uniform elongation and about 13% total elongation. The elongation level of this steel is comparable to DP 980, which is a cold-formable grade.

5.2.1 Metallurgy of Q&P steels

Q&P steels are a series of C-Si-Mn, C-Si-Mn-Al, or other similar compositions that are processed by the quenching and partitioning (Q&P) heat treatment. Q&P steels possess an excellent combination of strength and ductility with a final microstructure of ferrite (in the case of partial austenitization), martensite and retained austenite. This microstructure makes them suitable to use in the automotive industry as new-generation AHSS. They are suitable for cold stamping of various structures and safety parts having complicated shape to improve fuel economy and promoting passenger safety.

It is possible to change the amount of retained austenite at room temperature and its stability with alloying elements as carbon, manganese, nickel, etc. based on the knowledge gained by duplex stainless steels. However, it affects the cost and may be detrimental concerning the welding properties. The third generation of AHSS grades were developed to overcome these disadvantages; few of the good examples are those third-generation AHSS that are based partly on the quenching and partitioning process (Q&P steels) and on the properties of medium manganese steels. In this case, the composition of steel is not adequate for keeping the retained austenite at room temperature, but annealing, cooling, and thermal processes are optimized to change the austenite's composition and decrease its M_s temperature. For medium-Mn steels, where a relatively larger manganese amount (typically between 5 and 8 wt. %) is characteristic, the thermal treatment is slightly simplified. The intercritical annealing provides a chance to form austenite and to increase its carbon and manganese content; then the steel is cooled down to room temperature. The complex multiphase fine-grained microstructure together with the TRIP effect arising from the progressive transformation of the retained austenite during deformation provides the excellent mechanical behavior. By these processes, the UTS above 1200 MPa and uniform elongation larger than 12% can be achieved.

5.2.2 Microstructural properties and formability of Q&P steels

The concept of Q&P process for automotive materials was first published by Speer et al. [44]. In Q&P process, the material is quenched down below the M_s temperature, where austenite is not fully transformed. Due to the alloying concept of Q&P steels, this temperature usually is in the range of 200–350°C. It means that the microstructure is a mixture of martensite and austenite. Steel is then reheated and aging is done between 300 and 500°C; this is termed as the "partitioning step." During this treatment, carbon diffuses from the supersaturated martensite, providing the carbon enrichment of austenite, which increases its stability at room temperature; furthermore, it supports further TRIP effect during deformation. Besides these, tempering of martensite occurs, which improves its damage resistance properties, while keeping high strength.

This simplified scheme does not reveal all the complex evolution of the microstructure during partitioning, and the detailed mechanisms of Q&P evolution are still a matter of debate and not fully elucidated. For instance, the formation of bainite during partitioning cannot be completely excluded; it could explain the measured carbon enrichment in the retained austenite, as the partitioning temperatures are consistent with those for bainite formation.

Even if the detailed mechanisms are not fully revealed, the benefits of Q&P treatment by the improved mechanical properties have been clearly shown. The current range of strength that can be achieved with this new concept is between 1000 and 1500 MPa, with a total elongation of 20%. Moreover, as the matrix is a kind of tempered martensite, damage resistance is improved compared to DP or TRIP steels with the same strength level.

The development of such grades requires an important modification of the annealing line; quenching and reheating step was not possible until the recent years. The strong request from the automotive market toward third-generation Advanced High Strength Steels has led steel making companies to invest in the upgrading of their annealing lines to ensure processing of Q&P steel products.

5.3 TRIP assisted bainitic-ferritic (TBF) steels

TBF, a low-alloy grade like Q&P steels, can be produced by the existing heat treatment facilities. Stable retained austenite is its key component. Bainitic ferrite matrix with retained austenite inclusions may be regarded as the most common microstructure for TBF steels. It is produced by isothermal holding in the bainitic regions after fast cooling from fully austenitic microstructures. Typical chemical compositions of TBF steels contain C, Si, and Mn as major alloying elements. Alloy modifications include variations of the Al, Nb, and Cr content [47]. The cementite formation during bainitic transformation is suppressed by the Si constituent. The added Si enhances the C content in retained austenite, and it stabilizes the austenite. High Si contents of 1.5 wt% are used in these types of steels. Consequently, the transformation of retained austenite into martensite produced by either deformation or thermal processes during final cooling is prevented. Although Si has major importance to prevent carbide precipitation during annealing of the cold rolled material, it causes problems during processing via continuous annealing lines. Therefore, other alloying elements having similar effect of suppressing carbide formation have to be considered.

5.4 NanoSteel

NanoSteel®, a third class of third-generation AHSS, is still under development and not commercially available. In 2002 (following 6 years of research at Idaho National Laboratory), a NanoSteel Co. was established in the United States [48]. Trial production of NanoSteel sheets was started in 2012. The nanocrystalline structure was produced by special chemistry and heat treatment. After casting, the steel is mainly austenitic. Applying special heat treatment, the grain size of austenite is refined to nanometer scale. During plastic deformation, stress-induced nanoscale phase transformation increases strain hardening.

German company Engineering+Design AG (EDAG) recently published a design study in which the steel used in a 2011 Honda Accord® was replaced with NanoSteel products. The National Highway Traffic Safety Administration (NHTSA) sponsored a research study to compare the results to formerly applied conventional AHSS. The results showed further 8% weight reduction to conventional AHSS and 30% overall weight reduction to former model Honda Accord 2011 [49].

6. Conclusions

In this chapter, the recent developments and future trends in Advanced High Strength Steel production and application were overviewed. Considering both the customers' demand and the legal requirements, it was shown that some of these requirements are coinciding while others are contradictory. To fulfill these often contradictory requirements, the application of high strength steels may be regarded as one of the most promising developments. Among these developments, the application of new Advanced High Strength Steels (AHSS) is the most important one. In the last 45–50 years, different grades of AHSS were developed. They are classified as first-, second-, and third-generation AHSS. Some of these AHSS grades are already widely applied in the world automotive industry; some still are in the development phase. The main properties, the metallurgical background and the main processing routes of AHSS were discussed.

Acknowledgements

This work summarizes the results achieved within the project *Material developments for the Hungarian Automotive Industry* jointly financed by the European Union and the Hungarian Government (Grant No. AutoTech-4.2.2/A-11/1-KONV-2012-0029). Both financial supports are gratefully acknowledged.

Conflict of interest

The author declares no potential conflicts of interest with respect to the research, authorship, and/or publication of this article.

Author details

Miklós Tisza
Mechanical Engineering, University of Miskolc, Miskolc, Hungary

*Address all correspondence to: tisza.miklos@uni-miskolc.hu

References

[1] Tisza M, editor. Metal Forming in the Automotive Industry. 1st ed. Miskolc: University Press; 2015. 294 p. ISBN 978-963-358-082-0

[2] Lotus Engineering: An Assessment of Mass Reduction Opportunities for 2017-2020 Model Year Vehicle Programs. International Council on Clean Transportation; 2010. 308 p

[3] Wagener HW. New developments in sheet metal forming: Sheet materials, tools and machinery. Journal of Materials Processing Technology. 1997:342-357

[4] Chung JY, Kwon O. Development of high performance auto steels. In: Proceedings of ICTP 2008. Gyeongju-Korea, 7-11 September. 2008. pp. 3-6

[5] Matlock DK, Speer JG. Processing opportunities for new advanced high-strength sheet steels. Materials and Manufacturing Processes. 2010;**25**:7-13

[6] Lesch C, Kwiaton N, Frank B. Advanced High Strength Steels (AHSS) for automotive applications: Tailored properties by smart microstructural adjustments. Steel Research International. 2017;**88**:1-21. DOI: 10.1002/srin.201700210

[7] Nanda T, Singh V, Singh V, Chakraborty A, Sharma S. Third generation of advanced high-strength steels: Processing routes and properties. Journal of Materials: Design and Applications. 2019;**233**(2):209-238. DOI: 10.1177/1464420716664198

[8] Matlock DK, Speer JG, Moor ED, et al. Recent developments in advanced high strength sheet steels for automotive applications: An overview. Journal of Engineering Science and Technology. 2012;**15**:1-12

[9] International Iron and Steel Institute. UltraLight Steel Auto Body Concepts (ULSAB) Overview Report. 2002. Available from: www.worldautosteel.org

[10] WorldAutoSteel. UltraLight steel auto closures—ULSAC. Engineering Report. 2001. Retrieved from: http://www.worldautosteel.org/projects/ulsac-2/ulsac/

[11] WorldAutoSteel. UltraLight steel auto body—Advanced vehicle concepts—ULSAB-AVC. Engineering Report. 2002. Retrieved from: http://www.worldautosteel.org/projects/ulsab-avc-2/

[12] WorldAutoSteel: Future Steel Vehicle—FSV. Final Engineering Report. 2011. Retrieved from: http://www.autosteel.org/Programs/Future%20 Steel%20 Vehicle.aspx FutureSteelVehicle

[13] Keeler S, Kimchi M. AHSS Application Guidelines Version 6.0. WorldAutoSteel. World Steel Association; 2017. Available from: https://www.worldautosteel.org

[14] Rashid MS, Rao B. Tempering characteristics of a vanadium containing dual-phase steel. In: Kot RA, Bramfitt BL, editors. Fundamentals of Dual-Phase Steels. TMS-AIME, Warrendale, PA. 1981. pp. 246-264

[15] Hance B. Advanced high strength steel (AHSS) performance levels. In: SAE Technical Paper. 2018. DOI: 10.4271/2018-01-0629

[16] Li C, Li Z, Cen Y, et al. Microstructure and mechanical properties of dual phase strip steel in the overaging process of continuous annealing. Materials Science and Engineering A. 2015;**627**:281-289

[17] Adamczyk J, Grajcar A. Effect of heat treatment conditions on the

structure and mechanical properties of DP-type steel. Journal of Achievements in Materials and Manufacturing Engineering. 2006;**17**:305-308

[18] Mittal M. Effect of inter-critical annealing parameters on the recrystallization, austenite formation and stabilization in a dual phase steels [thesis]. Prem Nagar, Punjab, India: Thapar University; 2014

[19] Ramazani A, Ebrahimi Z, Prahl U. Study of the effect of martensite banding on the failure initiation in dual-phase steel. Computational Materials Science. 2014;**87**:241-247

[20] Meng Q, Li J, Wang J, et al. Effect of water quenching process on microstructure and tensile properties of alloy cold rolled dual-phase steel. Materials & Design. 2009;**30**:2379-2385

[21] Rana R, Liu C, Ray RK. Evolution of microstructure and mechanical properties during thermo-mechanical processing of a low-density multiphase steel for automotive application. Acta Materialia. 2014;**75**:227-245

[22] Kuziak R, Kawalla R, Waengler S. Advanced high strength steels for automotive industry: A review. Archives of Civil and Mechanical Engineering. 2008;**8**:103-117

[23] Skalova L, Divisova R, Jandova D. Thermo-mechanical processing of low-alloy TRIP-steel. Journal of Materials Processing Technology. 2006;**175**:387-392

[24] Shi W, Li L. Thermal stability of retained austenite in TRIP steel after different treatments. Journal of Iron and Steel Research International. 2008;**15**:61-64

[25] Basuki A, Aernoudt E. Influence of rolling of TRIP steel in the intercritical region on the stability of retained austenite. Journal of Materials Processing Technology. 1999;**89-90**:37-43

[26] Horvath CD. Advanced steels for lightweight automotive structures. In: Materials, Design and Manufacturing for Lightweight Vehicles. 2010. pp. 332-356. DOI: 10.1533/9781845697822.1.35

[27] Demeri MY. Advanced High-Strength Steels—Science, Technology, and Applications. Materials Park, Ohio: ASM International; 2013. 275 p

[28] Tisza M. Hot forming of boron alloyed Manganese steels. Materials Science Forum. 2015;**885**:25-30. DOI: 10.4028/www.scientific.net/MSF.885.25

[29] Banik J, Gerber T, Graff S, et al. Hot Forming in the Automotive Industry. Munich: Süddeutscher Verlag; 2013. 84 p

[30] Tisza M. Hot press forming of 22MnB5 steels. In: Proceedings of the International Conference on Material Science, 11-13 October 2015, Balatonalmádi. 2015

[31] Grassel O, Kruger L, Frommeyer G, et al. High strength Fe–Mn–(Al, Si) TRIP/TWIP steels development: Properties & application. International Journal of Plasticity. 2000;**16**:1391-1409

[32] Chung K, Ahn K, Yoo DH, et al. Formability of TWIP (twinning induced plasticity) automotive sheets. International Journal of Plasticity. 2011;**27**:52-81

[33] Allain S, Chateau JP, Bouaziz O, et al. A physical model of the twinning-induced plasticity effect in a high manganese austenitic steel. Materials Science and Engineering A. 2004;**384**:143-147

[34] Grajcar A, Borek W. Thermo-mechanical processing of

high-manganese austenitic TWIP-type steels. Archives of Civil and Mechanical Engineering. 2008;**8**:31-37

[35] Scott C, Remy B, Collet JL, et al. Precipitation strengthening in high manganese austenitic TWIP steels. International Journal of Materials Research. 2011;**102**:538-549

[36] Matlock DK, Speer JG. Third generation of AHSS: Microstructure design concepts. In: Haldar A, Suwas S, Bhattacharjee D, editors. Microstructure and Texture in Steels and Other Materials. London: Springer; 2009. pp. 185-205

[37] Chopra N. Integrated Computational materials engineering: A multi-scale approach. The Journal of the Minerals, Metals & Materials Society. 2015;**67**(1): 118-119. DOI: 10.1007/s11837-014-1260-0

[38] National Research Council. Integrated Computational Materials Engineering. A Transformational Discipline for Improved National Competitiveness and National Security. Washington, DC: The National Academies Press; 2008. Available from: http://www.nap.edu/catalog/12199.html. ISBN: 978-0-309-11999-3

[39] Joost WJ. Reducing vehicle weight and improving U.S. energy efficiency using integrated computational materials engineering. The Journal of the Minerals, Metals & Materials Society. 2012;**64**(9):1032-1038. DOI: 10.1007/s11837-012-0424-z

[40] Zamiri AR, Pourboghrat F. A novel yield function for single crystals based on combined constraints optimization. International Journal of Plasticity. 2010;**26**:731-746. DOI: 10.1016/j.ijplas.2009.10.004

[41] Savic V, Hector L, Singh H, Paramasuwom M, et al. Development of a lightweight third-generation advanced high-strength steel (3GAHSS) vehicle body structure. SAE International Journal of Materials and Manufacturing. 2018;**11**(4):303-313. DOI: 10.4271/2018-01-1026

[42] Schmitt JH, Iung T. New developments of advanced high-strength steels for automotive applications. C.R. Physique. 2018;**19**:641-656. DOI: 10.1016/j.crhy.2018.11.004

[43] Han HN, Oh CS, Kim G, Kwon O. Design method for TRIP-aided multiphase steel based on a microstructure-based modeling for transformation-induced plasticity and mechanically induced martensitic transformation. Materials Science and Engineering A. 2009;**499**:462-468

[44] Speer JG, Edmonds DV, Rizzo FC, Matlock DK. Partitioning of carbon from supersaturated plates of ferrite, with application to steel processing and fundamentals of the bainite transformation. Current Opinion in Solid State & Materials Science. 2004;**8**:219-237

[45] Arlazarov A, Gouné M, Bouaziz O, Hazotte A, Petitgand G, Barges P. Evolution of microstructure and mechanical properties of medium Mn steels during double annealing. Materials Science and Engineering A. 2012;**542**:31-39

[46] BaoSteel: Automotive Advanced High Strength Steels. Product Manual. BaoSteel Group, Corporate Publications; 2013. Available from: http://www.baosteel.com/group_en/contents/2875/39980.html

[47] Bachmaier A, Hausmann K, Krizan D, Pichler A. Development of TBF steels with 980 MPa tensile strength for automotive applications. In: Proceedings of International Conference on New Developments

in Advanced High Strength Steels; Colorado, June. p. 2013. DOI: 10.13140/RG.2.2.24907.80169

[48] Altan T. Three generations of advanced high-strength steels for automotive applications—The third generation. Stamping Journal. 2014:12-13. Available from: https://www.thefabricator.com/stampingjournal

[49] Singh H. Nanosteel intensive body-in-white. In: Research Study. EDAG Inc.; August 2013

Section 3

Science and Technology of High Entropy Materials

Chapter 7

Light-Weight and Flexible High-Entropy Alloys

Yasong Li and Yong Zhang

Abstract

The lightweight and flexible materials can improve people's quality of daily life; in addition, the materials can be widely used in aerospace, automotive, consumer electronics, etc. Recently, high-entropy alloys had become hot issues in materials science with many excellent properties; therefore, we can combine the design ideas of high-entropy alloys with lightweight materials and flexible materials, taking into account the advantages of two types of materials, and promoting the development and progress of new materials. In the chapter, we will elaborate on the relationship between the microstructure and properties of lightweight high-entropy alloys and the design ideas of high-entropy alloys with flexible materials that were investigated in recent years. Furthermore, as the microstructure and mechanical properties of the alloys exhibit the nonlinear behaviors with entropy on high-entropy alloys, we would like to define the lightweight high-entropy alloy as the density is lower than 6 g/cm^3, the mix-entropy of these alloys is higher than 1R (here, R is gas constant), and the number of components is four or more. Finally, it is expected to broaden the research field of high-entropy alloys and provide some new directions for the development of new materials.

Keywords: lightweight, high-entropy alloys, solid solution, alloy design, flexible materials

1. Introduction

Materials have always been a necessary progressive factor in human development; the progress of human society is often accompanied by advances in materials. From the Stone Age to the Bronze Age and then to the Iron Age, the emergence of each new materials has brought major changes in people's productivity. Nowadays, a series of materials have been used in various fields. The traditional structural materials such as steel, aluminum alloys, titanium alloys, magnesium alloys, etc. were still the most widely used materials. However, these materials cannot be applied in some specific areas. In addition, new materials have been developed such as composite material, nanostructure materials, carbon materials, bulk metallic glasses, high-entropy alloys, etc., as high-entropy alloys have been developed since 2004 by Yeh et al. [1] and Cantor et al. [2]. Due to the extremely complex composition of these alloys, the alloys also exhibit excellent properties that are difficult to achieve with many conventional alloys, such as high strength, high hardness, high fracture toughness, corrosion resistance, high temperature oxidation resistance, good low temperature performance, etc. In recent years, these high-entropy alloys such as AlCoCrFeNiCu, CoCrFeMnNi, CoCrFeNi (Ti, Al), NbMoTaW, CoCrNi (AlSi), etc. have been developed and studied [3, 4]. As these alloys also have large proportion of transition metal elements, they also show high density. However, lightweight materials in the aerospace, automotive (especially electric vehicles), consumer electronics, and other fields have become an important development direction. However, designing novel lightweight materials with the concept of high-entropy alloys has become a hot issue, which will promote the development and applications [5, 6].

In view of the excellent performance of high-entropy alloys, we firmly believe that the lightweight high-entropy alloys have superior performance than traditional lightweight materials such as aluminum alloys, titanium alloys, magnesium alloys, etc. The general definition of lightweight materials generally uses the density of titanium alloy as the limit. The existing elements with lower density than titanium (4.51 g/cm^3) are mainly lithium (0.53 g/cm^3), beryllium (1.85 g/cm^3), boron (2.46 g/cm^3), sodium (0.97 g/cm^3), carbon (2.26 g/cm^3), magnesium (1.74 g/cm^3), aluminum (2.70 g/cm^3), silicon (2.33 g/cm^3), potassium (0.86 g/cm^3), calcium (1.55 g/cm^3), yttrium (2.99 g/cm^3), rubidium (1.53 g/cm^3), strontium (2.64 g/cm^3), strontium (4.47 g/cm^3), barium (3.51 g/cm^3), etc., and most of these elements are main group elements, which tend to have a higher chemical activity, with larger atomic radius, also with large difference in melting point and boiling point (lower melt point such as rubidium 39.3°C and higher melt point as titanium 1668°C). Also as we design the lightweight high-entropy alloys, these elements are not exactly used for the new alloy systems. Therefore, the development of lightweight high-entropy alloys often shows more difficulty than that of traditional high-entropy alloys.

In addition, compared with rigid materials, flexible materials are also widely used, which include foils, fibers, films, ribbons, etc., and usually they are made of organic matter. The inorganic materials such as silica, bulk metallic glasses, and metal materials, etc. tend to exhibit the characteristics of rigid materials. However, after being made into fibers or films, such materials can often undergo bending deformation due to the size effect and can also exhibit the characteristics of flexible materials. Nowadays, there is an increasing demand for flexible electronic materials in the field of electronics, especially in the field of wearable electronics. High-entropy alloys have demonstrated excellent overall performance as a new class of alloy materials in the field of rigid materials. Combined with the design concept of high-entropy alloy, can high-entropy open up a new research field in terms of flexible materials?

Nowadays, some scholars have also carried out a lot of research works; therefore, we will give a brief review on the relevant research works (mainly based on the research works of our own research group) and put forward our own opinions on the design and preparation of lightweight high-entropy alloys and high-entropy flexible materials.

2. Lightweight high-entropy alloy systems

2.1 Al-Mg-Li lightweight high-entropy alloy systems

Nowadays, the most commonly used lightweight metal materials are aluminum alloys, titanium alloys, magnesium alloys, etc. As the lithium alloys is the lightest structural metal material, which magnesium and aluminum are the common lightweight metal materials, our group firstly design the two lightweight high-entropy alloys systems (AlLiMgZnCu and AlLiMgZnSn) by Yang et al. [7].

With the design concept of traditional high-entropy alloys, we hope to form a multicomponent solid solution by alloy design. In recent years, these factors such as ΔS_{mix}, ΔH_{mix}, δ, Ω, $\Delta\chi$, VEC, T_m, etc., have made significant effects on the formation of solid solution as the design of high-entropy alloys. Therefore, in order to

Alloy design elements	Al	Li	Mg	Zn	Cu	Sn
r (10^{-10} m)	1.43	1.56	1.60	1.39	1.28	1.55
A (g/mol)	26.98	6.94	24.31	65.39	63.55	118.7
Crystal structure	FCC	BCC	HCP	HCP	FCC	Tetragonal
χ	1.61	0.98	1.31	1.65	1.90	1.96
VEC	2	1	2	12	11	4
ρ (g/cm^3)	2.70	0.54	1.74	7.13	8.93	7.37
T_m (K)	933.5	453.7	922	692.7	1358	505.1

Table 1.
Atomic radius (r), standard atomic weight (A), crystal structure, electronegativity (χ), value electron concentration (VEC), density (ρ), and melting temperature (T_m) for constituent elements in present alloys [7].

enhance the formation of the solid solution in lightweight high-entropy alloys, we firstly considered these factors and made some relevant calculations, which are shown in **Table 1**.

We had designed six lightweight high-entropy alloys that were AlLiMgZnSn, $AlLi_{0.5}MgZn_{0.5}Sn_{0.2}$, $AlLi_{0.5}MgZn_{0.5}Cu_{0.2}$, $AlLi_{0.5}MgCu_{0.5}Sn_{0.2}$, $Al_{80}Li_5Mg_5Zn_5Sn_5$, and $Al_{80}Li_5Mg_5Zn_5Cu_5$, and the densities of these alloys are 4.23, 3.22, 3.73, 3.69, 3.05, and 3.08 g/cm^3, respectively; the density of these alloys is lower than that of titanium. The XRD pattern analysis of these alloys shows that the single-phase solid solution did not appear as the main phase under the condition of high mixing entropy; however, a large number of intermetallic compounds are produced during the smelting process. Only when the addition of aluminum reach to 80 at.%, the alloys show in a single face center cube(FCC) solid solution as the aluminum alloys. The SEM photos of these alloys which are shown in **Figure 1**, we can see a lot of intermetallic become the main phase of these alloys with high entropy, these show that the entropy did not victory when competition with the enthalpy, the solid solution did not form, also a lot of crack were found in the compounds, which cause the plasticity of these alloys are poor, also we when the aluminum become the main element of these alloy the α-Al (FCC) solid solution become the main phase in dendrite, some compounds which were rich in Cu or Sn in inter dendrite. The compression test of the alloys is shown in **Figure 2**, and the $Al_{80}Li_5Mg_5Zn_5Sn_5$ and $Al_{80}Li_5Mg_5Zn_5Cu_5$ alloys show good strength with higher than 800 MPa and yield strength higher than 400 MPa, with compressive plasticity better than 15%.

Also, the rare-earth elements lanthanum and cerium were added in these alloys to improve the solid solution formation ability of the alloys. In addition, Bridgeman directional solidification technology is also used in these alloys. However, these do not work in these systems. In order to further understand the formation law of solid solution of these lightweight high-entropy alloys, the δ, Ω, Δχ, and VEC of these low-density high-entropy alloys are shown in **Figure 3**. Comparing the formation regions of solid solutions and intermetallic compounds with the conventional high-entropy alloys, we find that for the lightweight high-entropy alloys, which tend to have higher mixing enthalpy and electronegativity with smaller Ω and VEC, the δ of the alloy is in an intermediate region, often close to the critical region where the solid solution phase forms. In addition, the δ-Δχ can be a better way to predict the phase formation ability of these alloys. When $\Delta\chi < 0.175$, the solid solution will become the main phase of these alloys. Mainly, we found that the Al-Mg-Li system low-density high-entropy alloys had high chemical activity, which made it easier to form intermetallic compounds with other elements. Finally, we found that with the study of composition design, microstructure performance, and phase

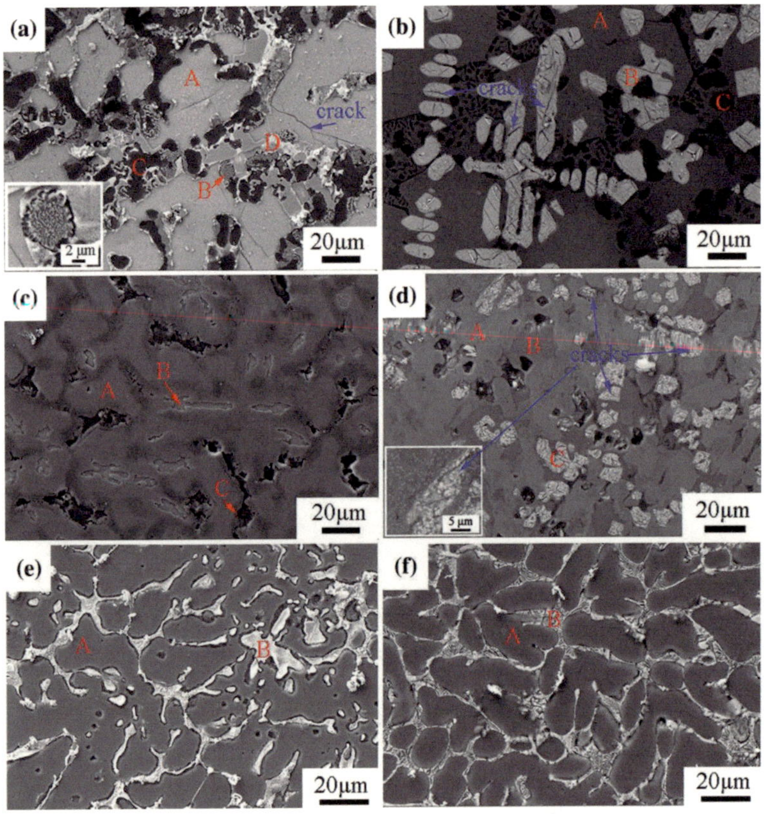

Figure 1.
SEM secondary electron images of low-density multicomponent alloys. (a) AlLiMgZnSn; (b) $AlLi_{0.5}MgZn_{0.5}Sn_{0.2}$; (c) $AlLi_{0.5}MgZn_{0.5}Cu_{0.2}$; (d) $AlLi_{0.5}MgCu_{0.5}Sn_{0.2}$; (e) $Al_{80}Li_5Mg_5Zn_5Sn_5$; and (f) $Al_{80}Li_5Mg_5Zn_5Cu_5$ alloys [7].

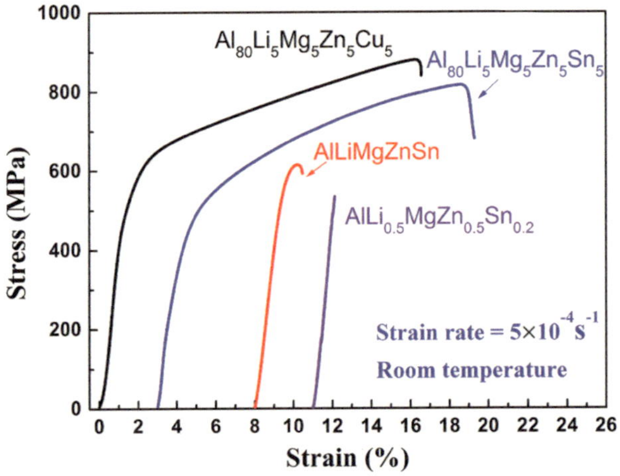

Figure 2.
Compressive engineering stress-strain curves of AlLiMgZnSn, $AlLi_{0.5}MgZn_{0.5}Sn_{0.2}$, $Al_{80}Li_5Mg_5Zn_5Sn_5$, and $Al_{80}Li_5Mg_5Zn_5Cu_5$ alloys at room temperature. The initial strain rate was 5×10^{-4} s^{-1}[7].

formation of multicomponent alloys, for low-density high-entropy alloys, high mixing entropy is not the key factor in the formation of solid solution structures of these alloys. Compared with the traditional high-entropy alloys (mostly composed of transition metal elements), the solid solution phase formation conditions of the

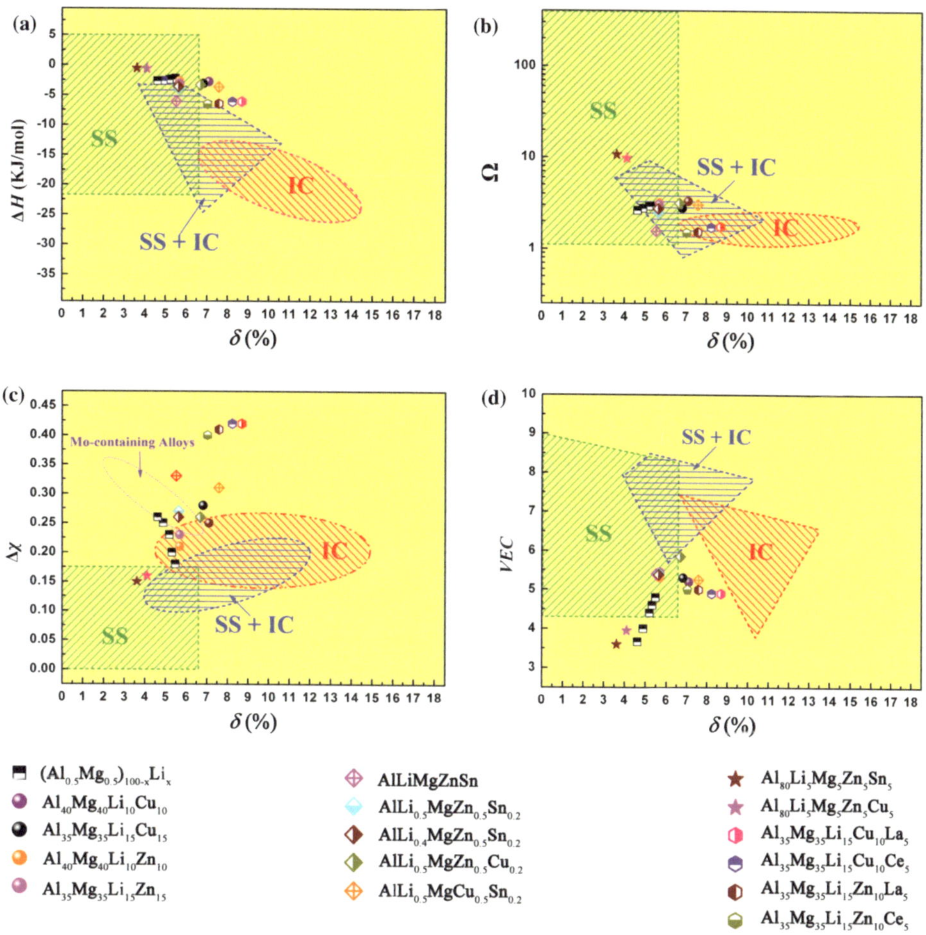

Figure 3.
Phase constituent prediction maps: (a) δ-Hmix, (b) δ-Ω, (c) δ-Δχ, and (d) δ-VEC plots for multicomponent alloys in this work overlaid on cross-hatched regions developed in previous HEA investigations. (For $(Al_{0.5}Mg_{0.5})_{100-x}Li_x$, x = 5, 10, 15, 25, and 33.33) [7].

Al-Mg-Li-based lightweight high-entropy alloys are more severe. The solid solution formation of these alloys can be predicted by electronegativity ($\Delta\chi$); when the $\Delta\chi < 0.175$, it is easier to form the solid solution; and as $\Delta\chi \geq 0.175$, it tends to form the intermetallic compounds.

Li et al. also studied the microstructure and properties of the Al-Mg-Li high-entropy alloy system by using super-gravity technology [8]. Under different conditions with the super-gravity experiments, which found that supergravity does not separate the heavy elements of the alloy from the light elements; however, the microstructure of the alloy changed, which caused different properties. The alloy structure is still composed of α-Al solid solution structure and intermetallic compounds, and with supergravity, the microstructure changes to the eutectic microstructure. As supergravity is one entopic force, there are a variety of entropic forces in the process of alloy during solidification. Since gravity increases with distance, there are pressure and viscosity gradients in the molten metal. Meanwhile, due to the high mixing entropy of the alloy and the combination of various factors, the microstructure of intermetallic compounds and solid solution will change during solidification. In addition, these effects also made the grain refinement of the alloy along the direction of gravity to a certain extent, resulting in an enhancement of the strength of the alloy. Nevertheless, the alloy still does not form a single-phase

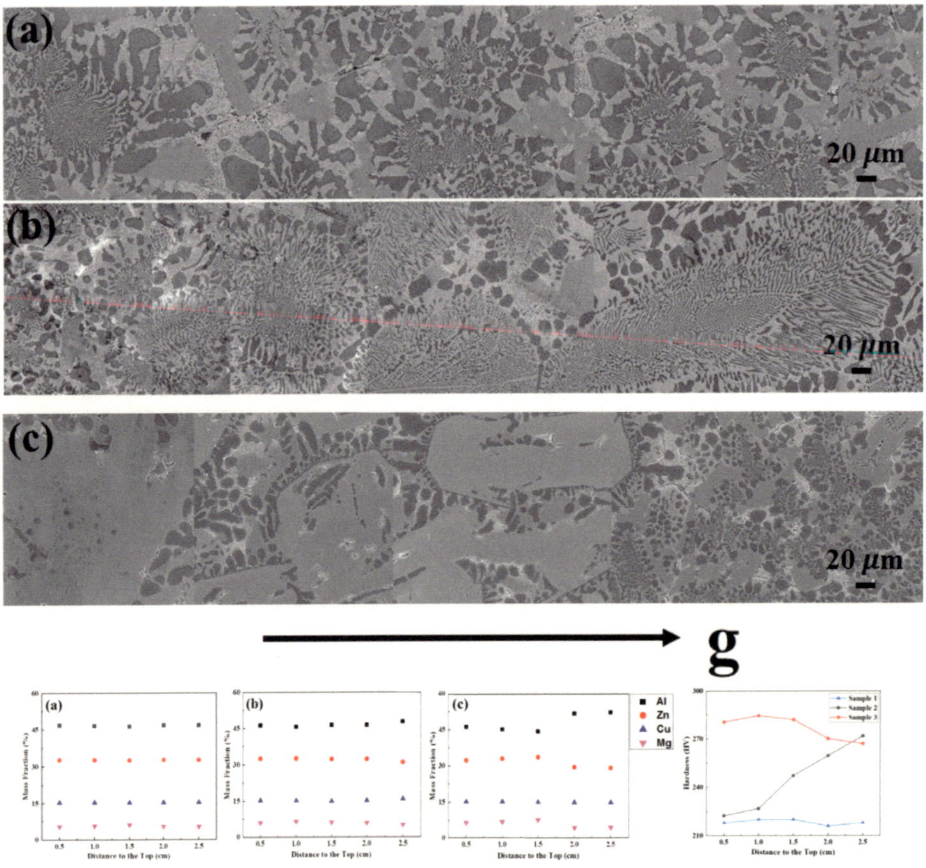

Figure 4.
Microstructure of the content of different element and hardness of the $AlZn_{0.4}Li_{0.2}Mg_{0.2}Cu_{0.2}$ *alloy with different supergravity experiments: (a) Sample 1; (b) Sample 2, and (c) Sample 3 [8].*

solid solution structure; therefore, the optimal structure of the alloy is the eutectic structure with the intermetallic compound and the solid solution with grain refine. **Figure 4** shows the microstructure, the composition of different elements by X-ray photoelectron spectroscopy, and the hardness of the alloy with the distance of gravity.

2.2 The Al-Mg-Zn-Cu-Si lightweight high-entropy alloy system

Based on the Al-Mg-Li study, our research group Shao et al. [9] used the Si exchange of Li, in order to reduce the cost of the alloy and expected to achieve lightweight, low-cost, high-entropy alloys; therefore, we studied the Al-Mg-Si system lightweight high-entropy alloys. Based on $\Delta\chi$, we designed the AlMgZnCuSi alloys, and these alloy samples were prepared by vacuum induction melting. We have found that the alloy forms a eutectic structure of solid solution and intermetallic compound when the content of Al is less than 80 at.%; however, these alloys show high strength with low ductility, and as the Al condition is higher than 80 at.%, they become α-Al face center cube solid solution. These alloys also have high strength with good compressive ductility. Which found that the $Al_{85}Mg_{10.5}Zn_{2.025}Cu_{2.025}Si_{0.45}$ alloy shows good toughness when the strength is higher than 800 MPa with ductility more than 20%. Currently, $\Delta\chi$ also predicts the phase formation of the alloy. We also found some serrated flow phenomena in the compressive strain curve of

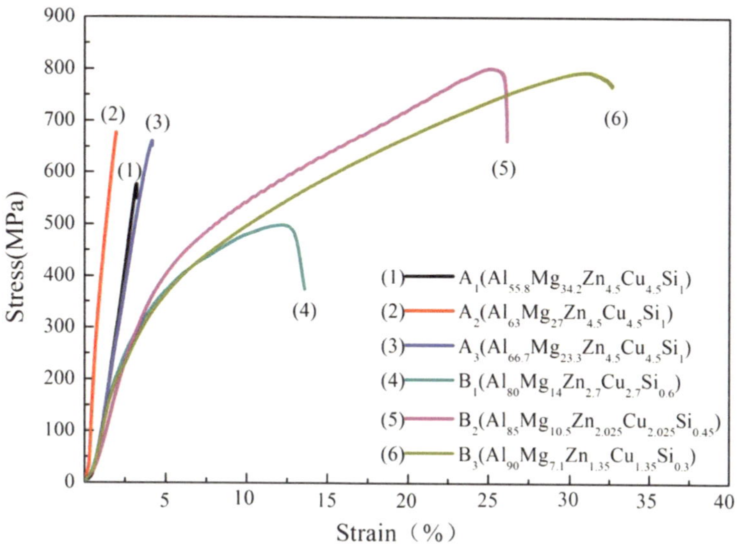

Figure 5.
The compressive stress-strain curves at room temperature [9].

the alloy and will do some further research on the mechanism of serration behavior with the alloy. This research shows that this inexpensive alloy system is the research direction of another high-strength lightweight high-entropy alloy. The compressive stress-strain curves of a series of alloys at room temperature with a strain rate of 10^{-3} s^{-1} are shown in **Figure 5**.

In addition, some other researchers have also studied similar alloy systems based on this type of lightweight high-entropy alloy system. Baek et al. [10] used the ultrasonic melt treatment to prepare lightweight $Al_{70}Mg_{10}Si_{10}Cu_5Zn_5$ alloy, this alloy also forms a large number of other precipitated phases in the aluminum matrix, and the effect of solution treatment of this alloy on the microstructure and properties of the alloy was investigated, which found that the alloy has an excellent performance at both room temperature and 350°C; however, the microstructure of the alloy is finely refined by the precipitation phase size by ultrasonic melt treatment technology, and mainly due to the introduction of trace amounts of Ti, the grain size is refined. In addition, mechanical properties of the alloy at room temperature have been improved with the solution treatment at 440°C, but the mechanical properties with high temperature (350°C) deteriorate. Through solution treatment, the Zn atoms redissolve into the second phases, which not only leads to the formation of fine super-saturated clusters in the matrix, but also spheroidizes the primary Si and Mg2Si phases, thereby improving the room temperature mechanical properties of the alloy. They also studied the effects of Al-6Mg-9Si-10Cu-10Zn-3Ni alloy aging treatment on properties and microstructure of alloys at different aging temperatures, and at 120°C, they found that the GP zone in the alloy with aging time was replaced by a Zn-rich metastable elliptical cluster to form a stable Zn precipitate containing a part of Cu atoms [11]. Besides, the aging precipitation behavior under different temperatures had also been studied [12]; as the aging temperature is below 70°C, a series of fine clusters and precipitates were formed, which greatly improves the strength of the composite. On the other hand, due to the coarsening of the precipitate, and the softening by the reduced volume fraction and the periodization of the second phase, a small strengthening effect was observed above 170°C. Sanchez et al. [13] have done some research on $Al_{65}Cu_5Mg_5Si_{15}Zn_5X_5$ and $Al_{70}Cu_5Mg_5Si_{10}Zn_5X_5$ systems and reported the effect of Fe, Ni, Cr, Mn, and Zr elements on the phase formation, microstructure, and properties of these alloys. These researches all showed that this kind of alloy system has a good prospect in foundry industry.

2.3 High temperature application of lightweight high-entropy alloy

The light-weight metal elements as beryllium, scandium, titanium, yttrium etc., and the light-weight non-metallic elements such as carbon, boron, silicon etc., in addition to aluminum have a higher boiling point; therefore, these elements were also used for the design of high temperature application of lightweight high-entropy alloy. Some researchers have done a series of research work on these alloys.

Tseng et al. [14] studied the $Al_{20}Be_{20}Fe_{10}Si_{15}Ti_{35}$ lightweight high-entropy alloy with a vacuum-arc-melting, and this alloy showed a single hexagonal close-packed (HCP) structure solid solution phase, with high hardness ~8.9 GPa, high strength ~2.976 GPa, with a density of ~3.91 g/cm^3; in addition, this alloy showed an excellent oxidation resistance at both 700 and 900°C, which is much better than the normal Ti-6Al-4 V alloy. Another way to prepare lightweight, high-temperature, high-entropy alloys is to reduce alloy density by adding lightweight elements Ti and Al to conventional alloys. These alloys usually have a higher density, however, lighter than the conventional superalloys, usually less than 6 g/cm^3. These lightweight high entropy alloys, such as $NbTiVTaAl_x$ [15], CrNbTiVZr [16], AlNbTiV [17], $Al_{1.5}CrFeMnTi$ [18, 19], AlTiVCr [20, 21] etc., which tend to have a single-phase solid solution structure with lower density, good plasticity, and high temperature properties. In addition, these are a powerful alternative to the next generation of superalloys, with great potential to replace existing superalloys. The application of such alloys will bring a big leap in materials for the aviation industry.

2.4 Other lightweight high-entropy alloy systems

Finally, we will briefly explain the existing research on other lightweight high-entropy alloys. Youssef et al. [22] made an investigation on $Al_{20}Li_{20}Mg_{10}Sc_{20}Ti_{30}$ lightweight high-entropy alloy with mechanical alloying. Since such alloys were prepared by mechanical alloying, the alloy structure exhibited an ultrafine grain structure with only 12 nm, and the alloy exhibited an ultra-high hardness of 5.9 GPa, and its density was only 2.67 g/cm^3; which shows a single face center cube (FCC) solid solution structure, when the power was milled without N, O, the alloy has a face centered cube (FCC) transformation into a hexagonal close-packed (HCP) structure with 500°C annealing treatment, however with N, O this transformation did not happen. Li et al. [23, 24] made an investigation on $Mg_x(MnAlZnCu)_{100-x}$ lightweight high-entropy alloys, the microstructure of $Mg_{20}(MnAlZnCu)$ alloy was consistent with HCP solid solution and Al-Mn icosahedral quasicrystal phase, and the compressive strength of these $Mg_x(MnAlZnCu)_{100-x}$ alloys were high; however, the plasticity of alloys was poor. In addition, the microstructure and properties of the $Mg_{20}(MnAlZnCu)$ alloy under different solidification conditions were also studied, and they found that with the faster cooling rate, the Al-Mn icosahedral quasicrystal phase was refined, which enhanced the strength of this alloy; however, with the brittleness of the HCP alloy, even the fast cooling rate can improve the plastic deformation ability of the alloy. The plasticity of this alloy is still poor, and with this work, they found that the high entropy can enhance the formation ability of icosahedral quasicrystal [24]. Du et al. [25] investigated the MgCaAlLiCu alloy, which shows a mainly single solid solution phase with tetragonal symmetry lattice structure, and the density of this alloy is ~2.2 g/cm^3, with high compressive strength ~910 MPa. Jia et al. investigated the AlLiMgCaSi high-entropy alloys and they found that the density of these alloys were 1.46 to 1.70 g/cm^3 and the strength was higher than 450 MPa, especially, as the $Al_{15}Li_{38}Mg_{45}Ca_{0.5}Si_{1.5}$ and $Al_{15}Li_{39}Mg_{45}Ca_{0}$

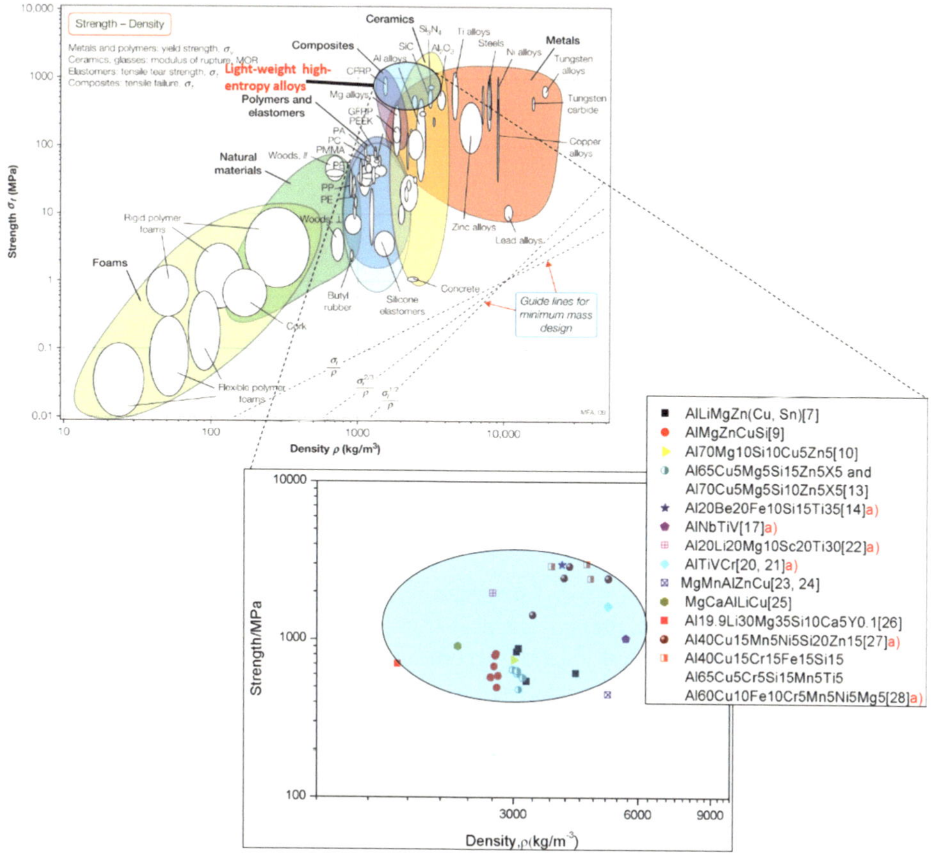

Figure 6.
The area of lightweight high-entropy alloys in the Ashby diagram of strength vs. density for structural materials [29], a) the strength of these alloys was equivalent as the hardness/3.

$_{.5}Si_{0.5}$ alloy exhibited good plasticity ~45 and ~60%, which is much higher than most of the lightweight high-entropy alloys [26]. Sanchez et al. investigated on the as-cast high-entropy aluminums, and they found these alloys showed high hardness than other lightweight alloys [27, 28]. **Figure 6** shows the area of lightweight high-entropy alloys in the Ashby diagram of strength vs. density for structural materials, which we found that the strength of lightweight high-entropy alloys was much higher and the density much lower than some ceramics such as the SiC, Al_3N, etc.; however, the ductility is better than that of ceramics.

There are still many problems in the existing lightweight high-entropy alloys to be solved. First, the formation conditions of conventional high-entropy alloy solid solution need to be corrected. In addition, lightweight high-entropy alloys tend to exhibit high strength and poor room temperature plasticity, and we need to improve the toughness of these alloys.

3. High-entropy flexible materials

High-entropy alloys tend to have a solid solution structure, which means that these alloys had good plastic deformation capacity. The face center cubic (FCC) high-entropy alloys such as CoCrFeMnNi, $Al_{0.3}$CoCrFeNi, CoCrFeNi, etc. [30–36], show the excellent tensile plasticity whish can exceed 50% at room temperature. Therefore, these alloys can be deformed by plastic deformation such as rolling,

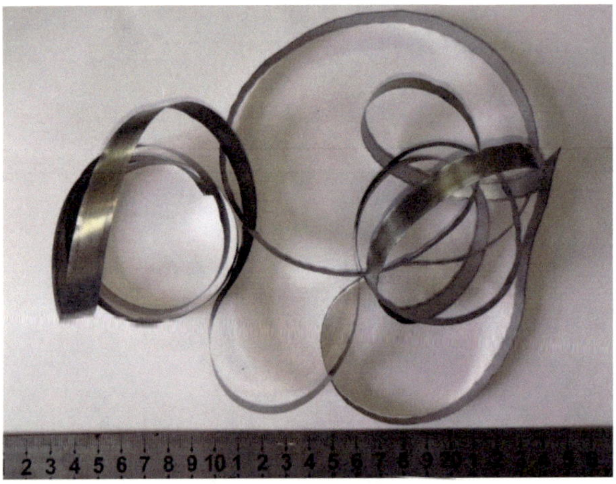

Figure 7.
High-entropy alloy ribbons by vacuum suspension quenching.

extrusion deformation, drawing deformation, etc., which can be made into foils, ribbon filaments, etc.; such materials tend to have the characteristics of flexible materials. In addition, further methods to obtain an alloy of a flexible material is the use of melt spinning method, or coating.

Ma et al. made the single-crystal structure the $Al0._3CoCrFeNi$ alloy with the Bridgman solidification which found that the elongation of this alloy ~80%, the alloy shows an excellent plastic deformation capacity [32]. Li et al. found that $Al0._3CoCrFeNi$ alloy shows the elongation more than ~60% with forging [33]. Based on these studies the Li et al. formed the fibers with this alloy [34]. In addition, the high-entropy alloy ribbons and fibers can also be prepared by vacuum suspension quenching system, which Zhao et al. [35] use this technology prepared the $CoFeNi(AlBSi)_x$ ribbons. High entropy alloy films can be prepared by chemical vapor deposition and physical vapor deposition, the thickness of these films tend to be 0.5 μm to 2 μm, which become two-dimensional materials [37], also such films after separation from the substrate will be a flexible materials. Xing et al. [38] corrected the phase formation of the film materials with the concept of cooling rate. **Figure 7** shows the $CoFeNi(AlBSi)_x$ high-entropy alloy ribbons by vacuum suspension quenching.

4. Conclusions

Summaries of the relevant properties and specificities of lightweight high-entropy alloys are as follows:

1. Lightweight high-entropy alloys are often limited by the addition of elements, and these elements tend to have high electronegativity. They are easy to form intermetallic compounds rather than solid solutions and it has been found that high mixing entropy does not promote the formation of solid solution phases.

2. Therefore, the concept of lightweight high-entropy alloy needs to be broadened, for which the mixing entropy $\Delta S_{mix} > 1R$ is a good choice.

3. Due to the excellent comprehensive mechanical properties of traditional high-entropy alloys, the lightweight high-entropy alloys are also supposed to have

great advantages. The density tends to locate between superalloys and titanium alloys, so we can broaden the density limit of lightweight superalloys, with a recommended density below 6 g/cm^3.

4. Lightweight high-entropy alloys have broad application prospects. However, the development of lightweight high-entropy alloys has great problems. It is expected to make breakthroughs in this area by using some advanced design concepts and preparation methods.

5. Using the concept of high-entropy alloys, there may be new breakthroughs in the development of flexible materials.

Acknowledgements

Y. Zhang would like to thank the financial support from National Natural Science Foundation of China (NSFC), Grant No. 51671020, and Yasong Li would like to thank Dongyue Li, Weiran Zhang, and Xuehui Yan for their help with accessing references.

Conflict of interest

The authors declare no conflict of interest.

Author details

Yasong Li and Yong Zhang*
State Key Laboratory for Advanced Metals and Materials, University of Science and Technology Beijing, Beijing, China

*Address all correspondence to: drzhangy@ustb.edu.cn

References

[1] Yeh JW, Chen SK, Lin SJ, Gan JY, Chin TS, Shun TT, et al. Nanostructured high-entropy alloys with multiple principal elements: Novel alloy design concepts and outcomes. Advanced Engineering Materials. 2004;**6**:299. DOI: 10.1002/adem.200300567

[2] Cantor B, Chang ITH, Knight P, Vincent AJB. Microstructural development in equiatomic multicomponent alloys. Materials Science and Engineering A. 2004;**375-377**:213. DOI: 10.1016/j.msea.2003.10.257

[3] Zhang Y, Zuo TT, Tang Z, Gao MC, Dahmen KA, Liaw PK, et al. Microstructures and properties of high-entropy alloys. Progress in Materials Science. 2014;**61**:1. DOI: 10.1016/j.pmatsci.2013.10.001

[4] Miracle DB, Senkov ON. A critical review of high entropy alloys and related concepts. Acta Materialia. 2017;**122**:448. DOI: 10.1016/j.actamat.2016.08.081

[5] Maulik O, Kumar D, Kumar S, Dewangan SK, Kumar V. Structure and properties of lightweight high entropy alloys: A brief review. Materials Research Express. 2018;**5**:52001. DOI: 10.1088/2053-1591/aabbca

[6] Shi R, Luo AA. Applications of CALPHAD modeling and databases in advanced lightweight metallic materials. Calphad. 2018;**62**:1. DOI: 10.1016/j.calphad.2018.04.009

[7] Yang X, Chen SY, Cotton JD, Zhang Y. Phase stability of low-density, multiprincipal component alloys containing aluminum, magnesium, and lithium. JOM-US. 2014;**66**:2009. DOI: 10.1007/s11837-014-1059-z

[8] Li R, Wang Z, Guo Z, Liaw PK, Zhang T, Li L, et al. Graded microstructures of Al-Li-Mg-Zn-Cu entropic alloys under supergravity. Science China Materials. 2019;**62**:736. DOI: 10.1007/s40843-018-9365-8

[9] Shao L, Zhang T, Li L, Zhao Y, Huang J, Liaw PK, et al. A low-cost lightweight entropic alloy with high strength. Journal of Materials Engineering and Performance. 2018;**27**:6648. DOI: 10.1007/s11665-018-3720-0

[10] Baek E, Ahn T, Jung J, Lee J, Cho Y, Euh K. Effects of ultrasonic melt treatment and solution treatment on the microstructure and mechanical properties of low-density multicomponent $Al_{70}Mg_{10}Si_{10}Cu_5Zn_5$ alloy. Journal of Alloys and Compounds. 2017;**696**:450. DOI: 10.1016/j.jallcom.2016.11.305

[11] Ahn T, Jung J, Baek E, Hwang SS, Euh K. Temporal evolution of precipitates in multicomponent Al-6Mg-9Si-10Cu-10Zn-3Ni alloy studied by complementary experimental methods. Journal of Alloys and Compounds. 2017;**701**:660. DOI: 10.1016/j.jallcom.2017.01.183

[12] Ahn T, Jung J, Baek E, Hwang SS, Euh K. Temperature dependence of precipitation behavior of Al-6Mg-9Si-10Cu-10Zn-3Ni natural composite and its impact on mechanical properties. Materials Science and Engineering A. 2017;**695**:45. DOI: 10.1016/j.msea.2017.04.015

[13] Sanchez JM, Vicario I, Albizuri J, Guraya T, Design AEM. Microstructure and mechanical properties of cast medium entropy aluminium alloys. Scientific Reports-UK. 2019;**9**:6792. DOI: 10.1038/s41598-019-43329-w

[14] Tseng K, Yang Y, Juan C, Chin T, Tsai C, Yeh J. A light-weight

high-entropy alloy $Al_{20}Be_{20}Fe_{10}Si_{15}Ti_{35}$. Science China Technological Sciences. 2018;**61**:184. DOI: 10.1007/s11431-017-9073-0

[15] Yang X, Zhang Y, Liaw PK. Microstructure and compressive properties of $NbTiVTaAl_x$ high entropy alloys. Procedia Engineering. 2012;**36**:292. DOI: 10.1016/j.proeng.2012.03.043

[16] Senkov ON, Senkov SV, Woodward C, Miracle DB. Low-density, refractory multi-principal element alloys of the Cr-Nb-Ti-V-Zr system: Microstructure and phase analysis. Acta Materialia. 2013;**61**:1545. DOI: 10.1016/j.actamat.2012.11.032

[17] Stepanov ND, Shaysultanov DG, Salishchev GA, Tikhonovsky MA. Structure and mechanical properties of a light-weight AlNbTiV high entropy alloy. Materials Letters. 2015;**142**:53. DOI: 10.1016/j.matlet.2014.11.162

[18] Feng R, Gao MC, Zhang C, Guo W, Poplawsky JD, Zhang F, et al. Phase stability and transformation in a light-weight high-entropy alloy. Acta Materialia. 2018;**146**:280. DOI: 10.1016/j.actamat.2017.12.061

[19] Feng R, Gao M, Lee C, Mathes M, Zuo T, Chen S, et al. Design of light-weight high-entropy alloys. Entropy-Switzerland. 2016;**18**:333. DOI: doi.10.3390/e18090333

[20] Qiu Y, Hu YJ, Taylor A, Styles MJ, Marceau RKW, Ceguerra AV, et al. A lightweight single-phase AlTiVCr compositionally complex alloy. Acta Materialia. 2017;**123**:115. DOI: 10.1016/j.actamat.2016.10.037

[21] Qiua Y, ST GMA, Fraser HL, Pohl K, Birbilis N. Microstructure and corrosion properties of the low-density single-phase compositionally complex alloy AlTiVCr. Corrosion Science. 2018;**133**:386. DOI: 10.1016/j.corsci.2018.01.035

[22] Youssef KM, Zaddach AJ, Niu C, Irving DL, Koch CC. A Novel low-density, high-hardness, high-entropy alloy with close-packed single-phase nanocrystalline structures. Materials Research Letters. 2015;**3**:95. DOI: 10.1080/21663831.2014.985855

[23] Li R, Gao JC, Fan K. Study to microstructure and mechanical properties of Mg containing high entropy alloys. Materials Science Forum. 2010;**650**:265. DOI: 10.4028/www.scientific.net/MSF.650.265

[24] Li R, Gao JC, Fan K. Microstructure and mechanical properties of MgMnAlZnCu high entropy alloy cooling in three conditions. Materials Science Forum. 2011;**686**:235. DOI: 10.4028/www.scientific.net/MSF.686.235

[25] Du XH, Wang R, Chen C, Wu BL, Huang JC. Preparation of a light-weight MgCaAlLiCu high-entropy alloy. Key Engineering Materials. 2017;**727**:132. DOI: 10.4028/www.scientific.net/KEM.727.132

[26] Jia Y, Jia Y, Wu S, Ma X, Wang G. Novel ultralight-weight complex concentrated alloys with high strength. Materials. 2019;**12**:1136. DOI: 10.3390/ma12071136

[27] Sanchez J, Vicario I, Albizuri J, Guraya T, Koval N, Garcia J. Compound formation and microstructure of As-cast high entropy aluminums. Metals-Basel. 2018;**8**:167. DOI: 10.3390/met8030167

[28] Sanchez JM, Vicario I, Albizuri J, Guraya T, Garcia JC. Phase prediction, microstructure and high hardness of novel light-weight high entropy alloys. Journal of Materials Research and Technology. 2018;**8**:795. DOI: 10.1016/j.jmrt.2018.06.010

[29] Ashby MF. Materials Selection in Mechanical Design. 4th ed. Burlington, MA, USA: Butterworth-Heinemann/Elsevier; 2011. p. 67. ISBN: 978-1-85617-663-7

[30] Kao Y, Chen T, Chen S, Yeh J. Microstructure and mechanical property of as-cast, -homogenized, and -deformed AlxCoCrFeNi ($0 \leq x \leq 2$) high-entropy alloys. Journal of Alloys and Compounds. 2009;**488**:57. DOI: 10.1016/j.jallcom.2009.08.090

[31] Vaidya M, Karati A, Marshal A, Pradeep KG, Murty BS. Phase evolution and stability of nanocrystalline CoCrFeNi and CoCrFeMnNi high entropy alloys. Journal of Alloys and Compounds. 2019;**770**:1004. DOI: doi. 10.1016/j.jmst.2018.09.068

[32] Ma SG, Zhang SF, Qiao JW, Wang ZH, Gao MC, et al. Superior high tensile elongation of a single-crystal $CoCrFeNiAl_{0.3}$ high-entropy alloy by bridgman solidification. Intermetallics. 2014;**54**:104. DOI: 10.1016/j.intermet.2014.05.018

[33] Li D, Zhang Y. The ultrahigh charpy impact toughness of forged $Al_xCoCrFeNi$ high entropy alloys at room and cryogenic temperatures. Intermetallics. 2016;**70**:24. DOI: 10.1016/j.intermet.2015.11.002

[34] Li D, Li C, Feng T, Zhang Y, Sha G, et al. High-entropy $Al0._3CoCrFeNi$ alloy fibers with high tensile strength and ductility at ambient and cryogenic temperatures. Act Mater. 2017;**123**:285. DOI: 10.1016/j.actamat.2016.10.038

[35] Zhao W, Miao D, Zhang Y, He Z. Big-data analysis of phase-formation rules in high-entropy alloys. Journal of Iron and Steel Research, International. 2017;**24**:358. DOI: 10.1016/S1006-706X(17)30053-5

[36] Liu J, Guo X, Lin Q, He Z, An X, et al. Excellent ductility and serration feature of metastable CoCrFeNi high-entropy alloy at extremely low temperatures. Science China Materials. 2019;**62**:853. DOI: 10.1007/s40843-018-9373-y

[37] Yan XH, Li JS, Zhang WR, Zhang Y. A brief review of high-entropy films. Materials Chemistry and Physics. 2018;**210**:12. DOI: 10.1016/j.matchemphys.2017.07.078

[38] Xing Q, Ma J, Wang C, Zhang Y. High-Throughput screening solar-thermal conversion films in a pseudobinary (Cr, Fe, V)-(Ta, W) system. ACS Combinatorial Science. 2018;**20**:602. DOI: 10.1021/acscombsci.8b00055

Chapter 8

Characterization and Testing of High-Entropy Alloys from AlCrFeCoNi System for Military Applications

Victor Geanta and Ionelia Voiculescu

Abstract

High-entropy alloys (HEAs) can be obtained using various metallurgical processes such as vacuum arc remelting (VAR), induction melting, powder metallurgy, additive manufacturing, plasma sintering of powders, etc. Among these methods, the obtaining process in the VAR plant provides superior homogeneity characteristics for metal matrices, simultaneously with advanced purity, duc to the high level of protection of the melts. The chapter presents a series of results on alloys with high entropy from the AlCrFeCoNi system, which can be used for various applications, including in the military field, for the realization of high-speed penetration protection panels. Experimental alloys were obtained by melting in electric arc under an argon atmosphere, using high-purity raw materials (greater than 99.5 wt%), and homogenization is ensured by successive five-times remelting of mini-ingots. The obtained alloys were subjected to microstructural analyses, mechanical tests, and also dynamic impact tests using incendiary perforation projectiles. At the same time, some tests were carried out on ballistic packages made of different materials, including high-entropy alloys. The results obtained in mechanical tests revealed high values of microhardness (over 600 $HV_{0.1}$) as well as compressive strengths above 2000 MPa. The mechanical characteristics of these alloys can undergo substantial changes by applying several heat treatments.

Keywords: high-entropy alloys, obtaining, characterization, dynamic testing

1. Introduction

High-entropy alloys (HEAs) are known for their special mechanical properties: high tensile strength resistance even at high temperature, high hardness, toughness exceeding that of most pure metals and alloys, comparable strength to that of structural ceramics and some metallic glasses, exceptional ductility and fracture toughness at cryogenic temperatures, and corrosion resistance. These specific properties were mainly attributed to complex concentrated solid solutions formed by the suppression of fragile intermetallic compounds as a result of high mixing entropy and enthalpy values [1]. HEAs generally tend to form single-phase solid solutions in the case of low mixing enthalpy and atomic size difference. Generally speaking, the formation of a single-phase solid solution corresponds to the mixing enthalpy values (DHmix) situated in domain of -15 kJ/mol < DHmix <5 kJ/ml and $0 < \delta < 5$ [2].

IntechOpen

As a result of these special features, HEA is currently an alternative to use as material for a number of special areas, such as structural applications; aerospace engineering and civil transportations; superconducting electromagnets such as magnetic resonance imaging, scanners, nuclear magnetic resonance machines, and particle accelerators; high-temperature applications such as gas turbines, rocket nozzles, and nuclear construction; cryogenic applications such as rocket casings, pipework, and liquid O_2 or N_2 equipment; refractory elements such as Nb, Mo, and Ta that can maintain their high strength even above 1200°C, superior to traditional super alloys such as Inconel 718 and Haynes 230; hardfacing applications; and military applications.

The specifications on the security of collective protection equipment and structures in the military field set forth enhanced requirements for the resistance of the protection panels/floors/elements against the penetration by various types of projectiles, due to the diversification of the types of interventions in the military activities.

The main characteristics of the materials intended for the manufacture of protection components are as follows: the highest possible breaking and yield strength values, the highest possible hardness and impact resistance, and the highest possible elongation at break and energy absorbed by a notched specimen while breaking under an impact load (the Charpy test) at temperatures down to minus 40°C. The current military specifications recommend hardness values of at least 540–600 BHN (Brinell hardness) or 55–60 HRC (Rockwell hardness) and, for strength characteristics, such as the yield strength, values above 1500 and 1700 MPa for the breaking strength. In the case of the impact fracture energy using the Charpy test, the values must be of approximately 13 J at −40°C, with elongations of at least 6% [3].

Such requirements have been met by designing metallic alloys of various compositions; the most widely used is the high-strength microalloyed steels, used to produce reinforcement elements of thicknesses between 8.5 and 30 mm. Some research papers in the military field [4] have shown that the hardness of the plating material is not a sufficient factor to provide maximum resistance against penetration by projectiles, considering that the values of the mechanical strengths (yield and breaking) are much more important in the behavior process under dynamic stress.

Accurate measurements of dynamic stresses during impact stress have detected mechanical stress values of 28 GPa in the case of bainitic microstructure steels, while in the case of static stress, the measured stresses were of no more than 2 GPa [5].

A ballistic performance index (BPI) [6, 7] was also proposed to estimate the ballistic strength of armored plates, containing data on steel density, the elastic modulus, yield and tensile strength, Poisson's coefficient, and the constriction or elongation during impact tests. There are terms that contain elastic and plastic deformation components, as well as terms that take into account the kinetic energy of the target-projectile system after impact. Therefore, the main characteristics that

Alloy	Yield strength, MPa	Compressive strength, MPa	Plastic deformation, %
AlCrFeCoNi	1250.96	2004.23	32.7
CrFeCoNiCuTi	1272	1272	1.6
AlCrMnFeCoNiCuTiV	1862	2431	0.95
$CrFeCoNiCuTi_{0.5}$	700	1650	21.26

Table 1.
Mechanical properties of some high-entropy alloys [8].

metallic materials used in special military applications need to feature for the best impact behavior are as follows:

- The highest possible hardness, as a measurement of the solid material's resistance against the penetration of its surface by various types of penetrators, with permanent shape changes, when a static or dynamic force is applied to them; the macroscopic hardness is generally characterized by the nature and strength of the intermolecular bonds, and the behavior of the solid material under the action of force is complex.

- Toughness, which describes the capacity of a metallic material to absorb the breaking energy, the strength of its metallic matrix when various cracks occur and propagate, and the energy of forming the breaking surfaces, considering that the breaking occurs by consuming the impact stress energy, with local plastic deformation.

- Impact strength, which is the relative susceptibility to breaking by the action of forces applied at high velocity.

High-entropy alloys from the AlCrFeCoNi system feature very good mechanical properties for military applications, as shown in **Table 1**. Thus, the yield stress, the compressive strength, and the plastic deformation of these alloys reach unexpected values [8–20], making them usable as composite structures, resistant to dynamic stresses with high deformation velocity, applicable in the field of collective protection [9–20].

2. Obtaining of $Al_xCr_yFe_zCo_vNi_w$ class of high-entropy alloys

The most commonly used method for successfully producing high-entropy alloys is the electric arc melting of the load materials in a vacuum arc remelting equipment (with current values of up to 500 A) in controlled atmosphere. The main technological operations pursued in this case are presented below.

2.1 Obtaining AlCrFeCoNi alloys in vacuum arc remelting furnace

Special type of materials chosen for making the ballistic target is experimental alloys AlCrFeCoNi, obtained in vacuum arc remelting equipment (MRF ABJ 900, in ERAMET Laboratory, University Politehnica of Bucharest, Romania). For calculating the metallic load, the theoretical degrees of assimilation of the elements in the melt and of the possible vaporization losses during the metallurgical process in vacuum or in argon-controlled atmosphere must be taken into account (**Figure 1**).

Mass losses are estimated on the basis of the literature data thereon, the degree of oxidation of the metallic load materials, the characteristics of the elements in the load, their positioning in the series of electrochemical potentials, the characteristics of the preparation aggregate, and the experience in the preparation process. It should be noted that the metallic load used to obtain high-entropy alloys must be of high quality, low in phosphorus and sulfur, and degreased and properly machined (in terms of granulometry). The degree of purity of the elements the alloy is over 99%. The elements dosed in equimolecular or quasi-equimolecular proportions are introduced following an order determined according to the type of alloy to be prepared, into the crucibles in the copper plate, water-cooled throughout the metallurgical process.

In order to produce high-entropy alloys under high-purity conditions, the working chamber must be suitably prepared by successive vacuuming and argon purging

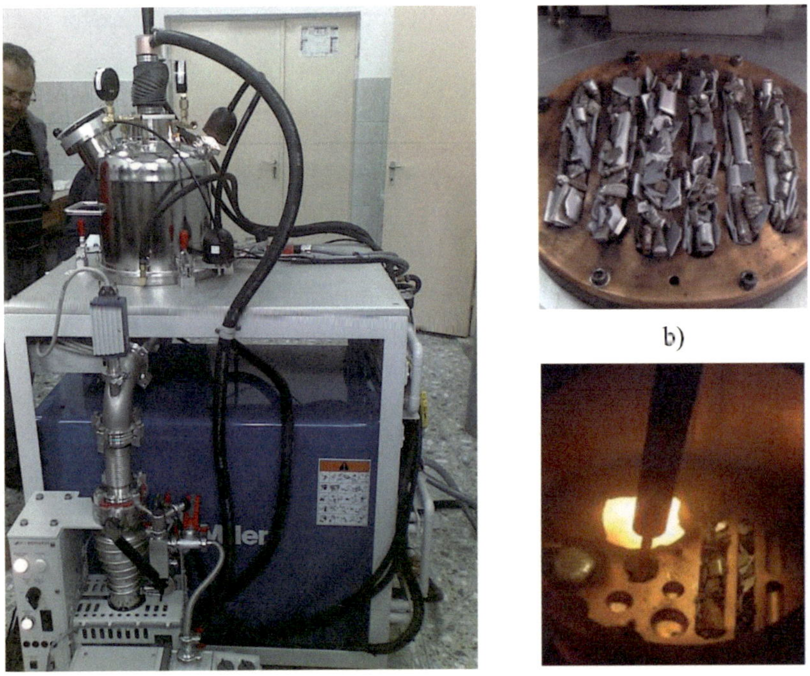

a) b) c)

Figure 1.
Vacuum arc remelting equipment MRF ABJ 900 and working chamber, in ERAMET Laboratory, University Politehnica of Bucharest, Romania.

operations performed at least three times. The vacuuming is performed using preliminary vacuum systems and diffusion pumps, which can provide pressure levels of about 3.5–4 × 10^{-4} mbar. Pure argon (Ar 5.3, 99.99%) is used for purging and melting. These operations provide a maximum oxygen content of 40–60 ppm in the working chamber. The final stage of the process is the argon purging of the working chamber and the setting of a working pressure level slightly above the atmospheric pressure.

The process of producing high-entropy alloys consists in melting the load materials, followed by the remelting of the samples for five to seven times, turning them on opposite sides to ensure a fully alloyed state and to improve the chemical homogenization of the mini-ingots. The entire preparation process is carried out by electric arc remelting in argon-controlled atmosphere.

The melting mode of the vacuum arc remelting furnace must be adapted to the type of alloy being prepared. These parameters vary during the preparation process, depending on the stages and the activities being carried out. The thorium tungsten electrode must be shifted during melting so that it is approximately 1/4" away from the copper plate electrode, and the electric arc formed sweeps the entire surface of the load for complete homogenization. Following melting and solidification, mini-ingots of weights almost constant compared to that of the load introduced into the VAR are produced.

Using this method, experimental high-entropy alloys have been instantaneously cooled by forced cooling of the water pumped at the base of the copper plate (the crystallizer of the furnace).

The melted material solidifies ultrafast in the water-cooled copper shell. The mini-ingots produced can have different shapes, depending on the shape of the cavity in the copper plate of the furnace. The samples were analyzed, according to the chemical composition, in order to define their physical, chemical, mechanical, etc. properties.

Figure2.
Metallic molds for casting $Al_{0.8}$CrFeCoNi alloy.

The chemical compositions of the metallic materials used are as follows:

- Extra soft steel, MK3 grade: C = 0.02 wt%; Si = 0.04 wt%; Mn = 0.21 wt%; S = 0.02 wt%; P = 0.015 wt%; Ni = 0.2 wt%; Cr = 0.15 wt%; Mo = 0.07 wt%; Cu = 0.14 wt%; Al = 0.12 wt%; Fe = ballance wt%
- Metallic chromium with 99 wt% Cr
- Electrolytic aluminum with 98.5 wt% Al
- Metallic cobalt with 99.5 wt% Co
- Electrolytic nickel with 99.5 wt% Ni

2.2 Obtaining of AlCrFeCoNi alloy in duplex system including induction and vacuum arc remelting furnace

A vacuum induction melting-vacuum arc remelting (VIM-VAR) duplex technology was selected to produce the high-entropy alloy AlCrFeCoNi in order to increase the purity of the alloy and to improve its mechanical properties. The experimental batches were prepared in the vacuum induction melting furnace Balzers, HU-40-25-40-04 type, with a capacity of 12 kg and in the MRF ABJ 900 vacuum arc remelting equipment, both from the ERAMET Laboratory, Bucharest.

The high-entropy alloy used to make the ballistic protection plates in the vacuum induction furnace was produced according to a classical preparation technology, using highly pure materials, and metallic molds made of iron Fc 250 (**Figure 2**) were used for casting to ensure the production of 150 × 100 × 10 mm rectangular plates.

The load calculation (based on the molar mass of the chemical elements, M_{mol}) for obtaining the $Al_{0.8}$CrFeCoNi alloy, taking into account the estimated elemental losses, is as follows (Eq. (1)):

$$M_{mol} = 27x0.8 + 52 + 56 + 59 + 59 = 247.6 \text{ g} \quad (1)$$

The concentration (% weight) of the alloying elements in the alloy was Al = 8.72 wt%; Cr = 21 wt%; Fe = 22.62 wt%; Co = 23.83 wt%; and Ni = 23.83 wt%. The alloy mass calculated based on the volume and density of the alloy cast in a plate and

Figure 3.
HEA $Al_{0.8}CrFeCoNi$ alloy plates obtained by VIM-VAR technology. (a) As-cast HEA samples obtained by VIM technology; (b) HEA sample positioned in copper plate of VAR equipment to be remelted; (c) sample after remelting procedure in VAR equipment.

related network was 1.111 and 0.574 kg, respectively, resulting in an alloy quantity of approx. 1.6855 kg per ingot.

Since alloy losses occur due to oxidation, due to the interaction with the furnace walls and to the casting ladle, a quantity of 8 kg of prepared alloy is considered per batch.

The quantities of materials used for each batch, also taking into account the oxidation losses, are as follows: Al = 0.74 kg; Cr = 1.70 kg; Fe = 1.85 kg; Co = 1.95 kg; and Ni = 1.93 kg. The total weight of the batch was of 8.17 kg. The $Al_{0.8}CrFeCoNi$ alloy plates obtained by casting are shown in **Figure 3**.

The production of the HEA $Al_{0.8}CrFeCoNi$ plates was completed by refining in the vacuum arc remelting furnace, for which, based on the requirements of the military field, a special copper plate was made, according to **Figure 2b**.

The working procedure in the VAR furnace was classical, seeking the most efficient homogenization of the alloy to ensure the best possible mechanical properties, by remelting each sample three times on each side (**Figure 3c**).

3. Mechanical tests

3.1 Impact fracture tests

The impact fracture tests were performed using a Charpy pendulum, to measure the energy absorbed in the process of dynamic fracture of standardized (notched) specimens. The values resulting from the impact fracture test for some AlCrFeCoNi alloys are shown in **Table 2**.

As can be seen from the values of the breaking energy shown in the table, the experimental materials have a tenacity corresponding to structural steels, while the hardness values are similar to those of tool steels.

3.2 Microhardness

The hardness of the experimental materials was determined using the Shimadzu HMV2T microhardness apparatus in Lamet Laboratories from *Politehnica* University of Bucharest. The measurements were made in line, with mark distances of about 500 μm, using the fingerprint force of 0.1 N and pressing time of 10 seconds [9–12]. For the $Al_xCr_yFe_zCo_vNi_w$ system, the entire spectrum of microhardness values in the x = y = z = v = w = 0.2 ... 2 at% range was analyzed, as shown in **Figure 4**.

The coding of the samples in **Figure 4** was based on the atomic proportions of the chemical elements; thus, HEA 1 = AlCrFeCoNi; HEA 2 = $Al_{1.5}CrFeCoNi$; HEA 3 = $Al_2CrFeCoNi$; HEA 5 = $Al_{0.8}CrFeCoNi$; HEA 6 = $Al_{0.6}CrFeCoNi$; HEA 7 = $Al_{0.4}CrFeCoNi$; HEA 8 = $Al_{0.2}CrFeCoNi$; HEA 9 = $Al_{1.2}CrFeCoNi$; and HEA 10 = $Al_{1.4}CrFeCoNi$. The maximum hardness value was obtained for the sample HEA 3, with maximum concentration of aluminum.

The coding of the samples in **Figure 5** was based on the atomic proportions of the chemical elements; thus, HEA 11 = $AlCrFeCoNi_{1.2}$; HEA 12 = $AlCrFeCoNi_{1.4}$; HEA 13 = $AlCrFeCoNi_{1.6}$; HEA 14 = $AlCrFeCoNi_{1.8}$; HEA 15 = $AlCrFeCoNi_2$; HEA 16 = $AlCrFeCoNi_{0.8}$; HEA 17 = $AlCrFeCoNi_{0.6}$; HEA 18 = $AlCrFeCoNi_{0.4}$; and HEA 19 = $AlCrFeCoNi_{0.2}$. The maximum hardness value was obtained for the sample HEA 11.

Test no.	Impact energy, J				
	HEA 1	HEA 5	HEA 6	HEA 12	HEA 14
1	62.1	67.1	67.0	66.0	62.4
2	62.7	67.2	66.9	65.8	62.3
3	62.1	67.0	67.1	65.9	62.1
4	62.2	67.3	68.9	66.1	62.1

Table 2.
Impact fracture energy values for some experimental HEAs.

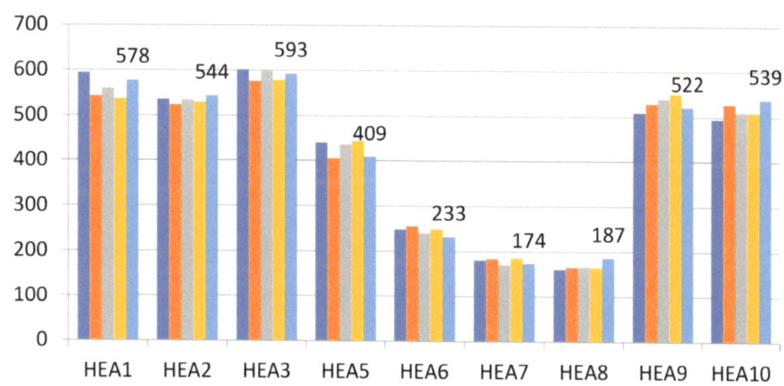

Figure 4.
Microhardness $HV_{0.1}$ values for experimental HEA 1 to HEA 10.

The coding of the samples in **Figure 6** was based on the atomic proportions of the chemical elements; thus, HEA 20 = $AlCrFeCo_{0.8}Ni$; HEA 21 = $AlCrFeCo_{0.6}Ni$; HEA 22 = $AlCrFeCo_{0.4}Ni$; HEA 23 = $AlCrFeCo_{0.2}Ni$; HEA 24 = $AlCrFeCo_{1.2}Ni$; HEA 25 = $AlCrFeCo_{1.4}Ni$; HEA 26 = $AlCrFeCo_{1.6}Ni$; HEA 27 = $AlCrFeCo_{1.8}Ni$; and HEA 28 = $AlCrFeCo_2Ni$. The maximum hardness value was obtained for the sample HEA 26.

The coding of the samples in **Figure 7** was based on the atomic proportions of the chemical elements; thus, HEA 29 = $AlCr_{0.2}FeCoNi$; HEA

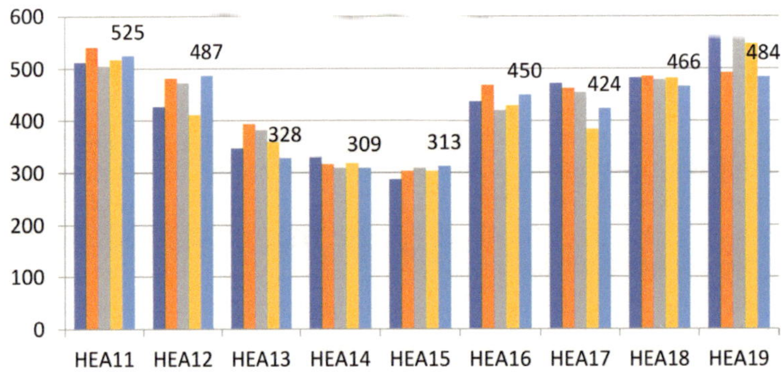

Figure 5.
Microhardness $HV_{0.1}$ values for experimental HEA 11 to HEA 19.

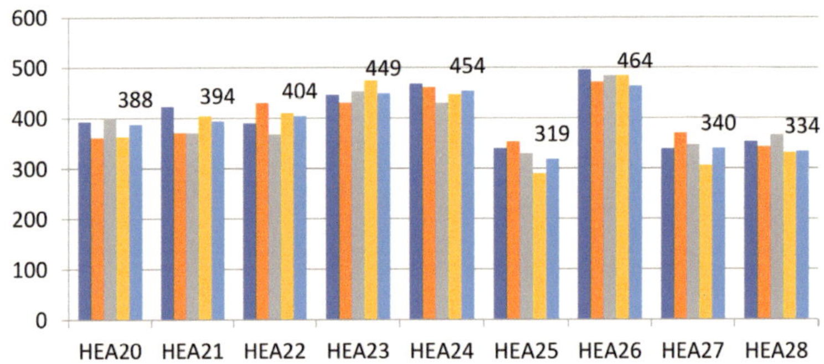

Figure 6.
Microhardness $HV_{0.1}$ values for experimental HEA 20 to HEA 28.

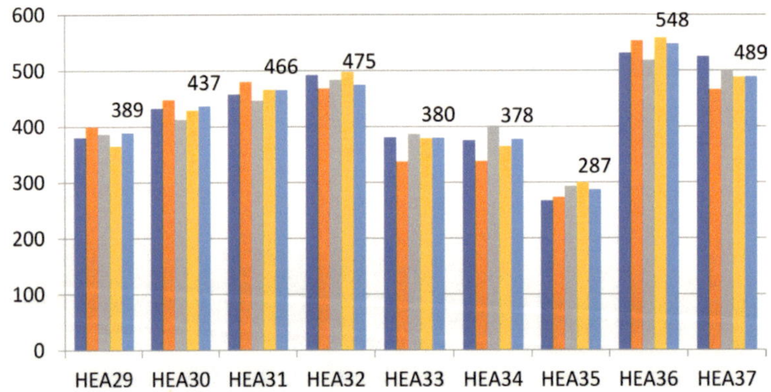

Figure 7.
Microhardness $HV_{0.1}$ values for experimental HEA 29 to HEA 37.

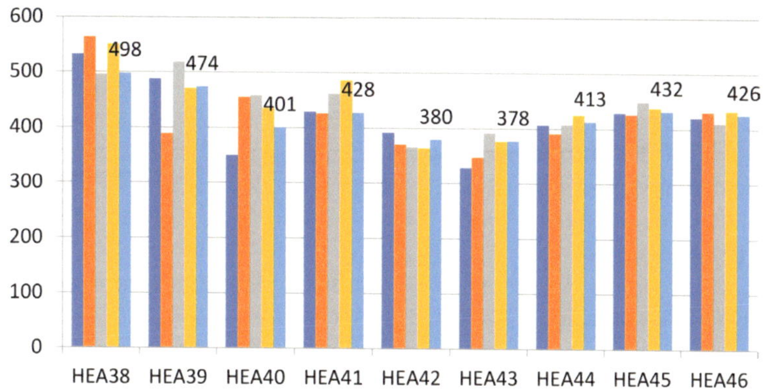

Figure 8.
Microhardness $HV_{0.1}$ values for experimental HEA 38 to HEA 46.

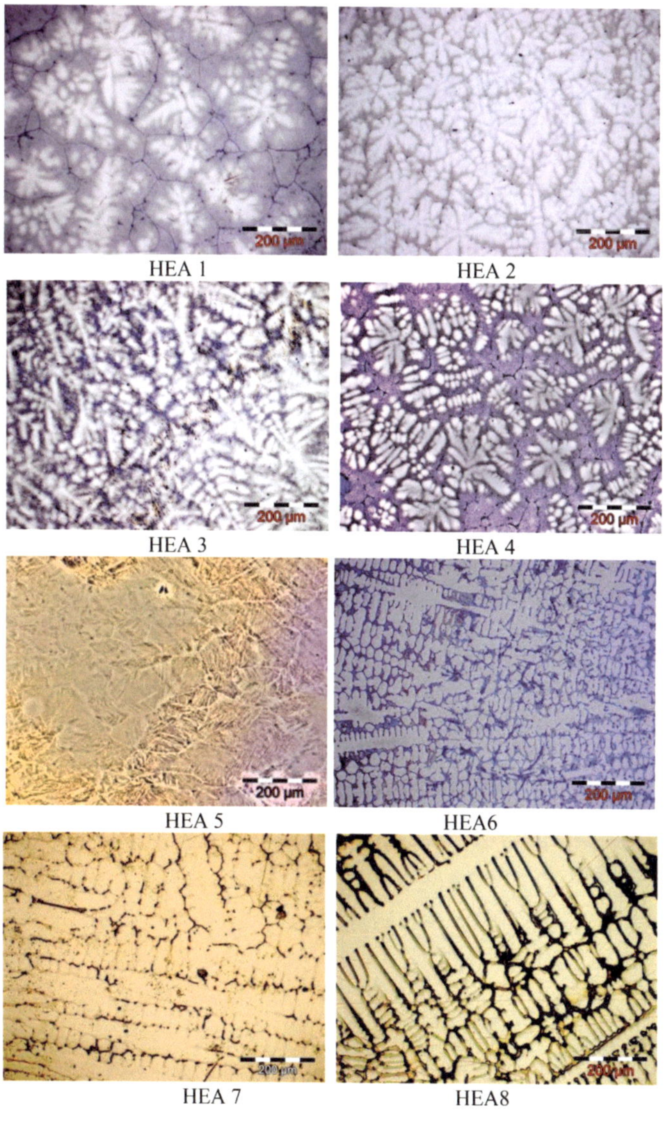

Figure 9.
Experimental as-cast $Al_xCrFeCoNi$ alloys.

30 = $AlCr_{0.4}FeCoNi$; HEA 31 = $AlCr_{0.6}FeCoNi$; HEA 32 = $AlCr_{0.8}FeCoNi$; HEA 33 = $AlCr_{1.2}FeCoNi$; HEA 34 = $AlCr_{1.4}FeCoNi$; HEA 35 = $AlCr_{1.6}FeCoNi$; HEA 36 = $AlCr_{1.8}FeCoNi$; and HEA 37 = $AlCr_2FeCoNi$. The maximum hardness value was obtained for the sample HEA 36.

The coding of the samples in **Figure 8** was based on the atomic proportions of the chemical elements; thus, HEA 38 = $AlCrFe_{0.2}CoNi$; HEA 39 = $AlCrFe_{0.4}CoNi$; HEA 40 = $AlCrFe_{0.6}CoNi$; HEA 41 = $AlCrFe_{0.8}CoNi$; HEA 42 = $AlCrFe_{1.2}CoNi$; HEA 43 = $AlCrFe_{1.4}CoNi$; HEA 44 = $AlCrFe_{1.6}CoNi$; HEA 45 = $AlCrFe_{1.8}CoNi$; and HEA 46 = $AlCrFe_2CoNi$. The maximum hardness value was obtained for the sample HEA 38, for minimum concentration of iron.

4. Microstructure

The microstructure of experimental high-entropy alloys was performed selectively by optical microscopy (Olympus GX51 reversed optical microscope) and scanning electron microscopy (Inspect SEM, FEI Company, scanning electron microscope equipped with EDAX Z2e detector) [9, 13–16]. The microstructural aspect of some experimental alloys is shown in **Figures 9–14**.

All materials show dendritic formations and interdendritic precipitations in experimental as-cast HEAs. Some of them (HEA 5) shows polyhedral grains and acicular phase growth from the grain boundary (**Figure 9**). The images have been performed for the same magnification (scale of 200 μm). The sample HEA 15 with the highest nickel concentration has the orientation and sequence of the alpha and gamma phases, in the form of parallel or perpendicular planes (**Figure 10**).

Changing the Co content does not make changes to the microstructure, which has a dendritic appearance (**Figure 11**). The increase in chromium content resulted

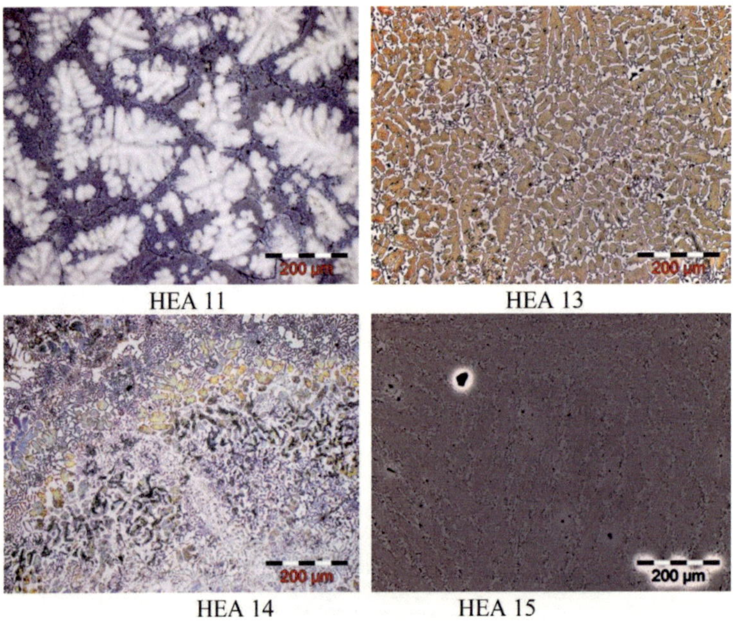

Figure 10.
Experimental as-cast AlCrFeCoNix alloys.

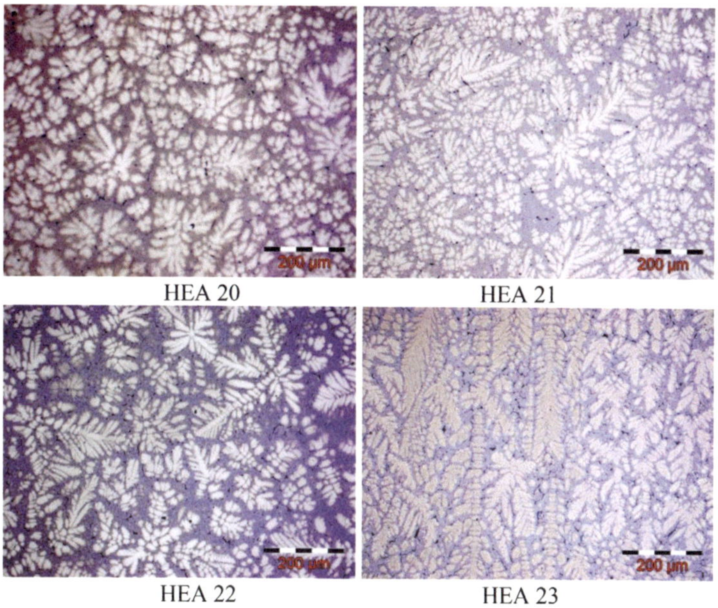

Figure 11.
Experimental as-cast AlCrFeCo$_x$Ni alloys.

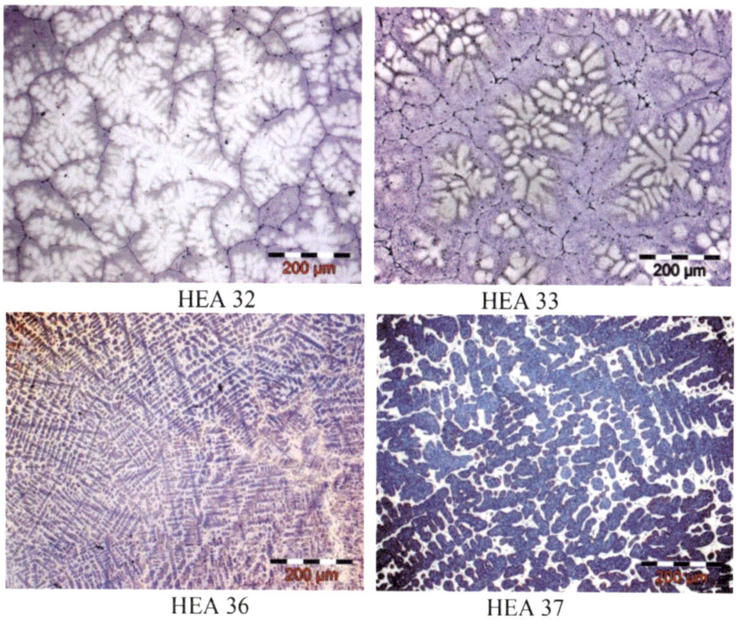

Figure 12.
Experimental as-cast AlCr$_x$FeCoNi alloys.

in an increased finishing of dendritic microstructure granulation for HEA 36 sample, which also led to an increase in hardness (**Figure 12**).

Scanning electron microscopy images revealed the fine and nanostructured microstructure with polyhedral grains and the layout of phases in quasi-parallel planes (**Figure 13**) [15].

In the case of high-Ni samples, higher magnification powers can be seen in phase configurations, in acicular or polymorphic form (**Figure 14**).

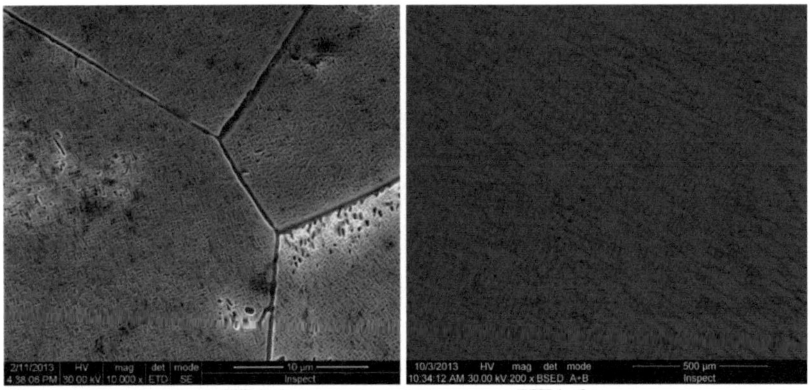

Figure 13.
SEM images of $Al_xCrFeCoNi$ alloys.

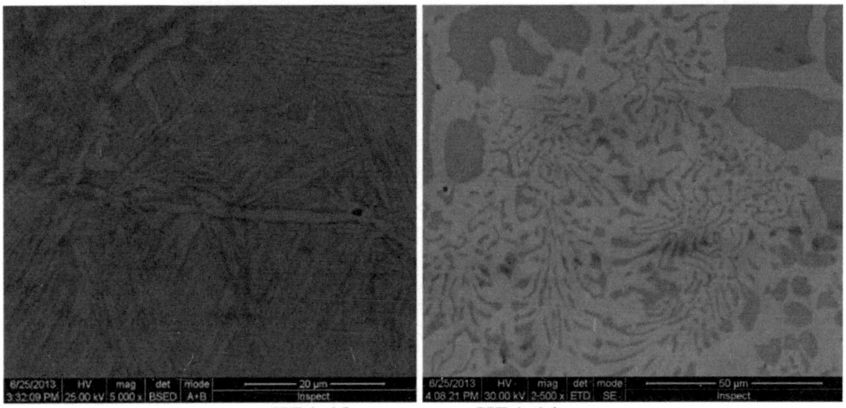

Figure 14.
SEM images of $AlCrFeCoNi_x$ alloys.

5. Dynamic testing

Several ballistic packages (**Figure 15**) were made for the testing of composite structures at high-velocity perforation, explosion, and high-velocity deformation. The packages have a sandwich structure, containing aluminum, steel, or ceramic plates or polymer HEA plate and, finally, aluminum plate again. The images from the firing room and the testing configuration are shown in **Figure 16**, and the experimental stand scheme for dynamic test is shown in **Figure 17**.

The ballistic packages undergoing tests for checking the bullet impact behavior were fastened to a wooden stand at a distance of 5 m from the machine-gun barrel's muzzle. The amount of powder in the cartridge shell was varied in order to obtain the desired initial velocity [15–19].

Two 7.62 × 39 mm incendiary armor-piercing bullets with initial velocities of 660 and 728 m/s, respectively, were fired at the targets. The ballistic package resisted punching for the first fire (**Figure 17a**) and then perforated at the second firing (**Figure 17b**).

The same procedure of dynamic testing, using 7.62 × 39 mm incendiary armor-piercing bullets at initial velocities of 723 and 728 m/s, respectively, was applied at the HEA-ceramic ballistic package. This type of ballistic package was perforated both at the first and second cases (**Figure 18**).

The last procedure was applied using 7.62 × 39 mm incendiary armor-piercing bullets at initial velocities of 720 and 725 m/s, respectively, for dynamic testing of HEA-ceramic ballistic package. This type of ballistic package was entirely perforated, in both cases (**Figure 19**).

a) b)

Figure 15.
Ballistic packages prepared for dynamic tests.

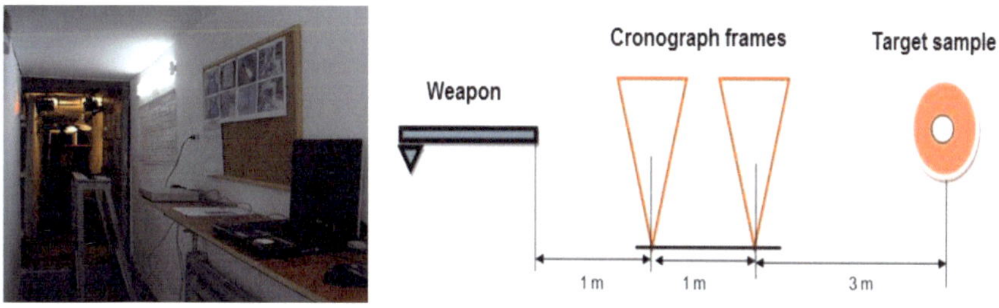

a) b)

Figure 16.
Firing room and experimental stand for dynamic test.

a) b)

Figure 17.
HEA-steel ballistic package during dynamic tests (a) after the first firing and (b) after the second firing.

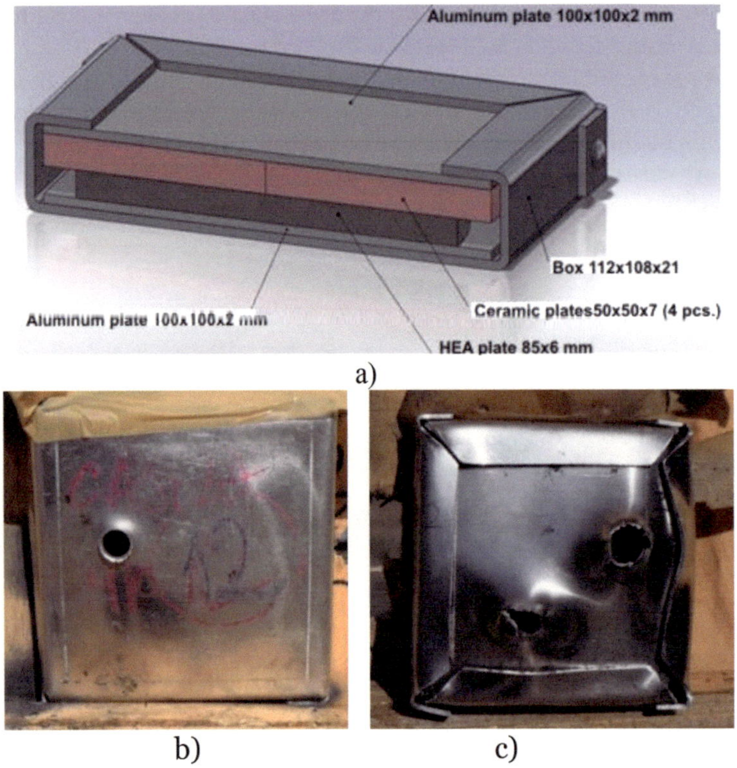

Figure 18.
HEA-ceramic ballistic package during dynamic tests (a) after the first firing (b) and after the second firing (c).

Figure 19.
HEA-polymer ballistic package during dynamic tests. (a) after the first firing and (b) after the second firing.

6. Conclusions

High-entropy alloys are well suited for production in vacuum arc remelting furnaces due to low material losses and to obtaining high purity by working in vacuum and argon. Moreover, they can also be produced in induction furnaces, for superior purity, using the VIM-VAR duplex process.

The hardness decrease of the AlCrFeCoNi class high-entropy alloys is proportional to the aluminum content decrease, from 500 HV_1 for HEA 1 (AlCrFeCoNi) at 400 HV_1 for HEA 5 ($Al_{0.8}CrFeCoNi$) and 224 HV_1 for HEA 6 ($Al_{0.6}CrFeCoNi$).

The hardness decrease can be explained by the reduction in the quantity and number of hard precipitates (Fe-Al compounds) in the metallic matrix.

The hardness further decreases by increasing the nickel content, which allows the formation of face-centered cubic (FCC) solid solutions of low hardness and high tenacity. In this type of solid solution the chemical elements have a good solubility, with a low tendency of separation of hard and brittle compounds.

The fracture energy values are in the range of 62–67 J for all five types of alloys, as the hardness oscillates in the range 200–500 $HV_{0.1}$. As a result, the hardening effect is not manifested by decreasing the metal matrix toughness in the case of high-entropy alloys in the analyzed alloy class.

The microstructure of high-entropy alloys is virtually "frozen" at the melt, with the solution retaining a conglomerate of chemical elements which are oftentimes very different (iron-related elements such as Cr, Ni, and Co, which together form solid solutions along with aluminum, a transition metal whose solubility greatly differs from that of Fe, Cr, Ni, and Co). The cooling condition creates entropy with very high values and explains the obtaining of completely different characteristics compared to the alloy cooled at usual rates.

Depending on the share of the alloying elements, one or two types of solid solutions are formed, partially embedding the other alloying elements. The dendritic microstructure predominates, separating into acicular compounds or globular precipitates, depending on the chemical composition and the cooling rate.

The tests on the behavior of high-entropy alloys at strong impacts were conducted under identical conditions. In the case of ballistic packages resistant at high-velocity penetration impacts, the best option is the HEA-steel system. High-entropy alloys, according to their composition, can be used in various sectors: medical engineering, earthwork equipment, and ballistic packages for individual and collective protection.

Acknowledgements

This paper was supported by the Romanian National Authority for Scientific Research CNDI-UEFISCDI, project number PN-III-P1-1.2-PCCDI-2017-0875 – PCCDI 20-2018 (HEAPROTECT).

Thanks

The authors want to thank the team of researchers coordinated by Professor PhD Tudor Chereche̦ş for collaborating in the dynamic tests and the processing of experimental data.

Author details

Victor Geanta and Ionelia Voiculescu*
Ionelia Voiculescu, University Politehnica of Bucharest, Bucharest, Romania

*Address all correspondence to: ioneliav@yahoo.co.uk

References

[1] Yeh J, Chen SK, Lin S, Gan JY, Chin TS, Shun TT, et al. Nanostructured high entropy alloys with multiple component elements: Novel alloy design concepts and outcomes. Advanced Engineering Materials. 2004;**6**:299-303

[2] Zhang Y et al. Microstructure and properties of high entropy alloys. Progress in Materials Science. 2014;**61**:1-93

[3] Maweja K, Strumpf W. The design of advanced performance high strength low-carbon martensitic armor steels microstructural considerations. Materials Science and Engineering A. 2008;**480**:160-166

[4] Maweja K, Strumpf W. Fracture and ballistic-induced phase transformation in tempered martensitic low-carbon armour steels. Materials Science and Engineering A. 2006;**432**:158-169

[5] R.I. Hammond, W.G. Proud. Does the pressure-induced α–ε phase transition occur for all low-alloy steels? Proceedings of the Royal Society of London A. 2004;**460**:2959-2974

[6] Srivathsan B, Ramakrishnan N. Ballistic performance maps for thick metallic armour. Journal of Materials Processing Technology. 1999;**96**:81-91

[7] Srivathsan B, Ramakrishnan N. A ballistic performance index for thick metallic armour. International Journal of Modeling, Simulation, and Scientific Computing. 1998;**3**:33-40

[8] Wang YP, Li BS, Ren MX, Yang C, Fu HZ. Microstructure and compresive properties of AlCrFeCoNi high entropy alloy. Materials Science and Engineering A. 2008;**491**:154-158

[9] Voiculescu I, Geanta V, Vasile IM, Ştefănoiu R, Tonoiu M. Characterisation of weld deposits using as filler metal a high entropy alloy. Journal of Optoelectronics and Advanced Materials. 2013;**15**(7-8):650-654

[10] Geantă V, Voiculescu I, Miloșan I, Istrate B, Mateș IM. Chemical composition influence on microhardness, microstructure and phase morphology of AlxCrFeCoNi high entropy alloys. Revista de Chimie (Bucharest). 2018;**69**(4):798-801

[11] Geantă V, Voiculescu I, Stefănoiu R, Chereches T, Zecheru T, Matache L, et al. Dynamic impact behaviour of high entropy alloys used in the military domain. Euroinvent ICIR 2018 IOP Publishing IOP Conference Series: Materials Science and Engineering 2018;**374**:012041. DOI: 10.1088/1757-899X/374/1/012041

[12] Geantă V, Voiculescu I, Istrate B, Vrânceanu D, Ciocoiu R, Cotruț C. The influence of chromium content on the structural and mechanical properties of AlCrxFeCoNi high entropy alloys. International Journal of Engineering Research in Africa, Trans Tech Publications, Switzerland. 2018;**37**:23-28. ISSN: 1663-4144

[13] Ştefănoiu R, Geantă V, Voiculescu I, Csaki I, Ghiban N. Researches regarding the influence of chemical composition on the properties of AlxCrFeCoNi alloys. Revista de Chimie. 2014;**65**(7):819-821

[14] Geantă V, Voiculescu I, Ştefănoiu R, Savastru D, Csaki I, Patroi D, et al. Processing and characterization of advanced multi-element high entropy materials from AlCrFeCoNi system. Optoelectronics and Advanced Materials—Rapid Communications. 2013;**7**:11, 874-12, 880

[15] Voiculescu I, Geantă V, Ştefănoiu R, Patroi D, Binchiciu H. Influence of the chemical composition on the

microstructure and microhardness of AlCrFeCoNi high entropy alloy. Revista de Chimie (Bucharest). 2013;**64**(12):1441-1444

[16] Geantă V, Voiculescu I, Chereches T, Zecheru T, Matache L, Rotariu A. Behavior to dynamic loads of composite multi-layer structures. Mats, Plastic (Bucharest). 2019;**6**(2):460-465

[17] Geantă V, Chereches T, Lixandru P, Voiculescu I, Stefănoiu R, Dragnea D, et al. Virtual testing of composite structures made of high entropy alloys and steel. Metals. 2017;**7**:496

[18] Geantă V, Chereches T, Lixandru P, Voiculescu I, Ștefănoiu R, Dragnea D, et al. Simulation of impact phenomena on the composite structures containing ceramic plates and high entropy alloys. ICIR Iasi, International Conference on Innovative Research-ICIR EUROINVENT 2017 IOP Publishing IOP Conference Series: Materials Science and Engineering. 2017;**209**:012043

[19] Csaki I, Ștefănoiu R, Geantă V, Voiculescu I, Sohaciu MG, Soare A, et al. Researches regarding the processing technique impact on the chemical composition, microstructure and hardness of AlCrFeCoNi high entropy alloy. Revista de Chimie (Bucharest). 2016;**67**(7):1373-1377

[20] Chereches T, Lixandru P, Geantă V, Voiculescu I, Dragnea D, Ştefănoiu R. Layered structures analysis, with high entropy alloys, for ballistic protection. IMANE Iași 2015. Applied Mechanics and Materials. 2015; **809-810**:724-729. ISSN: 1662-7482

Chapter 9

High-Entropy Ceramics

Sajid Ali Alvi, Hanzhu Zhang and Farid Akhtar

Abstract

High-entropy ceramics is an emerging class of high-entropy materials with properties superior to conventional ceramics. Recent research has been focused on the development of new high-entropy ceramic compositions. High-entropy oxides, carbides, borides, silicides, and boron carbides had been reported with superior mechanical, oxidation, corrosion, and wear properties. The research work on the processing and characterization of bulk high-entropy ceramics and coating systems has been summarized in this chapter. The composition design, structure, chemistry, composite processing of bulk high-entropy ceramics, and evolution of microstructure and properties are reported. The literature on the deposition of high-entropy ceramic coating and the influence of coating parameters have been discussed to produce high-entropy ceramic coatings with superior mechanical, oxidation, and wear properties.

Keywords: ceramics, spark plasma sintering, coatings, high temperature properties

1. Introduction

The concept of high-entropy alloy (HEA), an alloy with five or more constituents with a concentration between 5 and 35%, has triggered extensive research to develop new metallic and ceramic high-entropy materials. Like HEAs, high-entropy ceramics (HECs) consist of multicomponent ceramic oxides, borides, and nitrides. The HECs exhibit homogeneous crystalline single-phase despite the complex structure of ceramic compounds and offer superior properties over the constituents and conventional ceramics. The development of high-entropy ceramic coatings is an important route to design new and novel high-entropy ceramic materials with superior properties for practical applications. In this chapter, the recent work on HECs is summarized in the following two sections: bulk HECs and HEC coatings. Bulk HEC section is focused on the processing and the evolution of significantly enhanced mechanical properties of borides, oxides, and carbide ceramics and composites. Moreover, the potential of HECs in thermal-electrical, tribological, and high temperature applications are discussed. In the section on high-entropy coatings, recently reported researches on different HEC coating techniques and the properties of the HEC coatings have been summarized.

2. Bulk high-entropy ceramics

2.1 Fabrication routes

The fabrication of HEAs can be achieved through solid and liquid state routes. The processing of bulk HECs is preferred in solid state due to the high melting points of ceramics. The solid-state processing includes high energy ball milling, conventional solid state sintering, self-propagating high temperature synthesis (SHS), and spark plasma sintering (SPS) under the combined effect of heating using electric current and pressure. The SPS is a rapid and facile method to produce nearly dense components and suitable for sintering of high-entropy ceramics. Typically, the precursor ceramic powders are mixed and filled into a graphite die and rapidly sintered under high vacuum, pulsed direct electrical current, and uniaxial pressure.

Different types of the ceramic precursors have been used to fabricate HECs. The most commonly used precursors are commercial ceramic powders. The ceramic powders are mixed in desired stoichiometric ratio and homogenized using ball milling followed by spark plasma sintering to a target temperature and uniaxial pressure. The synthesis of high-entropy ceramic powders has been investigated as well. The precursor ceramic powders can be pre-synthesized via thermal reduction (TR) where metal oxide powders are used as reactants (**Figure 1**). The advantages of using oxide powders as raw materials include reducing the cost of the starting materials and possibilities of producing high purity ceramic powders with controlled grain size. Feng et al. [1] used metallic oxides and carbon black powder as starting material to produce high-entropy (Hf, Zr, Ti, Ta, Nb)C powder. The carbothermal reduction of oxide ceramics was completed at 1873 K under vacuum in 1 hour. Subsequent solid solution formation was achieved at a higher temperature of 2273 K, demonstrating a single-phase rocksalt structure of the solid solution. Ye et al. [2] reported the synthesis of high-entropy $(Zr_{0.25}Ta_{0.25}Nb_{0.25}Ti_{0.25})C$ powders of a single-phase rocksalt crystal structure by carbothermal reduction of metal oxides and graphite.

Liu et al. [3] demonstrated the synthesis of high-entropy $(Hf_{0.2}Zr_{0.2}Ta_{0.2}Nb_{0.2}Ti_{0.2})B_2$ powders with single-phase hexagonal structure through borothermal reduction from metal oxides and amorphous boron powders at 1973 K under argon atmosphere. The synthesized high-entropy metal diboride powders showed a fine particle size of 310 nm. Instead of boron, Zhang et al. [4] used the combination of B_4C, graphite, and metal oxides to synthesize $(Hf_{0.2}Zr_{0.2}Ta_{0.2}Nb_{0.2}Ti_{0.2})B_2$, $(Hf_{0.2}Zr_{0.2}Mo_{0.2}Nb_{0.2}Ti_{0.2})B_2$, and $(Hf_{0.2}Mo_{0.2}Ta_{0.2}Nb_{0.2}Ti_{0.2})B_2$ powders by boro-/carbothermal reduction. The reaction is reported to be more effective than borothermal reduction, with a lower reduction temperature 1873 K.

Wei et al. synthesized high-entropy carbide $(Ti_{0.2}Zr_{0.2}Nb_{0.2}Ta_{0.2}W_{0.2})C$ by employing different methods using commercial carbide powders and thermally reduced carbide powders and *in-situ* fabrication process from elemental powders [5]. All three compositions revealed single-phase face-centered cubic (FCC) structure according to X-ray diffraction (XRD) patterns. The comparison of the

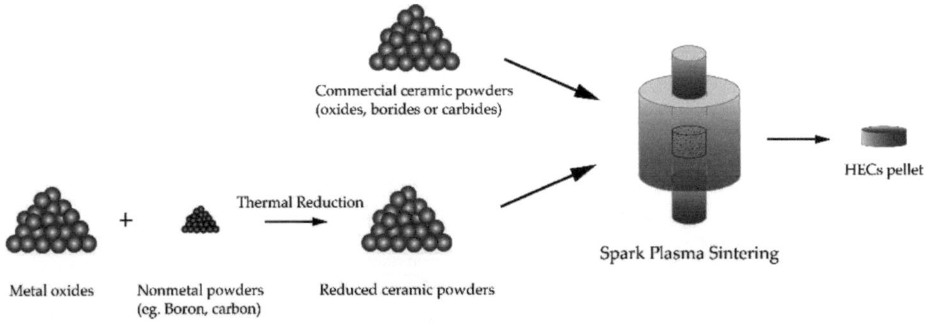

Figure 1.
The fabrication routes of dense high-entropy ceramics.

three methods suggested the microstructure inhomogeneity in case of elemental powders and TR approach.

2.2 Overview of current researches on bulk HECs

2.2.1 High-entropy oxides (HEOs)

The first high-entropy ceramic was reported by Rost et al. [6], on the production of entropy stabilized metal oxides with rocksalt crystal structure, synthesized from equimolar CoO, CuO, MgO, NiO, and ZnO in a tube furnace at temperatures above 850°C. Homogenous distribution of the cations in the crystal structure was observed. The system showed a reversible transformation between the high-entropy solid solution and multicomponent oxide mixture. Later on, detailed investigation using extended X-ray absorption fine structure (EXAFS) was performed to investigate the localize structure of the aforementioned high-entropy oxide [7], demonstrating that the anion lattice (O sublattice) has the most distortion to accommodate the size mismatch in the cation lattice (metal sublattice). The phase stability, as a result of configurational entropy, of the same HEO system was studied by Anand et al. [8].

The entropy stabilized oxide ((Mg, Ni, Co, Cu, Zn)O with colossal dielectric constant was discovered by Bérardan et al. [9]. With a Li^+ ionic conductivity of $>10^{-3}$ S/cm, which is much higher compared to lithium phosphorous oxynitride (LiPON) solid electrolyte (2×10^{-6} S/cm) [10], the produced high-entropy oxide was suggested to be an excellent substituent as superionic conductors for Li^+ and Na^+. The potential of high-entropy oxide in the lithium battery has been reported by Sarkar et al. [11] with improved storage capacity retention and cycling stability.

Jiang et al. [12] synthesized high-entropy perovskite oxides from multiple ABO_3 perovskite oxides. Djenadic et al. [13] utilized nebulized spray pyrolysis (NSP) to synthesize single-phase rare earth oxide powders from seven equiatomic rare earth oxides. In the REO system, the importance of selecting the starting component in the reported multicomponent oxide system was highlighted, with cerium (Ce^{4+}) addition being confirmed to improve the formation of single-phase solid solution. Gild et al. [14] fabricated high-entropy fluorite oxides from five fluorite oxides via high-energy ball milling and SPS, followed by various annealing treatments. Most of the fabricated HEOs revealed as nearly fully dense pellets with single-phase Fm-3 m crystal structure. Phonon scattering effect in the HEO system resulted in low thermal conductivity, making the synthesized HEOs desirable for thermal insulation applications.

2.2.2 High-entropy borides (HEBs)

The high-entropy borides are designed and fabricated from transition metal diborides, HfB_2, ZrB_2, TaB_2, etc., aiming at developing a new class of ultrahigh temperature metal diboride materials with superior mechanical properties over conventional diborides.

Gild et al. [15] sintered HEBs from five-component equimolar metal diborides via SPS. The fabricated HEBs from compositions like $(Hf_{0.2}Zr_{0.2}Ta_{0.2}Mo_{0.2}Ti_{0.2})B_2$ and $(Mo_{0.2}Zr_{0.2}Ta_{0.2}Nb_{0.2}Ti_{0.2})B_2$ possess single-phase hexagonal AlB_2 structure (**Figure 2**) as the parent metal diborides, with alternating hexagonal metal cations net and rigid boron net in the high-entropy structure. The authors reported that in case of using W_2B_5 in the starting precursors, single-phase solid solution was not successfully formed. Notably, W_2B_5 has a different crystal structure with other utilized metal borides, as well as limited solubility in other borides, which could

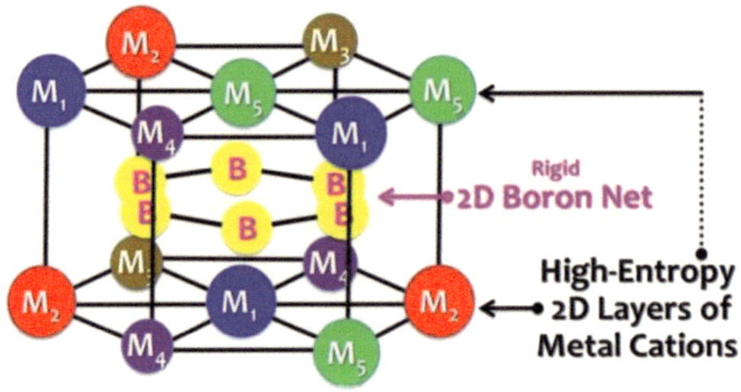

Figure 2.
The hexagonal AlB_2 structure of HEBs by Gild et al. [15] (licensed under CC BY 4.0, https://doi.org/10.1038/srep37946).

be potential factors of the formability of single-phase solid solution in the HEB system. The fabricated HEBs show improved hardness and oxidation resistance as compared to the average performances of the constituent borides.

The crystal structure, mechanical, and electronic properties of the HEB $(Hf_{0.2}Zr_{0.2}Ta_{0.2}M_{0.2}Ti_{0.2})B_2$ (M = Nb, Mo, Cr) were investigated by Sarkar et al. using ab initio calculations [16]. In the unique layered hexagonal lattice structure shown in **Figure 2**, the metal layer and 2D boron layer contain metallic bonds and covalent bonds, respectively, while both ionic and covalent bond were found between the two hexagonal layers. The high stability of the studied HEBs is contributed by strong boron-boron bond and metal-boron bond. Density-functional theory (DFT) calculations suggested negligible solid solution strengthening effect in elastic modulus, showing similar value with elastic modulus calculated by the mixing rule.

Tallarita et al. [17] synthesized bulk HEB $(Hf_{0.2}Mo_{0.2}Ta_{0.2}Nb_{0.2}Ti_{0.2})B_2$ by self-propagating high-temperature synthesis (SHS) from elemental metal and boron powders, followed by SPS at 2223 K. High fraction of high-entropy phase with hexagonal (space group P6/mmm) was obtained after the SHS process, along with minor amount of binary diborides and metal oxides. The SHS powders showed good sinterability with a relative density of 92.5% in the SPS product. The sintered HEB exhibits a single-phase hexagonal crystal structure which is similar to the hexagonal AlB_2 structure reported by Gild et al. [15], with improved hardness (22.5 ± 1.7 GPa) and oxidation resistance. The authors mentioned that the single-phase structure could not be obtained when performing one-step reactive SPS process with the same reactants.

Based on the developed HEB system, Zhang et al. [18] adopted borothermal reduction process to synthesize high-entropy borides from metal oxides and amorphous boron powder, aiming at producing ultra-fine HEB powder with improved sinterability. The obtained HEB powders $(Hf_{0.2}Zr_{0.2}Ta_{0.2}Cr_{0.2}Ti_{0.2})B_2$, $(Hf_{0.2}Mo_{0.2}Zr_{0.2}Nb_{0.2}Ti_{0.2})B_2$, and $(Hf_{0.2}Mo_{0.2}Ta_{0.2}Nb_{0.2}Ti_{0.2})B_2$ showed an average particle size smaller than 1 μm and led to final relative densities greater than 95% in the SPS products. The sintered HEBs showed ultrahigh hardness values of 28.3, 26.3, and 25.9 GPa, respectively. Compared to the HEBs with same compositions sintered from −325 mesh diboride powders by Gild et al. [15] (about 22 GPa), these HEBs have improved hardness properties as a result of utilization of ultrafine starting particles in SPS.

2.2.3 High-entropy carbides

The first reported high-entropy carbides were synthesized from transition metal carbides HfC, TaC, ZrC, NbC, and TiC by Castle et al. [19]. Four-component metal carbides were mixed in equiatomic ratio using high-energy ball milling and sintered in SPS with a maximum sintering temperature of 2573 K to form high-entropy (Hf-Ta-Zr-Ti)C and (Hf-Ta-Zr-Nb)C ultra-high temperature carbides. A single-phase solid solution with a rocksalt crystal structure was formed. More complete inter-diffusion in the formation of single-phase solid solution was found in the (Hf-Ta-Zr-Nb)C than (Hf-Ta-Zr-Ti)C system. As the atomic radius of Ti, as well as the lattice parameter of TiC is the smallest among the five starting components, the effect of lattice mismatch among the starting metal carbides on formability of high-entropy solid solution was highlighted. By considering the metal atomic radii, melting points and vacancy formation energy of the carbides, it is believed that TaC, which has the lowest metal vacancy formation energy, acted as the host lattice during the diffusion, while other metal atoms migrate into TaC lattice and occupy the cation positions. The nanoindentation of the fabricated high-entropy (Hf-Ta-Zr-Nb)C shows a hardness of 36.1 ± 1.6 GPa, which is 30% higher than the theoretical value calculated from the rule of mixture. Low thermal conductivity was observed in the reported SPS high-entropy carbides. Further investigation on the microstructure, atomic structure, and localized chemical disorder of the fabricated four-component high-entropy (Hf-Ta-Zr-Nb)C was reported by Dusza et al. [20]. Homogeneity of the chemical composition in both micro- and nanoscale was proved by various experimental approaches including scanning electron microscopy (SEM), transmission electron microscopy (TEM), and electron backscattered diffraction (EBSD).

The low thermal conductivity and diffusivity behavior were observed by Yan et al. [21] in the five-component high-entropy $(Hf_{0.2}Zr_{0.2}Ta_{0.2}Nb_{0.2}Ti_{0.2})C$ ceramic. The high-entropy carbide was synthesized in SPS at 2273 K and displayed rocksalt crystal structure with the metal atoms occupying the cation positions, while carbon occupied the anion positions in the lattice. The measured thermal diffusivity is reported to be lower than monocarbides and even some of the binary carbides. The increased number of principal elements in the structure led to lattice distortion, mainly through the carbon sublattice, which consequently caused severe phonon scattering. The same high-entropy carbide $(Hf_{0.2}Zr_{0.2}Ta_{0.2}Nb_{0.2}Ti_{0.2})C$ in powder form was synthesized by Zhou et al. [22] via pressureless sintering in SPS.

Entropy-forming-ability (EFA), the first systematic criterion to determine relative propensity of a multicomponent (metal carbide) system to form a high-entropy single-phase crystal structure was introduced by Sarker et al. [23], published on Nature Communication in November 2018. A total of 56 five-component systems from 12 metal carbides (Hf, Nb, Mo, Ta, Ti, V, W, and Zr) were investigated in the work by evaluating the energy distribution spectra of the structures that were generated by the AFLOW-POCC (automatic FLOW partial occupation) [24] algorithm. A narrow spectrum suggests low energy barrier to introduce more configurational disorder, thereby displays high EFA value. The calculation results were verified by synthesizing nine from the 56 compositions in SPS at 2473 K and comparing the experimental and calculation results. The phase identification in **Figure 3** suggested that six compositions $MoNbTaVWC_5$, $HfNbTaTiZrC_5$, $HfNbTaTiVC_5$, $NbTaTiVWC_5$, $HfNbTaTiWC_5$, and $HfTaTiWZrC_5$ that have higher EFA values than the rest three compositions reveals single-phase rocksalt crystal structure after SPS, while secondary phases were observed in compositions $HfMoTaWZrC_5$, $HfMoTiWZrC_5$, and $HfMoVWZrC_5$ with low EFA values.

The *ab initio* entropy descriptor invented by Sarker et al. [23] provides a new understanding of the formation of single-phase high-entropy materials and offers a systematic guide for researchers to design high-entropy carbides. Group IV and V carbides are known to have nearly complete mutual solubility; therefore, the formation of single-phase solid solution is relatively probable, attributed to enthalpy stabilization. Meanwhile, the incorporation of carbides with different

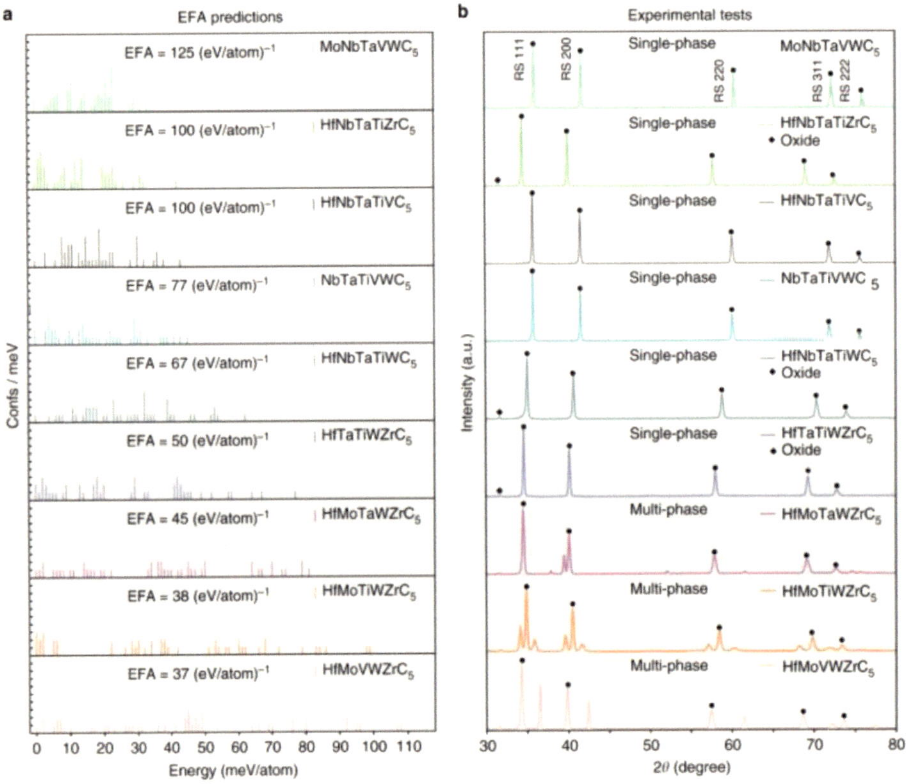

Figure 3.
(a) The energy distribution spectra of different configurations and the corresponding EFA value of the nine five-metal carbides. (b) The X-ray patterns of the sintered carbides with the same compositions, by Sarker et al. [23] (licensed under CC BY 4.0, https://doi.org/10.1038/s41467-018-07160-7).

crystal structures was commonly understood to impede the formation of the high-entropy phase because of the mismatch of lattice structure. It was observed that the addition of group VI elements (Cr, Mo, and W) is likely to reduce the chance of forming a single phase, as group VI metal monocarbides are generally demonstrated as non-cubic structure at room temperature. However, the $MoNbTaVWC_5$ composition that simultaneously contains tungsten carbide (W_2C) and molybdenum carbide (Mo_2C), which exhibit orthorhombic and hexagonal structure, respectively, shows the highest EFA value among the 56 analyzed compositions and demonstrates single-phase FCC structure after SPS. The mechanical properties measured by experiments showed that the Vickers hardness and elastic modulus have significant enhancement compared to the predicted value from the rule of mixture, contributed by mass disorder in the structure and solid solution hardening. Following the work of the pioneers, a systematic study focused on the phase stability and mechanical properties of high-entropy carbides synthesized from group IV, V, and VI metal carbides [25] was carried out. It was verified that carbide system that was stated to have EFA value lower than 45 cannot form single-phase solid solution via SPS.

2.2.4 Other high-entropy ceramics—silicide, borocarbide, sulfide, etc.

To date, the work of bulk high-entropy ceramics is mostly focused on oxide, boride, and carbide, and several other classes of high-entropy materials like silicide, sulfide, and borocarbide have been reported.

The synthesis of high-entropy silicide was reported by Gild et al. [26] and Qin et al. [27]. Gild utilized SPS and sintered high-entropy silicide $(Mo_{0.2}Nb_{0.2}Ta_{0.2}Ti_{0.2}W_{0.2})Si_2$ at 1923 K, resulting in a single hexagonal C40 crystal structure (space group $P6_222$). Hardness enhancement over the constituent silicide and low thermal conductivity similar to high-entropy carbide was reported. On the other hand, Qin produced $(Ti_{0.2}Zr_{0.2}Nb_{0.2}Mo_{0.2}W_{0.2})$ Si_2 from compositional elemental powders, leading to the formation of a hexagonal structure high-entropy silicide, with the same space group $P6_222$. Zhang et al. [28] developed a data-driven model to design high-entropy sulfides for thermoelectric applications. Compositions $Cu_5SnMgGeZnS_9$ and $Cu_3SnMgInZnS_7$ were picked to be synthesized via high energy ball milling and SPS. Single-phase high-entropy sulfide with homogeneous distribution of the elements is reported. By increasing the Sn content in the $Cu_5SnMgGeZnS_9$ system, a figure of merit zT value of 0.58 at 773 K was obtained.

By adding B_4C into the four-component HEC (HfMoTaTi)C, Zhang et al. [29] studied the capability of a high-entropy carbide system on accommodating one more nonmetal element, boron. The quaternary (HfMoTaTi)C contains FCC structures, while the addition of B_4C induced the formation of a small fraction of hexagonal phase. Similar with the diffusion process reported by Castle et al. [19], TaC is suggested to act as the host lattice in the borocarbide system. The effect of employing different particle sizes for host carbide TaC and other constituent carbides on the phase formation and mechanical properties of the HEC composites was discussed. A high-entropy $B_4(HfMo_2TaTi)C$ ceramic exhibiting hexagonal structure, with alternating metal and nonmetal layers in the lattice, was found when SiC whiskers are introduced to the borocarbide system [30]. The hexagonal HEC solid solution shows great improvement of the mechanical property with an ultrahigh hardness of 35 GPa.

3. High-entropy ceramic coatings

3.1 Preparation methods

3.1.1 Sputtering

Sputtering is a physical vapor deposition (PVD) technique where a solid surface (target) is bombarded with accelerated charged ions (Ar^+) resulting into backward scattering of surface atoms of solid. Sputtering technique is used to deposit thin coatings. There are several sputtering systems employed for deposition of thin coatings, such as direct current (DC) magnetron sputtering and radio frequency (rf) sputtering being the widely used techniques. DC-magnetron sputtering is a low pressure sputtering system for metal deposition and electrically conductive target coating materials. The magnetic field from magnetron lowers the sputtering gas pressure and increases the deposition rate of sputtered coatings. On the other hand, when metallic target is replaced by an insulator target, the plasma discharge cannot be maintained due to buildup of surface charge of positive ions on the target. Thus, an rf power supply is used instead of dc voltage power supply to sustain the glow discharge on an insulator target [31]. Another advantage of using sputtering for thin coating deposition is the use of reactive gases, substrate bias, and substrate temperature to vary the composition and properties of deposited coatings. In reactive sputtering, different reactive gases, such as oxygen, nitrogen, and CH_4 can be used to deposit ceramic coatings of oxides, nitrides, and carbides, respectively [32–34]. Furthermore, deposition of carbonitride and oxynitride coatings can be achieved by using a mixture of CH_4 + N_2 and O_2 + N_2, respectively [35, 36]. In reactive sputtering, a mixture of Ar and reactive gas (5–70%) is used for sputtering. The increasing bias between the target and substrate can affect the chemical composite, microstructure, and mechanical properties of the coating by giving compact coating structures. Lastly, substrate temperature can be varied in reactive sputtering to enhance the ion mobility and the interaction between the deposited ions, which can affect the microstructure, composition, and the properties of the coating. The following section will be used to highlight the use of reactive sputtering to deposit high-entropy nitride, carbide, and oxide ceramic coatings.

3.2 Structural evolution

3.2.1 High-entropy nitride coatings

High-entropy alloys based on multi-principle elements make it possible to design various nitride coatings. The use of nitride forming multi-principle elements can enhance the physical and mechanical properties that are not achievable in the conventional binary- and ternary-nitride coatings [37]. The high-entropy nitride coatings form amorphous and/or solid solution with face-centered cubic (FCC) structure. The severe lattice distortion and solid solution hardening help in developing high strength nitride coatings for various applications that require wear resistance, corrosion resistance, diffusion barrier, electrical resistivity, biocompatibility or light reflectivity [38]. The versatile sputtering deposition can be used to vary the N_2 flow, substrate temperature, and substrate bias to obtain nitride coatings with good physical, chemical, and mechanical properties, respectively.

3.2.2 Effect of N_2 flow rate

The high-entropy nitride (HEN) coatings developed by DC-magnetron sputtering were reported by research group of Yeh et al. using targets of FeCoNiCrCuAlMn, FeCoNiCrCuAl, AlxCoCrCuFeNi (x: 0.5 and 2), and AlCrNiSiTi [39]. In all the cases, the resulting HEN coatings showed formation of FCC solid solution at low N_2 flow rate (R_N) and became amorphous at higher N_2 flow rate (R_N) due to severe lattice distortion and presence of weak nitride forming elements, such as Al and Si, where $R_N = (N_2/(Ar + N_2))$. Following this work, Lai et al. developed (AlCrTaTiZr)N HEN coating, where the coating structure changed from amorphous phase metallic coating to FCC solid solution HEN with increasing N_2 flow rate [32]. The sputtering rate decreased from 35 nm/min (R_N = 0%) to 15 nm/min (R_N = 60%) with increasing N_2 flow ratio (R_N). The decreasing sputtering rate was attributed to a lower sputtering yield due to nitrogen absorption, nitridation of the target, and/or decreasing sputtering efficiency of reactive gases with increasing R_N as compared to argon ions. The increasing R_N ratio resulted in increasing the hardness from 9 GPa (0% R_N) to 32 GPa (15% R_N). In another study, Tsai et al. developed octonary principal element (AlMoNbSiTaTiVZr)N HEN coatings using magnetron sputtering with nitrogen flow ratio from 0 to 70% R_N [40]. The deposition rate decreased from 31 nm/min (R_N = 0%) to 8.3 nm/min (R_N = 67%) with increasing nitrogen flow ratio. In contrast, the hardness values increased from 13.5 GPa (0% R_N) to 37 GPa (50% R_N). Such high increase in hardness has been attributed to the stronger bonding between N and target elements. The coating morphology of (AlMoNbSiTaTiVZr)N HEN changed from coarser grain-like morphology (R_N = 0%; grain size: 30–100 nm) to reduced grain morphology (R_N = 11%; grain size: 10 nm) and then to rougher morphology (R_N = 33%). Similarly, the cross-section of the coating changed from a glass-like featureless morphology to fine columnar structure with increasing nitrogen flow ratio from R_N = 0–11% to R_N = 33%, respectively. Similarly, Xing et al. developed (NbTiAlSiZr)N HEN coating using RF sputtering with increasing nitrogen flow rate from 10 to 50% [41]. The coating thickness and deposition rate were found to decrease with increasing nitrogen flow rate, as shown in **Figure 4**.

The cross-section of (NbTiAlSiZr)N HEN coatings is shown in **Figure 5**. The coating thickness was found to decrease from 298.8 to 200 nm with increasing nitrogen flow rate. Furthermore, the hardness was found to increase from 9.5 to 12 GPa with the increasing nitrogen flow rate of R_N = 10% to R_N = 50%, respectively.

In another work, Chang developed a duodenary (TiVCrZrNbMoHfTaWAlSi)N HEN coatings with R_N from 0 to 50% using DC magnetron sputtering and similar structural evolution from amorphous to FCC solid solution was observed with increasing R_N [42]. The hardness after reactive sputtering increased from 13 GPa (0% R_N) to 34.8 GPa (50% R_N).

3.2.3 Effect of substrate bias

The changing substrate bias during coating deposition effects the chemical composition, microstructure, and mechanical properties of coatings. Chang et al. studied the effect of substrate bias from 0 to −200 V on (AlCrMoSiTi)N HEN coatings developed by DC magnetron sputtering at 50% R_N [43]. The coating showed FCC solid solution structure even though it contained immiscible nitrides, such as AlN, TiN, and Si_3N_4. However, the lattice parameter increased from 4.15 to 4.25 Å, and the grain size decreased from 16.8 to 3.3 nm with increasing substrate bias. This change in the lattice parameter was attributed to increase in adatom mobility, and the decreasing grain was due to increase in nucleation rate at ion-induced surface defects with changing substrate bias. The increasing substrate bias had a small effect on the hardness properties from 25 GPa with no bias to highest hardness of 32 GPa observed at −100 V bias. Following these findings, Huang et al. studied the effect of increasing substrate bias from 0 to −160 V on (AlCrNbSiTiV)N HEN coating using RF magnetron sputtering with R_N and substrate temperature kept constant at 28% and 300°C, respectively [44]. The XRD analysis of HEN coatings

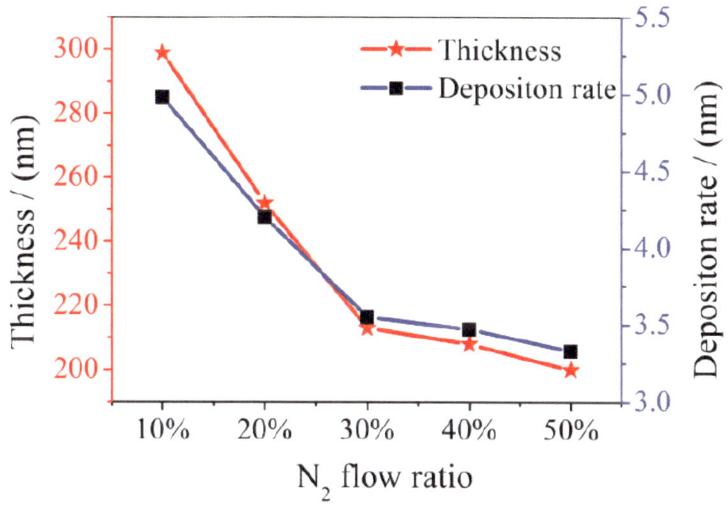

Figure 4.
Film deposition rate as a function of nitrogen flow ratio (R_N) in (NbTiAlSiZr)N HEN coatings [41].

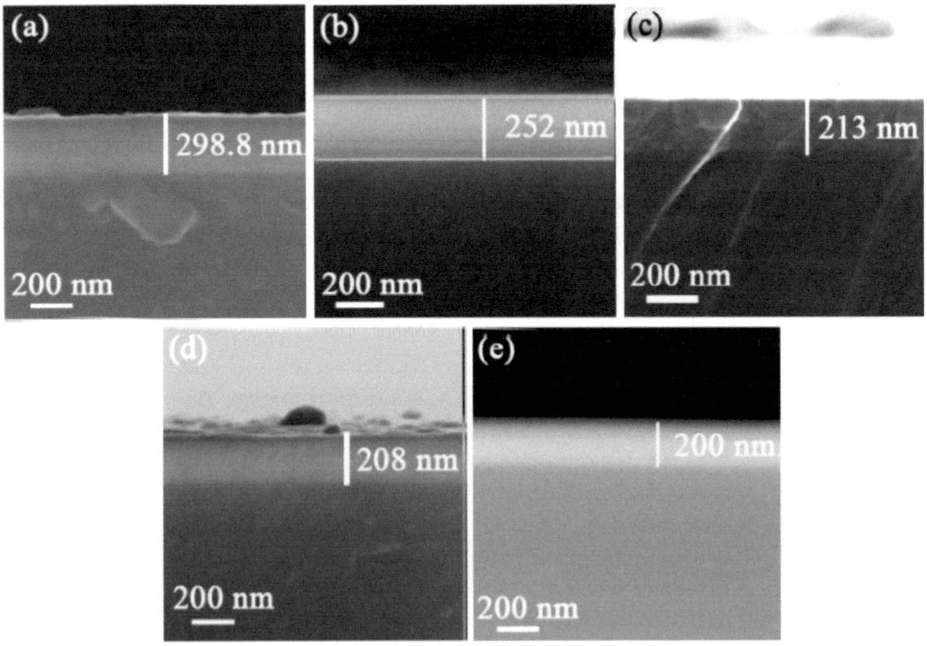

Figure 5.
Surface and cross-section SEM micrographs of (NbTiAlSiZr)N HEN coatings at various R_N: (a) R_N = 10%, (b) R_N = 20%, (c) R_N = 30%, (d) R_N = 40%, and (e) R_N = 50% [41].

showed FCC solid solution with a similar increasing and decreasing trends of lattice parameters and grain size, respectively. However, in this work, the hardness increased from 22 GPa with no bias to the highest hardness of 42 GPa achieved at −100 V bias. The (AlCrNbSiTiV)N HEN coating showed excellent thermal stability even after annealing at 800°C (5 h) and maintained a hardness of 40 GPa. Such increasing trend in hardness was attributed to changing grain size and residual stress with increasing substrate bias. Similar effect of substrate bias on the coating structure and properties were observed in (TiVCrZrHf)N [45], (AlCrTaTiZr)N [46], (TiHfZrVNb)N [47], and (TiZrHfNbTaY)N [48] HEN coatings.

3.2.4 Effect of substrate temperature

The growth of film during the deposition in sputtering system is dependent on the mobility of ions to the substrate. The substrate temperature can play a key role toward ion mobility and diffusion between deposited ions that affect the microstructure, composition, and properties of the coatings. Liang et al. investigated the effect of substrate temperature on the deposition of (TiVCrZrHf)N HEN coating using RF magnetron sputtering from room temperature (RT; 25°C) to 450°C with a fixed 4% R_N and −100 V substrate bias [49]. The XRD analysis showed formation of FCC solid solution without any significant phase separation at all substrate temperatures. However, grain size decreased from 10.7 nm at RT to 8 nm at 250°C and then increased to 9.7 nm at 450°C. The surface morphology of (TiVCrZrHf)N HEN coating became more smooth and dense with increasing substrate temperature. The cross-section morphology changed from an amorphous phase at coating-substrate interface to FCC phase toward coating surface. This phenomenon was reported to be due to higher stresses generated in the initial coating deposition and the greater lattice mismatch of 19% between the HEN coating and silicon substrate. The hardness of the coating increased from 30 to 48 GPa with increasing substrate temperature from RT to 450°C. The high hardness with increasing substrate temperature was attributed to the higher mobility of deposited atoms and reduction of growth void leading to denser coatings. Similar studies on the effect of substrate temperature on structure and enhancement of mechanical properties were studied on the deposition of (AlCrNbSiTiV)N [50–52] and (TiVCrAlZr)N [53] HEN coating.

3.3 High-entropy carbide coatings

Similar to high-entropy nitride coatings, high-entropy carbide coatings have been developed to obtain coatings for tribological and biomedical applications.

3.3.1 Effect of CH_4 flow rate ratio

Braic et al. performed the initial studies on development of high-entropy carbide coatings for tribological and biomedical applications. Their research group deposited (TiAlCrNbY)C high-entropy carbide coatings from co-sputtering of elemental targets using DC-magnetron sputtering with different CH_4 flow ratios (R_C), and at constant substrate temperature and substrate bias of 400°C and −100 V, respectively [54], where CH_4 flow ratio is given by: $R_C = CH_4/(CH_4 + Ar)$. The XRD data showed a change of structure from nanostructured broad FCC phase (0% R_C) to a single FCC carbide phase (10 and 17% R_C) and then to amorphous phase at higher carbon concentration (26 and 33% R_C). The coating morphology changed from slightly higher surface roughness of 7 nm (0% R_C) to fine grained surface roughness of 2 nm (33% R_C) with increasing CH_4 flow ratio. The hardness values increased from 8.2 GPa (0% R_C) to 22.6 GPa (26% R_C). Similarly, Braic et al. studied the effect of CH_4 flow ratio on (TiZrNbHfTa)C high-entropy carbide coating with elemental target co-sputtering using DC-magnetron sputtering on Ti6Al4V alloy substrate [55]. However, in this work, formation of only FCC solid solution was observed at R_C of 13 and 35% with hardness values of 22.4 and 32.1 GPa, respectively. Similar to earlier work, the surface roughness and crystallite size decreased, while the hardness increased with increasing CH_4 flow ratio. In another work, Braic et al. deposited (CuSiTiYZr)C high-entropy carbide coating using elements with large atomic radii differences and reported the effect of different CH_4 flow ratios on the structural and mechanical properties. In all the deposited high-entropy carbide coatings, the XRD showed the formation of amorphous phase irrespective of the amount of carbon. The higher lattice distortion in the high-entropy carbide coatings resulted in hardness values of 20.7 GPa (25% R_C), 27.2 GPa (35% R_C), and 29.5 GPa (50% R_C). Thus, proving the work of Zhang et al. [56] and Guo et al. [57] in the case of high-entropy alloys, a solid solution is formed when the constituent elements have a close atomic radius. Similarly, Jhong et al. developed (CrNbSiTiZr)C high-entropy carbide coatings and studied the effect of increasing CH_4 flow ratio on the structural evolution and mechanical properties [58]. In this system, the structure of high-entropy carbide coating changed from FCC solid solution phase at lower R_C of 3–10% to amorphous phase at higher R_C of 15–20%. Such structural change from FCC to amorphous phase resulted in reducing the hardness from 32.8 to 22.3 GPa.

3.3.2 Effect of substrate temperature

Braic et al. studied the effect of substrate temperature on the deposition of (CrCuNbTiY)C high-entropy carbide coating with DC co-sputtering of elemental targets with a constant substrate bias and two different CH_4 flow ratios. The substrate temperature was increased from 80 to 650°C, and its effect on the structural evolution and mechanical properties were reported. The high-entropy carbide coating with lower carbon concentration showed formation of FCC structures at all the deposition temperatures, while the coating with higher carbon concentration showed poor crystallinity and approaching toward an amorphous phase. The grain size and surface roughness increased with increasing substrate temperature in both coatings. In contrast, the hardness values increased with increasing substrate temperature from 13 GPa (80°C substrate temperature) to 30 GPa (650°C).

3.4 High-entropy oxide coatings

Most of the research on high-entropy ceramic coatings has been focused on high-entropy nitrides and carbide coatings. Few research works have been reported on high-entropy oxide (HEO) coatings. Initial work on HEO coatings was reported by Chen and Wong, where ($Al_xCoCrCuFeNi$)O (x: 0.5, 1 or 2) HEO coating by RF magnetron sputtering using different oxygen flow ratios (O_2/Ar) was developed, and the resulting structural evolution was characterized [59]. The structural evolution was observed to change from FCC, FCC + BCC or BCC (depending on the amount of Al content) to cubic spinal oxides with increasing oxygen flow ratio. The hardness increased from 5–8 GPa to 13–22.6 GPa with increasing O_2 flow ratio. In a subsequent work, Lin et al. used strong oxide forming elements to develop (AlCrTaTiZr)O HEO coating using DC magnetron sputtering with increasing O_2 flow ratio ($R_O = O_2/(O_2 + Ar)$) from 0 to 50%. The XRD analysis revealed a metastable amorphous structure irrespective of the O_2 flow ratio [33]. The tendency of forming amorphous phases at different O_2 flow ratios has been attributed to the large difference in the lattice parameters of oxides from each constituent element of Al, Cr, Ta, Ti, and Zr. The hardness values were reported to be in the range of 8–13 GPa. However, the hardness values increased in the range of 20–22 GPa after annealing at 900°C. The reported hardness values of HEO coatings were relatively higher than most of the reported oxide films, such as Al_2O_3 (10 GPa) [60], TiO_2 (18 GPa) [61], V_2O_5 (3–7 GPa) [62], and ZrO_2 (15 GPa) [63].

3.5 Properties of high-entropy ceramic coatings

3.5.1 Mechanical properties

Based on the reported high-entropy nitride (HEN) coatings in literature, it can be seen that the hardness of HEN coating is dependent on the selection of element that are strong nitride formers in multicomponent alloy. In the earlier work on (FeCoNiCrCuAlMn)N and ($FeCoNiCrCuAl_{0.5}$)N HEN coatings, a maximum hardness of 10.4–11.8 GPa was observed at higher N_2 flow ratio [39]. Consequently, HEN coatings containing strong nitride forming elements, such as (AlCrNbSiTiV) N [44], (TiVCrZrHf)N [49], and (TiZrNbAlYCr)N [64], were developed resulting in hardness increase of 40–48 GPa with increasing N_2 flow ratio. Furthermore, the mechanical properties have been found to increase with increasing substrate-bias and temperature. Some of the HEN coatings reported in literature with superior mechanical properties have been summarized in **Table 1**.

3.5.2 Tribological properties

The superior mechanical properties and high temperature stability of high-entropy nitride (HEN) and high-entropy carbide coating make them suitable toward tribological applications. Few research works have been reported on tribological studies of HEN and high-entropy carbide coating as a function of N_2/

HEN coating	Max. hardness (GPa)	Max. Young's modulus (GPa)	Ref.
(AlCrTaTiZr)N	32	368	[32]
(AlCrMoSiTi)N	35	325	[43]
(AlMoNbSiTaTiVZr)N	37	360	[65]
(AlCrNbSiTiV)N	42	350	[66]
(TiVCrZrHf)N	48	316	[67]
(TiZrNbHfTa)N	32.9	—	[68]
(TiVCrZrHf)N	33	276	[45]
(AlCrNbSiTi)N	36.7	425	[69]
(TiHfZrVNb)N	44.3	—	[47]
(AlCrMoTaTi)N	30.6	280	[70]
(AlCrMoTaTiSi)N	36	250	[71]
(TiVCrZrNbMoHfTa-WAlSi)N	34.8	276.5	[42]
(TiZrNbAlYCr)N	47	—	[64]
(TiZrHfNbTaY)N	40.2	—	[48]

Table 1.
Mechanical properties of high-entropy nitride coatings.

CH_4 gas flow ratio and substrate bias. Lai et al. studied the effect of substrate bias on tribological properties of (AlCrTaTiZr)N HEN coatings against steel counter ball [46]. The resulting wear test showed a high COF of 0.7, while the wear rate decreased from 6.4×10^{-6} to 3.6×10^{-6} mm^3/Nm with increasing substrate bias. The (AlCrTaTiZr)N HEN coating was found to be stable after the wear test of 70 m sliding distance. In another work, Cheng et al. studied the effect of N_2 flow rate on tribological properties of (AlCrMoTaTiZr)N HEN coatings against steel counter ball for a sliding distance of 90 m [72]. The resulting tribological test showed a low wear rate of 2.8×10^{-6} mm^3/Nm; however, the COF was found to be still high around 0.7. Similarly, Braic et al. studied the tribological behavior of (TiZrNbHfTa)N HEN coating and (TiZrNBHfTa)C high-entropy carbide coating on M2 steel substrate against sapphire counter ball for a sliding distance of 400 m at ambient conditions [68]. The wear test showed an average COF of 0.9 and an average wear rate of 2.9×10^{-6} mm^3/Nm for (TiZrNbHfTa)N HEN coating, and an average COF of 0.15 and an average wear rate of 8×10^{-7} mm^3/Nm for (TiZrNbHfTa)C high-entropy carbide coating. Following this work, wear test of (TiZrNbHfTa)N HEN coating and (TiZrNBHfTa)C high-entropy carbide coating was carried out in simulated body fluid (SBF) against sapphire counter ball for a sliding distance of 400 m [55]. The resulting wear test showed an average COF and wear rate of 0.17 and 2.9×10^{-7} mm^3/Nm for (TiZrNbHfTa)N HEN coating and an average COF and wear rate of 0.12–0.32 and $2–9 \times 10^{-7}$ mm^3/Nm for (TiZrNbHfTa)C high-entropy carbide coating, respectively. Furthermore, Braic et al. and Jhong et al. studied the tribological performance of (TiAlCrNbY)C [54], (CuSiTiYZr)C [34], (CrCuNbTiY)C [73], and (CrNbSiTiZr)C [58] high-entropy carbide coatings and showed high-entropy carbide coating possessing superior tribological properties with wear rate and COF values in the range of $0.12–12 \times 10^{-6}$ mm^3/Nm and 0.07–0.4, respectively.

Following the initial works on ball-on-disc sliding wear tests on HEN coatings, simulated tests have been performed for cutting tools application. Shen et al. studied the milling performance of (AlCrNbSiTi)N HEN coated WC-Co substrate

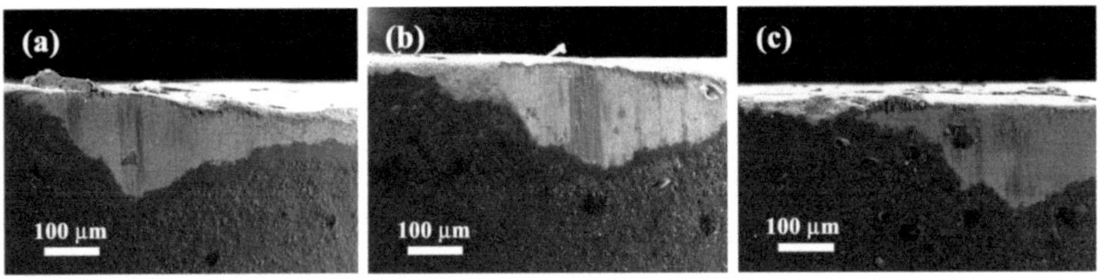

Figure 6.
SEM micrographs of flank wear morphology of cutting inserts coated with: (a) TiN, (b) TiAlN, and (c) AlCrNbSiTi HEN coating [71] (licensed under CC BY 4.0, DOI: 10.3390/coatings5030312).

against SKD11 steel for a sliding distance of 900 m and compared its performance to commercial TiN and TiAlN coatings [74]. The resulting milling tests showed a lower flank wear of 200 μm/min for (AlCrNbSiTi)N HEN coating as compared to 255 μm/min in TiN and 270 μm/min in TiAlN, as shown in **Figure 6**. Similarly, machining performance of (TiZrHfVNbTa)N [75], (AlCrNbSiTiV)N [51], and nanolaminate (TiAlCrSiY)N/(TiAlCr)N [76] HEN coatings showed better tribological properties than commercial nitride coatings.

3.5.3 Corrosion properties

The increase in lattice distortion with high number of principle elements results in formation of amorphous phase in high-entropy alloys, which in return gives better mechanical and electrochemical properties. The corrosion resistance of conventional alloys can be enhanced with amorphous HEA coatings by choosing the appropriate chemical compositions. Lin deposited (TiAlCrSiV)N HEN coatings on a mild steel substrate at different R_N flow ratios and studied its electrochemical properties in 3.5 wt.% NaCl solution at room temperature (RT; 22°C). The coating structure changed from amorphous phase at lower R_N flow ratio to FCC solid solution at higher R_N flow ratio. The highest polarization resistance of 11.36 kΩ/cm^2 was observed in the metallic TiAlCrSiV HEA coating, while the polarization resistance slightly decreased to 8.03–8.55 kΩ/cm^2 in its nitride coatings. Furthermore, the polarization resistance of HEN coatings was enhanced by developing an interlayer of metallic TiAlCrSiV HEA coating. In another work, Hsueh et al. deposited (AlCrSiTiZr)N HEN coating on 6061 aluminum alloy and mild steel substrates using DC magnetron sputtering at various R_N flow ratios, and studied the effect of R_N flow ratio and substrate bias during deposition on corrosion properties in 0.1 M H_2SO_4 aqueous solution at RT [77]. The (AlCrSiTiZr)N HEN coating changed from an amorphous structure to partially crystalline structure at higher R_N flow ratios. The resulting corrosion current density (i_{corr}) for 6061 aluminum alloy substrate decreased from 29.1 μA/cm^2 (uncoated) to 3.1 μA/cm^2 with (AlCrSiTiZr)N HEN coating; while for mild steel substrate, it decreased from 90.4 μA/cm^2 (uncoated) to 7.7 μA/cm^2. Similar increase in corrosion/oxidation resistance with HEN coating on conventional substrate was observed in (TiZrNbHfTa)N and (TiZrNbHfTa)C [55], (AlCrNbSiTi)N [69], nanolaminated AlCrMoNBZr/(AlCrMoNbZr)N [76, 78], and (NbTiAlSiZr)N [41] HEN coating.

4. Future possibility and commercialization

As compared to high-entropy alloys, fewer reports have been published on crystalline high-entropy ceramics (HECs). Due to the high melting point, high hardness, and good thermal and chemical stability, as well as excellent wear and oxidation resistance of ceramic materials, most of the bulk HECs such as high-entropy borides and carbides are designed to be the new type of ultra-high-temperature ceramics (UHTC) with increased high-temperature stability and superior mechanical properties over conventional ceramics. The current understanding of processing and evolution of properties of bulk HECs was built on and developed from the knowledge of HEAs. For bulk HECs, the preferred ceramic components are group IV, V, and VI metal ceramics (metal boride, carbide, silicide, etc.) due to their closely matched structure and properties. Compared to metallic HEAs system where only elemental materials are used, the combination of HEC composition is greatly limited. More experimental efforts on screening and investigation of HECs with different compositions are required to explore the potential of HECs. Furthermore, most of the reported bulk HECs is based on experimental observations, while a few are systematically discussed in combination with the modeling results. In order to give future researchers a clear guideline of designing HECs, further development on computational methods including corresponded thermodynamic ceramic database is needed to more effectively and accurately predict the structure and properties of HECs.

The research on the processing of HECs is progressing, and there have been considerable work performed in the industry on the coatings of HECs toward the next generation of nitride and carbide coatings. High-entropy carbide and nitride coatings have potential applications in biomedical industry, cutting tools, and hard facing die coatings due to their superior mechanical, corrosion, and oxidation resistant properties [55, 74]. HEC coatings of nitride with high thermal stability can be applied as diffusion barrier coatings in integrated circuits to inhibit the diffusion of adjacent materials (e.g., Cu and Si) [79]. Furthermore, high-entropy oxide coatings can be considered as a potential future material for cuprate superconductors, visible-light photocatalysts, and transparent field-effect transistors (TFETs) [59].

5. Summary

Researches of bulk high-entropy ceramic have been reported on metal oxides, refractory carbides, borides, and silicides. HECs with homogenous single-phase structure reveal superior mechanical performance and additional properties like thermal-electrical property of high-entropy oxides. Most scientific effort has been put in exploring various fabrication methods, characterizing the high-entropy structure and the remarkable physical and chemical properties. However, to fully understand HECs that involves a complex multicomponent ceramic system, the problem remains: what are the most significant phase formation rules in a HECs material? How can the potential applications of HECs be realized? A more systematic investigation of the material selection rules that already exists in a metallic high-entropy system is highly demanded in the ceramic system in order to optimize the material design and processing work.

The reported research in the past 10 years on high-entropy ceramic coatings has shown development of coatings with exceptional mechanical, high temperature, electrical, corrosion, and wear-resistant properties. High-entropy ceramic coatings have great potential in different applications, such as wear and corrosion resistant coatings, thermal barrier coatings, and electrical and biomedical applications. The control on variation of deposition parameters can be advantageous in achieving coating with extremely high strength values and highly densified structures, which can be corrosion resistant and biocompatible. Future work on development of nanocomposite and multilayer high-entropy ceramic coatings can possess super hardness properties. Furthermore, high temperature corrosion, mechanical, and tribological properties of high-entropy ceramic coatings need to be addressed.

Author details

Sajid Ali Alvi, Hanzhu Zhang and Farid Akhtar*
Division of Materials Science, Luleå University of Technology, Luleå, Sweden

*Address all correspondence to: farid.akhtar@ltu.se

References

[1] Feng L, Fahrenholtz WG, Hilmas GE, Zhou Y. Synthesis of single-phase high-entropy carbide powders. Scripta Materialia. 2019;**162**:90-93

[2] Ye B, Ning S, Liu D, Wen T, Chu Y. One-step synthesis of coral-like high-entropy metal carbide powders. Journal of the American Ceramic Society. 2019:jace.16514. 2019;**102**:6372-6378

[3] Liu D, Wen T, Ye B, Chu Y. Synthesis of superfine high-entropy metal diboride powders. Scripta Materialia. 2019;**167**:110-114

[4] Zhang AY, Jiang Z, Sun S, Chen Q, Qiu J, Plucknett K. Microstructure and mechanical properties of high-entropy borides derived from boro/carbothermal reduction. Journal of the European Ceramic Society. 2019;**39**(13):3920-3924

[5] Wei X-FF, Liu J-XX, Li F, Qin Y, Liang Y-CC, Zhang G-JJ. High entropy carbide ceramics from different starting materials. Journal of the European Ceramic Society. 2019;**39**(10):2989-2994

[6] Rost CM, Sachet E, Borman T, Moballegh A, Dickey EC, Hou D, et al. Entropy-stabilized oxides. Nature Communications. 2015;**6**:8485

[7] Rost CM, Rak Z, Brenner DW, Maria JP. Local structure of the MgxNixCoxCuxZnxO(x=0.2) entropy-stabilized oxide: An EXAFS study. Journal of the American Ceramic Society. 2017;**100**(6):2732-2738

[8] Anand G, Wynn AP, Handley CM, Freeman CL. Phase stability and distortion in high-entropy oxides. Acta Materialia. 2018;**146**:119-125

[9] Bérardan D, Franger S, Dragoe D, Meena AK, Dragoe N. Colossal dielectric constant in high entropy oxides. Physica Status Solidi Rapid Research Letters. 2016;**10**(4):328-333

[10] Bérardan D, Franger S, Meena AK, Dragoe N. Room temperature lithium superionic conductivity in high entropy oxides. Journal of Materials Chemistry A. 2016;**4**(24):9536-9541

[11] Sarkar A, Velasco L, Wang D, Wang Q, Talasila G, de Biasi L, et al. High entropy oxides for reversible energy storage. Nature Communications. 2018;**9**(1):3400

[12] Jiang S, Hu T, Gild J, Zhou N, Nie J, Qin M, et al. A new class of high-entropy perovskite oxides. Scripta Materialia. 2018;**142**:116-120

[13] Djenadic R, Sarkar A, Clemens O, Loho C, Botros M, Chakravadhanula VSK, et al. Multicomponent equiatomic rare earth oxides. Materials Research Letters. 2017;**5**(2):102-109

[14] Gild J, Samiee M, Braun JL, Harrington T, Vega H, Hopkins PE, et al. High-entropy fluorite oxides. Journal of the European Ceramic Society. 2018;**38**(10):3578-3584

[15] Gild J, Zhang Y, Harrington T, Jiang S, Hu T, Quinn MC, et al. High-entropy metal diborides: A new class of high-entropy materials and a new type of ultrahigh temperature ceramics. Scientific Reports. 2016;**6**:2-11

[16] Sarkar A, Velasco L, Wang D, Wang Q, Talasila G, de Biasi L, et al. Ab initio prediction of mechanical and electronic properties of ultrahigh temperature high-entropy ceramics ($Hf_{0.2}Zr_{0.2}Ta_{0.2}M_{0.2}Ti_{0.2})B_2$ (M = Nb, Mo, Cr). Scripta Materialia. 2018;**9**(8):328-333

[17] Tallarita G, Licheri R, Garroni S, Orrù R, Cao G. Novel processing route for the fabrication of bulk high-entropy metal diborides. Scripta Materialia. 2019;**158**:100-104

[18] Zhang Y, Guo WM, Bin JZ, Zhu QQ, Sun SK, You Y, et al. Dense

high-entropy boride ceramics with ultra-high hardness. Scripta Materialia. 2019;**164**:135-139

[19] Castle E, Csanádi T, Grasso S, Dusza J, Reece M. Processing and properties of high-entropy ultra-high temperature carbides. Scientific Reports. 2018;**8**(1):1-12

[20] Dusza J, Švec P, Girman V, Sedlák R, Castle EG, Csanádi T, et al. Microstructure of (Hf-Ta-Zr-Nb)C high-entropy carbide at micro and nano/atomic level. Journal of the European Ceramic Society. 2018;**38**(12):4303-4307

[21] Yan X, Constantin L, Lu Y, Silvain JF, Nastasi M, Cui B. $(Hf_{0.2}Zr_{0.2}Ta_{0.2}Nb_{0.2}Ti_{0.2})C$ high-entropy ceramics with low thermal conductivity. Journal of the American Ceramic Society. 2018;**101**(10):4486-4491

[22] Zhou J, Zhang J, Zhang F, Niu B, Lei L, Wang W. High-entropy carbide: A novel class of multicomponent ceramics. Ceramics International. 2018;**44**(17):22014-22018

[23] Sarker P, Harrington T, Toher C, Oses C, Samiee M, Maria J-PP, et al. High-entropy high-hardness metal carbides discovered by entropy descriptors. Nature Communications. 2018;**9**(1):4980

[24] Yang K, Oses C, Curtarolo S. Modeling off-stoichiometry materials with a high-throughput ab-initio approach. Chemistry of Materials. 2016;**28**(18):6484-6492

[25] Harrington TJ, Gild J, Sarker P, Toher C, Rost CM, Dippo OF, et al. Phase stability and mechanical properties of novel high entropy transition metal carbides. Acta Materialia. 2019;**166**:271-280

[26] Gild J, Braun J, Kaufmann K, Marin E, Harrington T, Hopkins P, et al. A high-entropy silicide: $(Mo_{0.2}Nb_{0.2}Ta_{0.2}Ti_{0.2}W_{0.2})Si_2$. Journal of Materiomics. 2019;**5**(3):337-343

[27] Qin Y, Liu J-X, Li F, Wei X, Wu H, Zhang G-J. A high entropy silicide by reactive spark plasma sintering. Journal of Advanced Ceramics. 2019;**8**(1):148-152

[28] Zhang R, Gucci F, Zhu H, Chen K, Reece MJ. Data-driven design of ecofriendly thermoelectric high-entropy sulfides. 2018;**57**(20):13027-13033

[29] Zhang H, Akhtar F. Processing and characterization of refractory quaternary and quinary high-entropy carbide composite. Entropy. 2019;**21**(5):474

[30] Zhang H, Hedman D, Feng P, Han G, Akhtar F, Akhta F. A high-entropy $B4(HfMo_2TaTi)C$ and SiC ceramic composite. Dalton Transactions. 2019;**48**:5161-5167

[31] Wasa K. Fundamentals and applications for functional thin films. In: Handbook of Sputter Deposition Technology. Elsevier; 2012. 660 p

[32] Lai CH, Lin SJ, Yeh JW, Chang SY. Preparation and characterization of AlCrTaTiZr multi-element nitride coatings. Surface and Coatings Technology. 2006;**201**(6):3275-3280. DOI: 10.1016/j.surfcoat.2006.06.048

[33] Lin MI, Tsai MH, Shen WJ, Yeh JW. Evolution of structure and properties of multi-component (AlCrTaTiZr)ox films. Thin Solid Films. 2010;**518**(10):2732-2737. DOI: 10.1016/j.tsf.2009.10.142

[34] Braic M, Balaceanu M, Vladescu A, Zoita CN, Braic V. Deposition and characterization of multi-principal-element (CuSiTiYZr)C coatings. Applied Surface Science. 2013;**284**:671-678. DOI: 10.1016/j.apsusc.2013.07.152

[35] Lv CF, Zhang GF, Cao BS, He YY, Hou XD, Song ZX. Structure and mechanical properties of a-C/(AlCrWTaTiNb)CxNy composite films. Surface Engineering. 2016;**32**(7):541-546

[36] Yu R, Huang R, Lee C, Shieu F. Synthesis and characterization of multi-element oxynitride semiconductor film prepared by reactive sputtering deposition. Applied Surface Science. 2012;**263**:58-61

[37] Li W, Liu P, Liaw PK. Microstructures and properties of high-entropy alloy films and coatings: A review. Materials Research Letters. 2018;**6**(4):199-229. DOI: 10.1080/21663831.2018.1434248

[38] Yeh JW, Chen YL, Lin SJ, Chen SK. High-entropy alloys—A new era of exploitation. Materials Science Forum. 2007;**560**:1-9

[39] Chen TK, Wong MS, Shun TT, Yeh JW. Nanostructured nitride films of multi-element high-entropy alloys by reactive DC sputtering. Surface and Coatings Technology. 2005;**200**(5-6):1361-1365

[40] Tsai MH, Lai CH, Yeh JW, Gan JY. Effects of nitrogen flow ratio on the structure and properties of reactively sputtered (AlMoNbSiTaTiVZr)Nx coatings. Journal of Physics D: Applied Physics. 2008;**41**(23)

[41] Xing Q, Wang H, Chen M, Chen Z, Li R, Jin P, et al. Mechanical properties and corrosion resistance of NbTiAlSiZrNx high-entropy films prepared by RF magnetron sputtering. Entropy. 2019;**21**:396

[42] Chang Z. Structure and properties of duodenary (TiVCrZrNbMoHfTaWAlSi)N coatings by reactive magnetron sputtering. Materials Chemistry and Physics. 2018;**220**:98-110. DOI: 10.1016/j.matchemphys.2018.08.068

[43] Chang HW, Huang PK, Yeh JW, Davison A, Tsau CH, Yang CC. Influence of substrate bias, deposition temperature and post-deposition annealing on the structure and properties of multi-principal-component (AlCrMoSiTi)N coatings. Surface and Coatings Technology. 2008;**202**(14):3360-3366

[44] Huang P-K, Yeh J-W. Effects of substrate bias on structure and mechanical properties of (AlCrNbSiTiV)N coatings. Journal of Physics D: Applied Physics. 2009;**42**(11):115401. Available from: http://stacks.iop.org/0022-3727/42/i=11/a=115401?key=crossref.6aa81f793403c9956100753c5880ec88

[45] Tsai DC, Liang SC, Chang ZC, Lin TN, Shiao MH, Shieu FS. Effects of substrate bias on structure and mechanical properties of (TiVCrZrHf)N coatings. Surface and Coatings Technology. 2012;**207**:293-299. DOI: 10.1016/j.surfcoat.2012.07.004

[46] Lai CH, Cheng KH, Lin SJ, Yeh JW. Mechanical and tribological properties of multi-element (AlCrTaTiZr)N coatings. Surface and Coatings Technology. 2008;**202**(15):3732-3738

[47] Pogrebnjak AD, Yakushchenko IV, Bagdasaryan AA, Bondar OV, Krause-Rehberg R, Abadias G, et al. Microstructure, physical and chemical properties of nanostructured (Ti-Hf-Zr-V-Nb)N coatings under different deposition conditions. Materials Chemistry and Physics. 2014;**147**(3):1079-1091. DOI: 10.1016/j.matchemphys.2014.06.062

[48] Beresnev VM, Sobol' OV, Andreev AA, Gorban' VF, Klimenko SA, Litovchenko SV, et al. Formation of

superhard state of the TiZrHfNbTaYN vacuum–arc high-entropy coating. Journal of Superhard Materials. 2018;**40**(2):102-109

[49] Liang SC, Chang ZC, Tsai DC, Lin YC, Sung HS, Deng MJ, et al. Effects of substrate temperature on the structure and mechanical properties of (TiVCrZrHf)N coatings. Applied Surface Science. 2011;**257**(17):7709-7713. DOI: 10.1016/j.apsusc.2011.04.014

[50] Huang PK, Yeh JW. Effects of substrate temperature and post-annealing on microstructure and properties of (AlCrNbSiTiV)N coatings. Thin Solid Films. 2009;**518**(1):180-184. DOI: 10.1016/j.tsf.2009.06.020

[51] Chang KS, Chen KT, Hsu CY, Da HP. Growth (AlCrNbSiTiV)N thin films on the interrupted turning and properties using DCMS and HIPIMS system. Applied Surface Science. 2018;**440**:1-7. DOI: 10.1016/j.apsusc.2018.01.110

[52] Chang CH, Li PW, Wu QQ, Wang MH, Sung CC, Hsu CY. Nanostructured and mechanical properties of high-entropy alloy nitride films prepared by magnetron sputtering at different substrate temperatures. Materials and Technologies. 2019;**34**(6):343-349

[53] Chang ZC, Liang SC, Han S. Effect of microstructure on the nanomechanical properties of TiVCrZrAl nitride films deposited by magnetron sputtering. Nuclear Instruments and Methods in Physics Research B. 2011;**269**(18):1973-1976. DOI: 10.1016/j.nimb.2011.05.027

[54] Braic M, Braic V, Balaceanu M, Zoita CN, Vladescu A, Grigore E. Characteristics of (TiAlCrNbY)C films deposited by reactive magnetron sputtering. Surface and Coatings Technology. 2010;**204**(12-13):2010-2014. DOI: 10.1016/j.surfcoat.2009.10.049

[55] Braic V, Balaceanu M, Braic M, Vladescu A, Panseri S, Russo A. Characterization of multi-principal-element (TiZrNbHfTa)N and (TiZrNbHfTa)C coatings for biomedical applications. Journal of the Mechanical Behavior of Biomedical Materials. 2012;**10**:197-205. DOI: 10.1016/j.jmbbm.2012.02.020

[56] Zhang BY, Zhou YJ, Lin JP, Chen GL, Liaw PK. Solid-solution phase formation rules for multi-component alloys. Advanced Engineering Materials. 2008;**10**(6):534-538

[57] Guo S, Ng C, Lu J, Liu CT. Effect of valence electron concentration on stability of fcc or bcc phase in high entropy alloys. Journal of Applied Physics. 2011;**109**(10)

[58] Jhong YS, Huang CW, Lin SJ. Effects of CH_4 flow ratio on the structure and properties of reactively sputtered (CrNbSiTiZr)cx coatings. Materials Chemistry and Physics. 2018;**210**:348-352. DOI: 10.1016/j.matchemphys.2017.08.002

[59] Chen TK, Wong MS. Structure and properties of reactively-sputtered AlxCoCrCuFeNi oxide films. Thin Solid Films. 2007;**516**(2-4):141-146

[60] Edlmayr V, Moser M, Walter C, Mitterer C. Thermal stability of sputtered Al_2O_3 coatings. Surface and Coatings Technology. 2010;**204**(9):1576-1581 Available from: http://www.sciencedirect.com/science/article/pii/S0257897209008019

[61] Bendavid A, Martin PJ, Jamting Å, Takikawa H. Structural and optical properties of titanium oxide thin films deposited by filtered arc deposition. Thin Solid Films. 1999;**355-356**:6-11. Available from: http://www.

sciencedirect.com/science/article/pii/S0040609099004368

[62] Fateh N, Fontalvo GA, Mitterer C. Structural and mechanical properties of dc and pulsed dc reactive magnetron sputtered V_2O_5 films. Journal of Physics D: Applied Physics. 2007;**40**(24):7716-7719

[63] Bernard O, Huntz AM, Andrieux M, Seiler W, Ji V, Poissonnet S. Synthesis, structure, microstructure and mechanical characteristics of MOCVD deposited zirconia films. Applied Surface Science. 2007;**253**(10):4626-4640. Available from: http://www.sciencedirect.com/science/article/pii/S0169433206013080

[64] Pogrebnjak AD, Beresnev VM, Smyrnova KV, Kravchenko YO, Zukowski PV, Bondarenko GG. The influence of nitrogen pressure on the fabrication of the two-phase superhard nanocomposite (TiZrNbAlYCr)N coatings. Materials Letters. 2018;**211**:316-318

[65] Tsai MH, Yeh JW, Gan JY. Diffusion barrier properties of AlMoNbSiTaTiVZr high-entropy alloy layer between copper and silicon. Thin Solid Films. 2008;**516**(16):5527-5530

[66] Huang P-K, Yeh J-W. Effects of nitrogen content on structure and mechanical properties of multi-element (AlCrNbSiTiV)N coating. Surface and Coatings Technology. 2009;**203**(13):1891-1896. DOI: 10.1016/j.surfcoat.2009.01.016

[67] Liang SC, Tsai DC, Chang ZC, Sung HS, Lin YC, Yeh YJ, et al. Structural and mechanical properties of multi-element (TiVCrZrHf)N coatings by reactive magnetron sputtering. Applied Surface Science. 2011;**258**(1):399-403. DOI: 10.1016/j.apsusc.2011.09.006

[68] Braic V, Vladescu A, Balaceanu M, Luculescu CR, Braic M. Nanostructured multi-element (TiZrNbHfTa)N and (TiZrNbHfTa)C hard coatings. Surface and Coatings Technology. 2012;**211**:117-121. DOI: 10.1016/j.surfcoat.2011.09.033

[69] Hsieh MH, Tsai MH, Shen WJ, Yeh JW. Structure and properties of two Al-Cr-Nb-Si-Ti high-entropy nitride coatings. Surface and Coatings Technology. 2013;**221**:118-123. DOI: 10.1016/j.surfcoat.2013.01.036

[70] Tsai DC, Chang ZC, Kuo BH, Shiao MH, Chang SY, Shieu FS. Structural morphology and characterization of (AlCrMoTaTi)N coating deposited via magnetron sputtering. Applied Surface Science. 2013;**282**:789-797. DOI: 10.1016/j.apsusc.2013.06.057

[71] Tsai DC, Chang ZC, Kuo BH, Chang SY, Shieu FS. Effects of silicon content on the structure and properties of (AlCrMoTaTi)N coatings by reactive magnetron sputtering. Journal of Alloys and Compounds. 2014;**616**:646-651

[72] Cheng KH, Lai CH, Lin SJ, Yeh JW. Structural and mechanical properties of multi-element (AlCrMoTaTiZr)Nxcoatings by reactive magnetron sputtering. Thin Solid Films. 2011;**519**(10):3185-3190. DOI: 10.1016/j.tsf.2010.11.034

[73] Braic V, Parau AC, Pana I, Braic M, Balaceanu M. Effects of substrate temperature and carbon content on the structure and properties of (CrCuNbTiY)C multicomponent coatings. Surface and Coatings Technology. 2014;**258**:996-1005. DOI: 10.1016/j.surfcoat.2014.07.044

[74] Shen W-J, Tsai M-H, Yeh J-W. Machining performance of sputter-deposited ($Al_{0.34}Cr_{0.22}Nb_{0.11}Si_{0.11}Ti_{0.22}$)50N50 high-entropy nitride coatings. Coatings. 2015;**5**(3):312-325. Available from: http://www.mdpi.com/2079-6412/5/3/312/

[75] Bushlya V, Johansson D, Lenrick F, Ståhl J, Schultheiss F. Wear mechanisms of uncoated and coated cemented carbide tools in machining lead-free silicon brass. Wear. 2017;(376-377):143-151. DOI: 10.1016/j.wear.2017.01.039

[76] Zhang W, Tang R, Yang ZB, Liu CH, Chang H, Yang JJ, et al. Preparation, structure, and properties of high-entropy alloy multilayer coatings for nuclear fuel cladding: A case study of AlCrMoNbZr/(AlCrMoNbZr)N. Journal of Nuclear Materials. 2018;**512**:15-24. DOI: 10.1016/j.jnucmat.2018.10.001

[77] Hsueh H-T, Shen W-J, Tsai M-H, Yeh J-W. Effect of nitrogen content and substrate bias on mechanical and corrosion properties of high-entropy films (AlCrSiTiZr)100−xNx. Surface and Coatings Technology. 2012;**206**(19-20):4106-4112. Available from: http://linkinghub.elsevier.com/retrieve/pii/S0257897212002873

[78] Zhang W, Wang M, Wang L, Liu CH, Chang H, Yang JJ, et al. Interface stability, mechanical and corrosion properties of AlCrMoNbZr/(AlCrMoNbZr)N high-entropy alloy multilayer coatings under helium ion irradiation. Applied Surface Science. 2019;**485**:108-118. DOI: 10.1016/j.apsusc.2019.04.192

[79] Chang S, Chen M, Chen D. Multiprincipal-element AlCrTaTiZr-nitride nanocomposite film of extremely high thermal stability as diffusion barrier for Cu metallization. Journal of the Electrochemical Society. 2009;**156**(5):37-42

Section 4

Novel Applications of High Entropy Alloys

Chapter 10

High Entropy Alloys for Medical Applications

Victor Geanta, Ionelia Voiculescu, Petrica Vizureanu and Andrei Victor Sandu

Abstract

A wide variety of metallic biomaterials have been developed so far, including various types of alloys. However, there is a strong need in the medical field for new solutions in what concerns metallic biomaterials with superior biocompatibility and mechanical properties in order to meet future requirements, including the recently developed high entropy alloys (HEAs). This chapter presents some characteristics of high entropy biocompatible metallic alloys produced in an electric-arc remelting furnace in argon inert atmosphere. The effects of the chemical elements used, the microstructural features, and some mechanical characteristics, both in the cast state or after some heat treatments, are highlighted.

Keywords: high entropy alloys, biocompatible, microstructure, microhardness

1. Introduction

1.1 Background

HEAs are defined as alloys comprising more than five main elements mixed in an equiatomic or near-equiatomic fraction [1–14]. Many HEAs have been reported to have: superior mechanical properties, such as ultrahigh fracture even at high temperatures, high hardness, toughness exceeding that of most pure metals and alloys, excellent comparable strength to that of structural ceramics and some metallic glasses, exceptional ductility, and fracture toughness at cryogenic temperatures [1, 3], and good physical properties, such as superconductivity, supermagnetism, and significant resistance to corrosion [5].

By replacing one or several elements in the composition of high entropy alloys, properties that significantly differ from the initial ones can be obtained. Furthermore, the decrease or increase in the ratio of additional elements can generate different metallographic structures with significant influences on the properties of alloys [6–10]. While high-strength conventional alloys are based mainly on the controlled distribution of one or two high-hardness phases at most, in high entropy alloys, the exceptional properties are based on the quenching effect of the supersaturated solid solution and on the suppression of the intermetallic phases [1, 4–12]. The complex distribution of the various chemical elements within the crystalline network of high entropy alloys appears to be the main cause of their special characteristics when compared to the classical or bi-component alloys. Choosing the combination of chemical elements could allow to simultaneously cumulate superior mechanical properties as well as to ensure special corrosion resistance and biocompatibility, making their use as a new class of biocompatible metallic materials suitable [10, 12].

According to the most recent evaluations, the criteria for forming simple solid solutions in high entropy alloys must comply with the following conditions [13, 14]:

- Configuration entropy (ΔS_{am}), which in cases of high entropy alloys must be higher than 11 J/mol K. The entropy of mixing [Eq. (1)] is calculated using Boltzmann's formula:

$$\Delta S_{mix} = -R\sum c_i \ln c_i \tag{1}$$

where R is the gas constant (8.314 J/mol K) and c_i is the molar fraction of element i.

- The enthalpy of mixing (ΔH_{am}) of the alloy must be between −11.6 and 3.2 kJ/mol, and it is calculated using the derived formula [Eq. (2)] from Miedema's macroscopic model:

$$\Delta H_{am} = \sum c_i c_j \, \Delta H_{ij} \tag{2}$$

where ΔH_{ij} is the binary enthalpy of the formation of the elements i and j.

- The atomic radius difference criterion (δ), which claims that the phases that contain predominantly solid solutions are formed for values lower than 6.6%, and at values lower than 4%, only solid solutions are formed. The calculation formula Eq. (3) for δ was defined as follows:

$$\delta = 100\sqrt{\sum c_i \left(1 - \frac{r_i}{\overline{r}}\right) 2} \tag{3}$$

where r_i is the atomic radius of element i and $\overline{r}$ is the average atomic radius.

- The derived parameter Ω includes ΔSam and ΔHam and is taken into consideration only together with δ. If $\Omega > 1.1$ and $\delta < 3.6\%$, only solid solutions are formed. If $1.1 < \Omega < 10$ and $3.6\% < \delta < 6.6\%$, only solid solutions and intermetallic compounds are formed, and if $\Omega > 10$, only solid solutions are formed. The calculation formula for Ω Eq. (4) is:

$$\Omega = T_{top} \, \Delta S_{am} / |\Delta H_{am}| \tag{4}$$

where T_{top} is the melting temperature calculated using the expression Eq. (5):

$$T_{top} = \Sigma c_i T_{top\ i} \tag{5}$$

where $T_{top\ i}$ is the melting temperature of element i.

- The difference in electronegativity Δχ (according to Allen) of the various components of the alloy must be comprised between 3 and 6% in order to form only solid solutions. The calculation formula was deduced in a similar way to the one used to calculate the difference in atomic radius Eq. (6):

$$\Delta\chi = 100\sqrt{\Sigma c_i \left(1 - \frac{\chi_i}{\overline{\chi}}\right) 2} \tag{6}$$

where χ_i is the electronegativity of the element i, and $\bar{\chi}$ is the average electronegativity.

- The critical correlation ratio to obtain only solid solutions is determined by the expression Eq. (7):

$$k_1^{cr} = 1 - \frac{T_A \Delta S_{am}}{\Delta H_{am}} (1 - k_2) > \frac{\Delta H_{IM}}{\Delta H_{am}} \tag{7}$$

where IM index refers to intermetallic compounds, while T_A is the homogenization temperature; k_2 index is considered to be 0.6 and represents the ratio between the entropy of the formation of the compounds and that of the formation of solid solutions.

Even though some inconsistencies can be noticed between the above-mentioned criteria, they are useful to evaluate the conditions under which solid phases are formed in multi-component alloys [13, 14].

1.2 State of the art

High entropy alloys with different compositional characteristics have attracted a lot of attention due to their potentially interesting properties for special fields. Meanwhile, this area provides vast opportunities for new compositions and microstructures, especially for complex alloys [15, 16].

After developing refractory high entropy alloys composed of a single BCC phase in W-Nb-Mo-Ta and W-Nb-Mo-Ta-V alloy systems, it was necessary to produce alloys that comprised transitional metals such as Nb-Mo-Ta-W, V-Nb-Mo-Ta-W, Ta-Nb-Hf-Zr-Ti, Hf-Nb-Ta-Ti-Zr, Mo-Nb-Ta-V-W and the equiatomic Hf-Mo-Nb-Ta-Ti-Zr alloys [3, 15–18]. From a biocompatibility perspective, it is interesting to note that the majority of these elements are biocompatible, with the exception of vanadium. By combining the HEA concept with the need to ensure alloy biocompatibility, biocompatible high entropy alloys have been designed from the above-mentioned systems, with potential use for orthopedic implants. Results are reported as regards the production and testing of high entropy alloys in the system TiNbTaZrMo, TiNbTaZrFe, TiNbTaZrW, TiNbTaZrCr, and TiNbTaZrHf having deformability and biocompatibility characteristics superior to pure titanium, considered to be the least cytotoxic of all metals.

The high entropy alloys in the CrCoFeMoMnNiNb system microalloyed with Ta, Ti or Zr have special mechanical characteristics (compression strength above 2000 MPa, good deformation capacity under severe conditions and good dynamic impact behavior), excellent passive film chemical stability (corrosion potential in simulated biological environment), and reduced cytotoxicity (determined within the MTT test, ISO 10993) compared with the classical alloys used in highly demanding medical devices (CoCr or CoCrMo alloys), which have recorded side effects (tissue necrosis and release of metal ions in the body exceeding the acceptable limits) [15].

The data for the VEC parameter (valence atom concentration) from the TiNbTaZr and TiNbTaZrX alloys (where X was the element replaced, in turn, with Cr, V, Mo, W, and Fe) were located around value 5, indicating the formation of a BCC structure as well as the tendency to form a solid solution phase. In accordance with the above-mentioned criteria, the TiNbTaZr and TiNbTaZrX1 alloys (where X1 can be one of the Mo or W elements) show a reduced possibility of solid solution formation due to the high value of the δ parameter (difference of the atomic radius) [18].

Refractory element Mo is preferred in manufacturing metallic biomaterials due to the fact that it can be found in several conventional metallic biomaterials, such as Ti-15Mo-5Zr-3Al and Co-Cr-Mo [15–19]. Cast and quenched HEA TiNbTaZrMo had breaking strength values exceeding 1000 MPa, higher than those of TiNbTaZrHf and Ti6Al4V refractory alloys, but also good deformability. Quenching led to the improvement of TiNbTaZrMo deformability, which was attributed to coarse granulation and/or redistribution of constituent elements in the dendritic and interdendritic regions [15, 18]. The distribution of cells formed on different types of substrates plays a significant role in cellular functions that involve protein migration, proliferation, and synthesis. In osteointegration tests, the osteoblasts formed on cast and quenched HEA TiNbTaZrMo surfaces showed a widespread morphology, fairly similar to the morphology of the cells on the CP-Ti titanium alloy. On the other hand, the osteoblasts formed on the 316 L austenitic stainless steel were smaller and had a less widespread morphology. The results obtained indicated that osteoblasts had a better tendency to develop, contributing significantly to forming the bone matrix in the case of HEA TiNbTaZrMo, with or without heat quenching treatments, the effects being similar to CP-Ti alloys [15, 16].

Further research is necessary in order to clarify the origin of the excellent biocompatibility of HEA TiNbTaZrMo. These results clearly indicate that these alloys are a new class of metallic biomaterials with exceptional characteristics. To conclude, the new TiNbTaZrMo equiatomic biocompatible alloy contains two BCC solid solution-type phases with a fine equiaxial dendritic structure and an excellent biocompatibility compared to pure Ti, together with superior mechanical properties, indicating the possibility of being used as a new class of metallic biomaterials.

Over the last few years, there has been an increased need to manufacture stents to take over blood vessel functions. Alongside the classical titanium alloys used in this respect, some high entropy alloys from the CoCrFeNiMn and $Al_{0.1}CoCrFeNi$ systems were also investigated [16, 20].

CoCrFeNiMn alloy is equiatomic and was developed for the first time by Cantor [17]. He found the alloy to be very stable, to have higher configuration entropy than melting entropy, being made up of a single phase solid solution with FCC crystalline structure [17, 18]. Moreover, it possesses remarkable mechanical properties, such as high plasticity and ductility, as well as significant tear resistance. This alloy's microstructure is dendritic, and its diffusion rates are very slow. This conclusion was reached after carrying out analyses on heat treated specimens at 700 or 900°C for one-hour periods, when it was discovered that the diffusion of elements was extremely low regardless of the value of the holding temperature [20].

The heat treatments led to a moderate increase of the breaking strength from 447 to 515 MPa (in the case of homogenization to 900°C, maintaining the elongation at break at the same value of 51%) for AlCrFeNiMn alloys [17, 18, 20]. The increase of the homogenization time to 48 hours and of the temperature to 1000°C did not produce major changes in the mechanical characteristics, there even being a slight decrease in the breaking strength to 475 MPa and in elongation to 50%. This behavior led to the conclusion that these alloys are not substantially consolidated by precipitating intermetallic compounds during heat treatments. However, different results were obtained by applying a combined treatment that consisted of annealing + rolling + annealing. The heat and mechanical processing parameters were: annealing at 1000°C for 4 hours, cold rolling with a 50% thickness reduction, from 5.8 to 2.9 mm, and a reduction speed of 0.2 mm per passing, followed by a new annealing at 1000°C for 4 hours.

Following the initial annealing treatment, the material loses its hardness, acquires even greater plasticity, and partial diffusion phenomena occur, while the dendritic microstructure suffers from a rolling finish, reducing interdendritic spaces and eliminating pores and casting defects [17, 20, 21]. The $Al_{0.1}CoCrFeNi$ alloy contains 2.44 at % Al and 24.4 at % of elements Co, Cr, Fe, and Ni, respectively, being single-phased with FCC structure. While casting, the mechanical characteristics of this alloy are modest; the yield strength is 140 MPa, the tensile strength is 370 MPa, and the elongation reaches 65%. By applying combined thermo-mechanical treatments (cold rolling with 60% reduction and homogenization at 1000°C for 24 hours) increases in mechanical properties and microstructure modification from dendritic to polygonal are obtained [17, 18].

Combining a Ti oxide layer with a Zr oxide layer in a TiNbTaZr alloy demonstrated an excellent biocompatibility. From this viewpoint, obtaining a high corrosion resistance of the implanted metals, which work under physiological corrosion conditions over long periods of time, became a major concern. The excellent corrosion resistance of HEA TiZrNbTaMo in corrosive environments, comparable to that of the Ti6Al4V, was due to the passivation effects of the surface and to achieve high stability to the pitting phenomenon [21–26]. In general, the biomaterial-made implantable elements are aimed at improving and extending the patients' lives. After using orthopedic prostheses made from bio inert materials for a long time, the emphasis is now laid on using materials that can activate tissue repair mechanisms, called bioactive materials, through phenomena aimed at increasing proliferation and differentiation of osteoblasts, resulting in in situ reformation of bone architecture [27, 28]. Regarding the biocompatibility of implantable alloys, it is mandatory to ensure an increased corrosion resistance in the corrosive physiological environment compared to the classical alloys [18].

This chapter presents a series of results regarding the obtaining and characterization of high entropy biocompatible alloys from the CrFeMoNbTaTiZr, CrFeMoNbTaTi, CrFeMoNbTaZr, CrFeMoTaTiZr, CrFeTaNbTiZr, CrTaNbTiZrMo, and FeTaNbTiZrMo alloying systems. All the alloys were produced at laboratory scale, in an electric-arc remelting furnace in an inert argon atmosphere. Given that the alloys contain easily fusible elements, some of the metal components were not completely melted during the primary processing. The microstructural characteristics and the microhardness of the alloys suffered changes following the application of homogenization heat treatments. Some of these experimental alloys underwent corrosion resistance tests in simulated biological environments, the results obtained being encouraging. The microscopic investigation of cell viability in direct contact with FeMoTaTiZr alloy in a 1:1:1:1:1 ratio, in which the in vitro cultivation of mesenchymal stem cells isolated from human bone tissue was carried out, demonstrated the biocompatibility of this type of alloy [28]. The conditions of adhesion to the implanted metallic material can be improved by depositing hydroxyapatite-based layers on its surfaces using the magnetron sputtering method. This method allows controlling the deposition parameters so as to obtain thin, flawless, uniform layers, with a very good adhesion to the substrate, low roughness, resistant to corrosion and wear, low stress layers, which are essential properties to be used in medical applications [29].

2. Experimental results

2.1 Obtaining of biocompatible high entropy alloys in the RAV MRF ABJ 900 furnace

High entropy alloys can be obtained in optimum conditions in RAV furnaces working in high purity argon-controlled environments. The concept for the design of experimental alloy recipes from the CrFeMoNbTaTiZr system was based on the choice of chemical elements having extremely low biotoxicity, currently used as an alloying base for classical alloys used to manufacture medical devices. In order to produce high entropy alloys in the CrFeMoNbTaTiZr system in the MRF ABJ 900 vacuum arc remelting equipment within the **ERAMET**—SIM, UPB Laboratory, seven classes of different alloys were chosen in which the chemical composition varied, maintaining the equiatomic proportion in each of them, as follows: HEAB 1—CrFeMoNbTaTiZr; HEAB 2—CrFeMoNbTaTi; HEAB 3—CrFeMoNbTaZr; HEAB 4—CrFeMoTaTiZr; HEAB 5—CrFeTaNbTiZr; HEAB 6—CrTaNbTiZrMo; HEAB 7—FeTaNbTiZrMo. Raw materials consisting of elements with purity greater than 99.5 wt % were mechanically processed to be introduced into the RAV equipment, then weighed, and dosed in equiatomic reports (**Table 1**).

For each experimental alloy sample, a metal load constant of approx. 30 g was maintained. The raw materials were deposited on the copper plate of the RAV equipment in an order meant to ensure the quickest possible formation of a metal bath under the action of the electric arc (**Figure 1**). This mode of operation is very important when working with refractory chemical elements. After coupling to the cooling system, the process continued with successive suctions until a pressure of 5×10^{-3} mbar was obtained in the working area.

Argon atmosphere (5.3 purity levels) was then introduced, and approx. 8–10 homogenization melts were performed, by rotating the samples, in order to ensure a uniform distribution of the chemical elements in the alloys produced. The high

Alloy	Element, g							Production efficiency, g
	Cr	Fe	Mo	Nb	Ta	Ti	Zr	
HEAB 1 CrFeMoNbTaTiZr	2.55	2.73	4.68	4.54	8.77	2.37	4.38	29.85
HEAB 2 CrFeMoNbTaTi	2.97	3.19	5.47	5.30	10.32	2.74	—	29.91
HEAB 3 CrFeMoNbTaZr	2.74	2.95	5.06	4.90	9.54	—	4.80	29.42
HEAB 4 CrFeMoTaTiZr	2.98	3.20	5.50	—	10.36	2.75	5.21	29.69
HEAB 5 CrFeTaNbTiZr	3.00	3.22	—	5.36	10.42	2.76	5.24	29.72
HEAB 6 CrTaNbTiZrMo	2.78	—	5.13	4.98	9.68	2.56	4.87	29.94
HEAB 7 FeTaNbTiZrMo	—	2.97	5.10	4.94	9.61	2.55	4.83	29.95

Table 1.
Mass and production efficiency of biocompatible high entropy alloy batches.

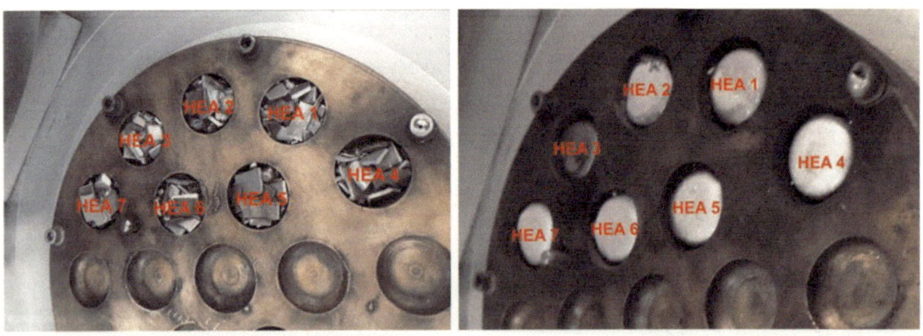

a) b)

Figure 1.
HEAB 1 to HEAB 7 batches prepared for melting on the copper plate of the RAV equipment (a) and after melting (b).

number of successive remelting was imposed by the fact that the load contained high melting temperature elements, such as: Fe—1538°C; Ti—1668°C; Zr—1855°C; Cr—1907°C; Nb—2477°C; Mo—2623°C; Ta—3017°C.

The buttons made from biocompatible HEA alloys (**Figure 1b**) were weighed in order to determine the production coefficient. The losses were due to the fact that, during the process of producing the alloys, there were small drips in the working area, under the action of the electric arc, without their excessive vaporization.

2.2 Microstructure

Samples were taken from the biocompatible high entropy alloys produced in order to carry out the microstructural analysis. The samples taken by abrasive disk cutting under cooling liquid jet were then subjected to the metallographic preparation procedure, using abrasive paper with grit sizes ranging between 360 and 2500, followed by polishing using alpha alumina suspension with grit sizes ranging between 3 and 0.1 μm. The experimental alloys were not chemically etched using metallographic reagents so as to also perform localized chemical composition analyses using the EDAX detector. The microstructural analysis was carried out by optical and scanning electron microscopy, using an Olympus GX 51 optical microscope and a SEM Inspect S electron microscope equipped with an EDAX type Z2e detector from the LAMET, UPB laboratory.

Following the metallographic examination, a significant part of the chemical elements included in the matrix of the HEAB 1 (CrFeMoNbTaTiZr alloy) and HEAB 2 (CrFeMoNbTaTi alloy) was dissolved, forming a solid homogeneous solution (**Figure 2a**, **b**).

Elements such as Cr, Zr, Fe, and Nb formed a common solid solution in which intermetallic compounds precipitated. Nonetheless, a series of compounds of hard-to-melt elements (Mo and Nb), dispersed relatively uniformly in the base matrix (**Figure 3**), were visible. High melting temperature chemical elements, such as Ta and Mo, were not completely melted, the rounded grains being partially adherent to the solid solution existing between the other chemical elements of the alloy (**Figure 3a**, **b**). In the case of the HEAB 3 (CrFeMoNbTaZr alloy), the tantalum was impossible to melt within the volume of the designed batch, which was too small. The big difference between the melting temperatures of the chemical elements constituting the alloy, corroborated with the rapid cooling in the water cooled copper base plate, also generated fracture effects in the metal matrix at the interface with the Ta or Mo grains (**Figure 3a**, **b**).

Even though the HEAB 4 (CrFeMoTaTiZr alloy) contained only two hard-to-melt elements (Mo and Ta), it was still impossible to completely dissolve its metallic grains (**Figure 4a**). Furthermore, between the hard-to-melt metal grain (Ta) and the embedded metal matrix, some micro-cracks were formed perpendicular to the interface, as a result of solidification stresses (in other words, as a result of the inability of the high-hardness metal matrix to disperse the stresses generated by rapid cooling).

The HEAB 5 (CrFeTaNbTiZr alloy) and HEAB 6 (CrTaNbTiZrMo alloy) ran a similar course with the other alloys presented above, the dissolution of the Ta particles being impossible in this case as well (**Figure 4**). A possible cause could be the volume of the grains (having a mass of approx. 10.42 g), which was too big in relation to the total volume of the batch (having a total mass of approx. 29.22 g), which was too small. The very short time in which the alloy was in liquid state did not allow for the complete dissolution of larger grains.

In the case of the HEAB 7 (FeTaNbTiZrMo alloy), the issue of dissolving hard-to-melt metal particles (Ta and Nb) could not be solved during the melting stage

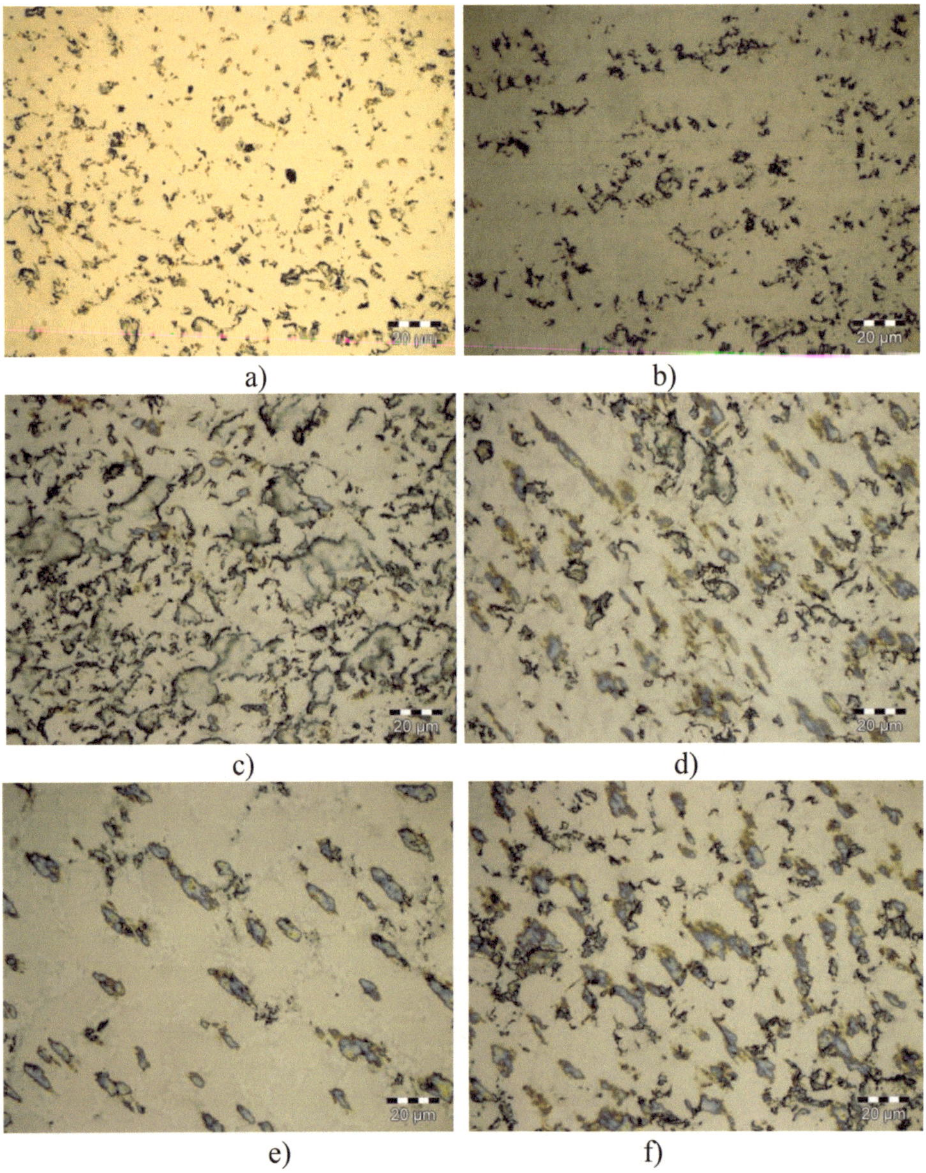

Figure 2.
Cross-section of biocompatible HEABs upon the same magnification. Optical microscopy (1000 magnification) (a) CrFeMoNbTaTiZr; (b) CrFeMoNbTaTi; (c) CrFeMoNbTaZr; (d) CrFeMoTaTiZr; (e) CrFeTaNbTiZr; (f) CrTaNbTiZrMo.

(**Figure 5**). The FeTaNbTiZrMo alloy shares characteristics similar to the ones previously mentioned. The Ta particle failed to dissolve in this case as well, a fracture being developed from its interface with the base matrix (**Figure 6**).

The remaining elements formed a fairly homogenous solid solution, with a dendritic appearance, in which a series of intermetallic compounds were dispersed and uniformly distributed. The dendritic appearance of the metallic matrix was also highlighted on the breaking surfaces of the mini ingots (**Figure 7a**). The images obtained with the help of SEM electron microscopy show micro-dendrites alternating with surfaces of cleavage fractures and micro-fractures, which explain the high brittleness of these alloys in cast state. The localized chemical composition analyses

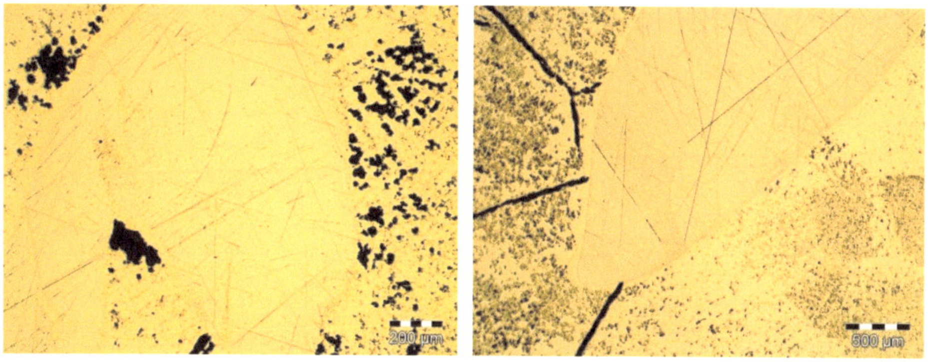

a) b)

Figure 3.
Undissolved blocks present in the experimental HEAB alloys. (a) CrFeMoNbTaTiZr; (b) CrFeMoNbTaZr.

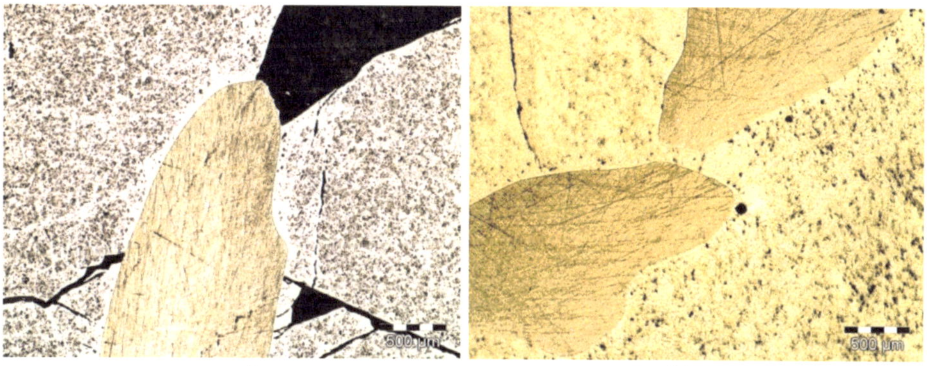

a) b)

Figure 4.
Undissolved tantalum fragments in the high entropy alloys. (a) CrFeTaNbTiZr; (b) CrTaNbTiZrMo.

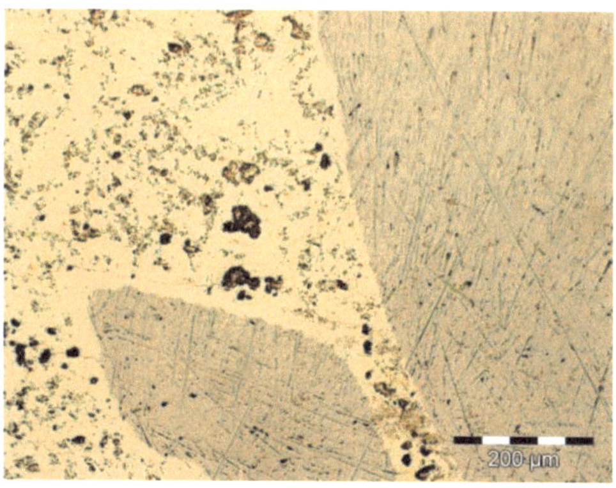

Figure 5.
Undissolved Ta and Nb fragments in the FeTaNbTiZrMo alloy.

(**Table 2**) highlighted the distribution of elements into the dendritically metallic matrix and the segregation of some elements (Nb, Ti, and Ta) near the interface with undissolved blocks of Nb and Ta (**Figure 8**).

Figure 6.
Undissolved tantalum fragment in the FeTaNbTiZrMo alloy.

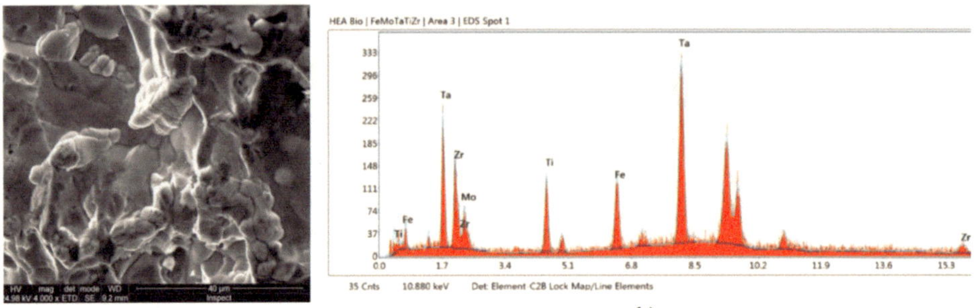

a) b)

Figure 7.
Microstructure (a) and semi-quantitative composition spectrum (b) of FeTaNbTiZrMo alloy on the breaking surface.

Spot	Chemical elements, wt%						Location
	Fe	Ta	Nb	Ti	Zr	Mo	
1	—	—	100	—	—	—	Nb block
2	—	24.74	67.99	7.28	—	—	Interface with Nb
3	24.19	10.26	14.30	14.25	28.81	8.20	Matrix
4	3.66	25.7	30.00	7.98	5.49	26.90	Matrix
5	—	42.14	48.27	9.59	—	—	Interface with Ta
6	—	100	—	—	—	—	Ta block

Table 2.
*The chemical composition of micro-zones which corresponds to **Figure 8**.*

2.3 Heat treatments

Given the fact that HEAs are metallic materials with a high degree of chemical heterogeneity resulted from the association of chemical elements with significant differences in atomic diameters and different mutual solubility, the majority of the researchers in the field resorted to applying heat treatments after producing the alloys [20, 21]. The homogenization heat annealing treatments can reduce or eliminate the segregation effects of chemical elements which take place during casting in the case of high entropy alloys [20]. Microstructures near-equilibrium thus results, either by dissolving the metastable phases or by nucleating the equilibrium phases,

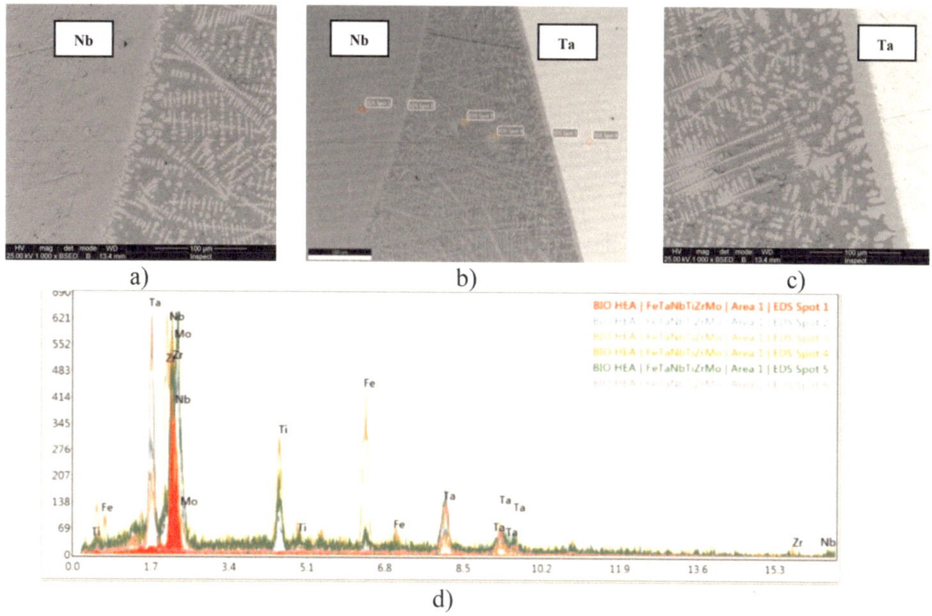

Figure 8.
SEM image of interface between two undissolved blocks and dendritically metallic matrix (Nb, a), (Nb and Ta, b), (Ta, c) and semi-quantitative composition spectrum for the FeTaNbTiZrMo alloy (d) for the micro-zones (spot 1 to spot 6).

the formation of which is suppressed during rapid cooling. The final result of annealing treatments is to reduce the level of micro- or macroscopic residual stresses [3, 4]. Thus, homogenization annealing combined with rapid cooling was shown to allow for increased values of mechanical characteristics. In some cases, depending on the chemical composition of the alloy, hardness values decreased if successive heat treatments were applied at different temperatures [1, 22]. Some of the heat treatments applied to high entropy alloys can either lead to hardening effects or to an increased plasticity and toughness, depending on the value of the heating temperature, on the actual duration of holding at those temperatures as well as on the cooling mode. The effects of heat treatments on macro- and microstructure depending on temperature are very interesting and suggestive [22–30]. The dendritic morphology specific to cast alloys is unaffected if the holding temperature does not exceed 1040°C. On the contrary, when the temperature exceeds 1200°C, phases rich in certain chemical elements occur (e.g., Cu) [20]. The formation of HC-rich in BC-Ta-Nb and hexagonal-packed close (HCP) was highlighted during the heat treatments, depending on the annealing time at 700°C [31].

When it comes to the experimental alloys developed in this chapter, a series of heat treatments were applied in order to achieve the microstructural homogenization and the dissolution of hard-to-melt particles. The samples underwent heat treatment for aging at 600°C for 4 hours and then at 900°C for 6 hours. The heat treatments were performed in the Nabertherm LT 15/12/P320 furnace with a programmable chart for the heat regime. The heating speed was of 20°C/min, and the samples kept for 4 hours were cooled in air, while the samples kept for 6 hours were cooled in the furnace. The evolution of the new alloy microstructure was highlighted by optical and electron microscopy. Thus, before applying heat treatments, the hard-to-melt elements (Ta, Nb, and Mo) did not completely dissolve in the metal melt; they were only diffused at the level of the separation boundaries, on short distances. After the heat treatment, an increased tendency toward oxidation was noticed in the elements located in the superficial layer of the samples, as well as

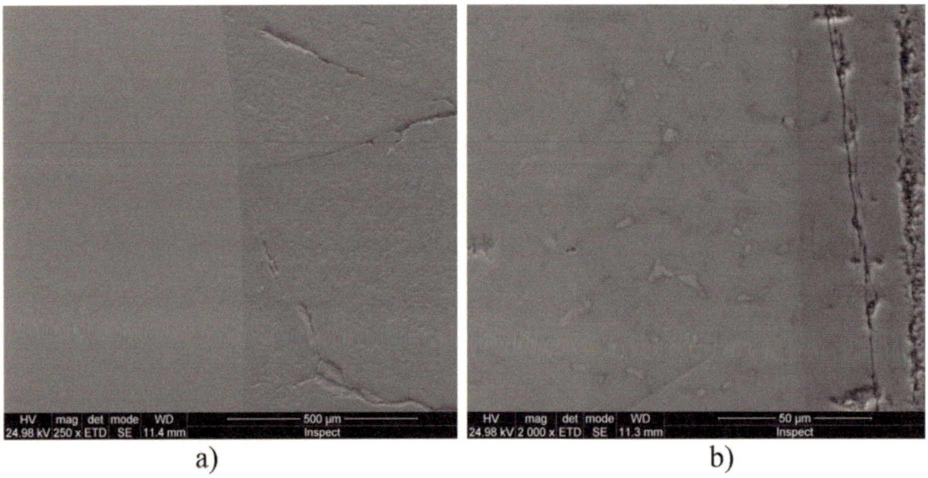

Figure 9.
The oxidized layer of CrFeMoTaTiZr alloy after heat treatment at 800°C/24 hours/slow cooling in the furnace with fractures (a) and peeling (b).

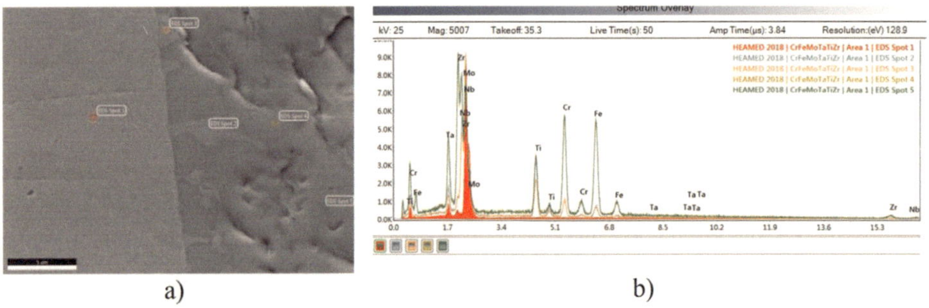

Figure 10.
The interface with undissolved Ta particle (a) and semi-quantitative composition spectrum (b) of the heat-treated CrFeMoTaTiZr alloy.

the formation of a homogenized alloy band, with a dendritic microstructure, located immediately below the complex oxide layer (**Figure 9**).

The effects of the heat treatment are also highlighted by means of an EDS analysis performed with an AMETEC Z2e analyzer on micro-zones located both at the center of the CrFeMoTaTiZr samples as well as at the edges, in order to quantify the oxidation and diffusion effects. An image of the boundary between an undissolved Ta particle and the embedding metal matrix is illustrated in **Figure 10**.

3. Microhardness

The heat treatment effects on mechanical hardness characteristics are highlighted by the different microhardness values determined with the help of the Shimadzu HMV 2 T microhardness tester, presented in **Table 3**.

The analysis of the microhardness values resulted from applying a homogenization treatment showed that hardness increases in the highly alloyed metal matrix up to the average value of 1290 HV0.2 in case of the CrFeMoTaTiZr alloy. The diffusion of the chemical elements during the treatment determined a reduced hardness in the marginal zone adjacent to the surface (located at a distance of approx. 200

Alloy	Microhardness values in points of HV0.2/10	Average value, HV 0.2	Variation coefficient of hardness
CrFeMoNbTaTiZr	749; 739; 700; 743; 747	736	2.76
CrFeMoNbTaTi	750; 732; 747; 804; 736	754	3.85
CrFeMoNbTaZr	775; 748; 719; 725; 803	754	4.66
CrFeMoTaTiZr	697; 764; 795; 790; 712	752	5.97
CrFeMTaNbTiZr	749; 700; 748; 736; 705	728	3.24
CrTaNbTiZrMo	581; 567; 591; 590; 544	575	3.42
FeTaNbTiZrMo	626; 686; 655; 657; 651	655	3.26
CrFeMoNbTiZr Central zone Marginal zone	791; 832; 817; 775; 772 801; 809; 822; 798; 815	797 809	3.30 1.22
Heat-treated CrFeMoTaTiZr Central zone Marginal zone homogenized layer	974; 1349; 1406; 1286; 1434 906; 876; 893; 893; 840 1296; 1429; 1305; 1339; 1316	1290 882 1337	14.38 2.88 4.03

Table 3.
Microhardness values for biomedical HEA samples.

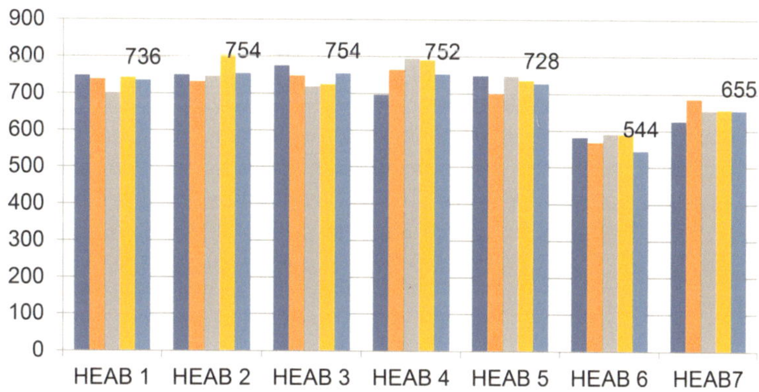

Figure 11.
Evolution of microhardness for experimental biocompatible HEA.

microns), where the average hardness value was 882 HV0.2, that is, approx. 66 HRC, as well as the formation of a homogenization band of approx. 45 microns, where the hardness increased to 1337 HV0.2. The microstructural aspects are in accordance with the microhardness values.

As can be seen from the data presented in **Table 3** and **Figure 11**, HEAB 1–HEAB 5 have very close values of microhardness, with the average being around 750 HV0.2. A notable difference was obtained with the HEAB 6 alloy, which had the lowest hardness value (544 HV0.2) before heat treatment. Also, for the HEAB 7 alloy, the microhardness was between the minimum and maximum values of the other materials (655 HV 0.2).

After heat treatment, the HEAB 6 alloy recorded a considerable increase in hardness in some areas (under the oxidized layer), which came to about 1337 HV0.2. This development, which represents an increase of over 150%, can be exploited in terms of obtaining high-wear-resistant surfaces for medical instruments.

4. Cell viability preliminary tests

In order to evaluate the cell viability of the new alloys with high entropy in simulated biological environments, a series of experiments on bone fragments taken from bone extractions from human patients were performed. Bone fragments were transferred to minimally modified Dulbecco essential medium (DMEM) (Sigma cat. No. D6046) supplemented to the final concentration of 1% with penicillin/streptomycin (Sigma cat.no.P4333) and 10% fetal bovine serum (FBS) (Sigma F7524) and centrifuged at 4000 rpm for 10 min to dislodge stem cells. Bone fragments were cultured as explants in a vial with a surface of 75 cm^2 (Eppendorf No. 0030711122) at 37°C, in a humid atmosphere, with 5% CO_2. After 7–10 days, the first cells that migrated from the explants were observed under the microscope. Cells directly in contact with the alloy samples were analyzed in the Leica DMi8 inverted microscope, using FITC and Rhodamine fluorescence cubes, after incubation with Calcein AM and Propidium Iodide.

5. Conclusion

The high entropy alloys in the CrFeMoNbTaTiZr, CrFeMoNbTaTi, CrFeMoNbTaZr, CrFeMoTaTiZr, CrFeTaNbTiZr, CrTaNbTiZrMo, and FeTaNbTiZrMo alloying systems can be produced in the RAV furnace provided that the batch volume is large enough to allow the dissolution of hard-to-melt elements. Raw materials need to be introduced in the form of smaller grain particles (below 1 mm in diameter) in order to favor the dissolution phenomena by diffusion in the common molten metal bath. Moreover, the liquid state holding time must be increased so as to achieve the complete dissolution of particles from hard-to-melt elements. Heat treatments produce superficial oxidation effects, forming oxide layers of approx. 42 microns thick, partly fractured or peeled. The hardness value in case of the CrFeMoTaTiZr alloy increased from 800 HV0.2 to approx. 1290 HV0.2 after applying the heat treatment. The epifluorescence microscopy technique showed that mesenchymal stem cells adhered to the surface of the alloy.

Acknowledgements

The researches were financed by the Executive Agency for Higher Education, Research, Development and Innovation (CNCS CCDI—UEFISCDI), under the grant project no. PN-III-P1-1.2-PCCDI-2017-239/60PCCDI 2018—Obtaining and expertise of new biocompatible materials for medical applications—MedicalMetMat within PNCDI III.

Thanks

The authors would like to thank their colleagues and collaborators Mirela Codescu, Alina Vlădescu, Hajnal Kelemen, Sanda Manu, and Dumitru Mitrică for their contribution to the testing and characterization of experimental materials.

Author details

Victor Geanta[1], Ionelia Voiculescu[1*], Petrica Vizureanu[2] and Andrei Victor Sandu[2]

1 University Politehnica of Bucharest, Bucharest, Romania

2 Gheorghe Asachi Technical University of Iasi, Romania

*Address all correspondence to: ioneliav@yahoo.co.uk

References

[1] Yeh J, Chen SK, Lin S, Gan JY, Chin TS, Shun TT, et al. Nanostructured high entropy alloys with multiple component elements: Novel alloy design concepts and outcomes. Advanced Engineering Materials. 2004;**6**:299-303

[2] Zhang Y et al. Microstructures and properties of high-entropy alloys. Progress in Materials Science. 2014;**61**:1-93

[3] Tsai M-H, Yeh J-W. High-entropy alloys: A critical review. Materials Research Letters. 2014;**2**(3):107-123. DOI: 10.1080/21663831.2014.912690

[4] Miracle DB, Miller JD, Senkov ON, Woodward C, Uchic MD, Tiley J. Exploration and development of high entropy alloys for structural applications. Entropy. 2014;**16**:494-525. DOI: 10.3390/e16010494

[5] Tang Z, Huang L, He W, Liaw PK. Alloying and processing effects on the aqueous corrosion behavior of high-entropy alloys. Entropy. 2014;**16**: 895-911. DOI: 10.3390/e16020895

[6] Geantă V, Voiculescu I, Miloșan I, Istrate B, Mateș IM. Chemical composition influence on microhardness, microstructure and phase morphology of AlxCrFeCoNi high entropy alloys. Revista de Chimie. 2018; **69**(4):798-801

[7] Geantă V, Voiculescu I, Stefănoiu R, Chereches T, Zecheru T, Matache L, et al. Dynamic impact behaviour of high entropy alloys used in the military domain Euroinvent ICIR 2018. IOP Conference Series: Materials Science and Engineering. 2018;**374**:012041. DOI: 10.1088/1757-899X/374/1/012041

[8] Geantă V, Voiculescu I, Istrate B, Vrânceanu D, Ciocoiu R, Cotruț C. The influence of chromium content on the structural and mechanical properties of AlCrxFeCoNi high entropy alloys. International Journal of Engineering Research in Africa. 2018;**37**:23-28

[9] Voiculescu I, Geanta V, Vasile IM, Ştefănoiu R, Tonoiu M. Characterisation of weld deposits using as filler metal a high entropy alloy. Journal of Optoelectronics and Advanced Materials. 2013;**15**(7–8):650-654

[10] Voiculescu I, Geantă V, Ştefănoiu R, Patroi D, Binchiciu H. Influence of the chemical composition on the microstructure and microhardness of AlCrFeCoNi high entropy alloy. Revista de Chimie. 2013;**64**(12):1441-1444

[11] Geantă V et al. Virtual testing of composite structures made of high entropy alloys and steel. Metals. 2017;7: 496

[12] Csaki I et al. Researches regarding the processing technique impact on the chemical composition, microstructure and hardness of AlCrFeCoNi high entropy alloy. Revista de Chimie. 2016; **67**(7):1373-1377

[13] Fazakas E et al. Microstructure, thermal, and corrosion behavior of the AlAgCuNiSnTi Equiatomic multicomponent alloy. Materials. 2019; **12**(6):926

[14] Popescu G., et al. New TiZrNbTaFe high entropy alloy used for medical, IOP Conference Series: Materials Science and Engineering 2018;**400**:022049. DOI: 10.1088/1757-899X/400/2/022049

[15] Todai M, Nagase T, Hori T, Matsugaki A, Sekita A, Nakano T. Novel TiNbTaZrMo high-entropy alloys for metallic biomaterials. Scripta Materialia. 2017;**129**:65-68

[16] Saini M, Singh Y, Arora P, Arora V, Jain K. Implant biomaterials: A comprehensive review. World Journal of Clinical Cases. 2015;**3**(1):52-57

[17] Cantor B, Chang ITH, Knight P, Vincent AJB. Microstructural development in equiatomic multicomponent alloys. Materials Science and Engineering A. 2004; **375–377**:213-218

[18] Wang S-P, Xu J. TiZrNbTaMo high-entropy alloy designed for orthopedic implants: As-cast microstructure and mechanical properties. Materials Science and Engineering C. 2017;**73**: 80-89

[19] Minciună MG, Vizureanu P, Geantă V, Voiculescu I, Sandu AV, Achiței DC, et al. Effect of Si on the mechanical properties of biomedical CoCrMo alloy. Revista de Chimie. 2015; **66**(6):891-894

[20] Voiculescu I, Geanta V, Ionescu M. Effects of heat treatments on the microstructure and microhardness of AlxCrFeNiMn alloys. Annals of "Dunarea de Jos" University of Galati, Fascicle XII, Welding Equipment and Technology. 2015;**26**:5-11

[21] Munitz A, Kaufman MJ, Nahmany M, Derimow N, Abbaschian R. Microstructure and mechanical properties of heat treated Al1.25CoCrCuFeNi high entropy alloys. Materials Science and Engineering A. 2018;**714**:146-159

[22] Kambic HE. Changing strategies for biomaterials and biotechnology. In: Kambic HE, Yokobory AT Jr, editors. Biomaterials Mechanical Properties, ASTM STP 173. Philadelphia: American Society for Testing and Materials; 1994. pp. 293-301

[23] Hildenbrand H. Biomaterials—A history of 7000 years. BioNanoMaterials. 2013;**14**(3–4): 119-133. DOI: 10.1515/BNM-2013-0014

[24] Balaban DP. Biomaterials. Constanța, Romania: Ovidius University Press; 2005

[25] Williams DF. Definition in biomaterials. In: Progress in Biomedical Engineering. Amsterdam: Elsevier; 1987. p. 67

[26] Williams DF. On the nature of biomaterials. Biomaterials. 2009;**30**: 5897-5909

[27] Floroian L, Badea M, țamotă I. Biomaterials with applications in medicine. JMB. nr. 1. 2015. Project POSDRU/159/1.5/S/134378

[28] Geanta V, Voiculescu I, et al. Grant project no. PN-III-P1–1.2-PCCDI-2017-239/60PCCDI 2018. Obtaining and expertise of new biocompatible materials for medical applications. MedicalMetMat within PNCDI III, Romania

[29] Gao L, Liao W, Zhang H, Utama Surjadi J, Sun D, Lu Y. Microstructure, mechanical and corrosion behaviors of CoCrFeNiAl0.3 high entropy alloy (HEA) films. Coatings. 2017;7:156. DOI: 10.3390/coatings7100156

[30] Munitz A, Salhov S, Guttmann G, Derimow N, Nahmany M. Heat treatment influence on the microstructure and mechanical properties of AlCrFeNiTi0.5 high entropy alloys. Materials Science and Engineering A. 2019;**742**:1-14

[31] Chen SY et al. Phase transformations of HfNbTaTiZr high-entropy alloy at intermediate temperatures. Scripta Materialia. 2019; **158**:50-56

Chapter 11

Multicomponent Alloys for Biomedical Applications

Lucien Reclaru, Lavinia Cosmina Ardelean, Alexandru Florian Grecu and Catalin Adrian Miu

Abstract

Titanium alloys are considered to be the most advanced materials for orthopedic implants due to the favorable combination of mechanical properties, low density, tissue tolerance, high strength-to-weight ratio, good resistance to corrosion by body fluids, biocompatibility, low density, nonmagnetic properties, and the ability to join with the bone. This is the reason why we decided to assess the resistance of two titanium alloys currently used for orthopedic implants, namely, Ti6Al7Nb and Ti6Al4V, as reference, to cyclic fatigue by dynamic tests with crevice corrosion stimulation. According to the results obtained, the examined electrochemical quantities, the visual and SEM observations, and EDX analysis reveal better corrosion behavior of the prostheses made of Ti6Al4V—anodized series compared to prostheses made of Ti6Al7Nb. The further comparison of two explanted proximal modules, made of Ti6Al7Nb and Ti6Al4V, to the same type of prostheses evaluated by cyclic fatigue dynamic tests with crevice corrosion stimulation reveals that there are significant similarities, in particular with regard to the electrolyte diffusion, deposition of products and corrosion. Cation extraction tests which were carried out for Ti6Al7Nb prostheses that have undergone particular surface treatments show significant differences depending on the surface treatment and demonstrate that orthopedic implant materials are not "inert."

Keywords: titanium alloys, Ti6Al4V, Ti6Al7Nb, orthopedic implant, cyclic dynamic test, static test, fatigue corrosion, crevice corrosion, localized corrosion, tribocorrosion, biocompatibility

1. Introduction

Starting with the twentieth century, a wide range of alloys have been used in medical applications as surgically implanted medical devices or denture materials, aimed at providing improved physical and chemical properties, such as strength, durability, and corrosion resistance [1, 2].

The classes of alloys used in medical devices and denture materials include stainless steels, cobalt-chromium, and titanium (alloyed and unalloyed) [3].

Orthopedic implants are subjected to heavy and cyclic load bearing, and they work in a bioactive environment. Most parts of hip joint implants are made of metallic materials. Titanium alloys are considered to be the most advanced materials in this type of application. However, Co-Cr-Mo alloys and austenitic stainless steels as biosteels paired with appropriate metallic alloys, ceramics, and polymers are also being used for hip implant components [4].

The alloys currently accepted for orthopedic implant applications are:

- Stainless steels: 18Chromium-14Nickel-2.5Molybdenum; Nitrogen Strengthened 21Chromium-10Nickel-3Manganese-2.5Molybdenum; Nitrogen Strengthened 23Manganese-21Chromium-1Molybdenum Low-Nickel; Stainless Steel Forgings

- Cobalt chromium alloys: Cobalt-28 Chromium-6 Molybdenum; Cobalt-20Chromium-15Tungsten 10Nickel; Cobalt-28Chromium-6Molybdenum; Cobalt -28Chromium-6Molybdenum Powder

- Cobalt chromium nickel alloys: 40Cobalt-15Nickel-20Chromium-7 Molybdenum-16Iron; 35Cobalt-35Nickel-20Chromium-10Molybdenum; 35 Cobalt-35Nickel-20Chromium-10Molybdenum Forgings

- Titanium & titanium alloys: Unalloyed Titanium; Titanium Alloy in the Alpha Plus Beta Condition; Titanium-6Aluminum-4Vanadium; Titanium and Titanium-6 Aluminum-4Vanadium Alloy Powders; Titanium-6Aluminum-4 Vanadium Casting; Titanium-3Aluminum-2.5Vanadium; Titanium-6 Aluminum -7Niobium; Titanium-13Niobium-13Zirconium; Titanium-12 Molybdenum-6 Zirconium-2 Iron; Titanium-15 Molybdenum; Nickel-Titanium Shape Memory

- Zirconium: Zirconium-2.5Niobium Stabilized Zirconium (Mg-PSZ)

- Tantalum: Unalloyed Tantalum

The biocompatibility of any material in contact with a living tissue is part of the general context of chemical toxicity effects on the human body [5–7]. A biocompatible material may be defined as inert, nontoxic, non-mutagenic, non-recognizing, nonirritating, and non-allergenic [8–9].

Since the early twenty-first century, Registration, Evaluation, Authorization and Restriction of Chemicals (REACH) [7, 10] regulates Europe's approach regarding chemical toxicology to humans, aiming to list the substances present in Europe (manufactured or imported in volumes exceeding one ton) and to control the high-risk substances while reconciling public health and environmental protection.

According to the European Chemicals Agency (ECHA) [11], the list of preregistered substances contains around 143,000 chemicals. Substances that may have serious effects on human health and the environment can be identified as substances of very high concern (SVHCs):

1. CMR group: substances which are carcinogenic, mutagenic, or toxic for reproduction.

2. ED group: substances with endocrine-disrupting (ED) properties.

3. Sensitizers and other equivalent level of concern (ELoC) substances.

4. PBT group: substances that are persistent, bioaccumulative, and toxic, whereas vPvB substances are very persistent and very bioaccumulative [12].

Among the SVHCs incriminated by ECHA, about 4000 substances which can cause a contact allergy are listed. It is estimated that 15–20% of Europe's population are sensitized to allergens. Allergic reactions are a significant and growing health problem affecting large parts of the European population [7, 13].

According to the International Agency for Research on Cancer, metals are classified into three classes:

Class 1: The agent is carcinogenic to humans—Ni derivatives, Cr^{6+}, Cd and its derivatives, and Be and its derivatives.

Class 2: The agent is possibly carcinogenic to humans—metallic Ni and Co.

Class 3: The agent is not classifiable as to its carcinogenicity to humans but reveals mutagenic properties—Sn^{2+}, Cu^{2+}, and Fe^{2+}.

Metals with none or limited references as mutagens are Cu^{1+}, Sn^{4+}, Au, Pt, Ag, Pd, In, and G [8, 9, 14].

Corrosion is an electrochemical reaction characteristic to all metals in contact with biological systems, and its consequence is the formation of metal ions which may trigger hypersensitivity reactions and affect the immune response system [15]. It characterizes the chemical reactivity of metals and alloys, which results in a visible alteration of the material and affects the function of a metallic component or of the entire ensemble [16]. Crevice corrosion is the localized corrosion of a metal surface at, or immediately adjacent to, an area that is shielded from full exposure to the environment because of the close proximity between the metal and the surface of another material (ASTM G15-97). Tribocorrosion refers to all the mechanical and chemical interactions that cause the degradation of solids in relative displacement with or without contact lubricant.

A major characteristic that concerns any metallic material used for medical applications is good resistance to corrosion, and this is the most relevant property when it comes to its biologic safety [8]. The tendency of a metal to corrode is given by its electrode potential.

Conceived for a biological environment, alloys for medical use should essentially be integrated without developing adverse effects, maintaining their function without degrading within an acceptable time limit [17].

The potential systemic and local toxicity, allergy, and carcinogenicity result from releasing elements during the corrosion process. Elements such as Ni and Co are known for their high allergic potential, and prudence dictates that alloys containing these elements should be avoided as much as possible. Several elements are known mutagens, and a few, such as Be and Cd, are known carcinogens in different chemical forms [17].

The potential negative effects on the tissues or on the body at cellular level are precisely induced by the presence of certain components released as degradation products, especially metal cations in solution, due to surface corrosion [9]. The degradation of metallic devices in a biological environment is accompanied by the release of cations such as Cr, Co, Ni, and Ti. Therefore, we deal with a cumulative effect: allergic, irritating, mutagenic, and toxic [7].

Metal ions and debris have been shown to be released from orthopedic implants which are made of stainless steel and Co-Cr alloys [18].

Cr, Mo, Si, Fe, and Mn are the ions released from stainless steel implants, while Ti, Al, V, and Nb are released from titanium alloy implants [19]. The in situ degradation of an implant decreases its structural integrity and also releases products which may trigger an adverse biological reaction [15]. Biological risks associated with the released metal ions have been identified to include those from wear debris, colloidal organometallic complexes, free metal ions, and inorganic metal salts or resulting oxides [20].

The exposure to a variety of chemicals is known as the "cocktail effect" and expresses the way in which different chemicals are released from different sources and affect humans [7, 21]. Individual chemicals can become more dangerous when mixed together and act as an aggravating factor [22].

About 10%–15 female European adults and 1–3% male European adults suffer from Ni contact allergy. Because this is considered an important health problem, the European Union (EU) legislated this matter as follows:

a. "the Ni release from parts in direct contact with the skin must be lower than 0.5 mg/cm^2/week" (European Parliament and council directive 94/27/EC of June 30, 1994) [23] and

b. "all metallic parts that are inserted into pierced ears and other parts of the human body must not have a nickel release rate greater than 0.2 mg/cm^2/week" (Commission directive 2004/96/EC of September 27, 2004) [24].

In dentistry, the percentage of females allergic to Ni is reported to vary from 9 to 20%, and in case of orthodontic patients with pierced ears, 30% are allergic to Ni, Cu, and Cr [2]. In certain countries, nickel-based, cheaper alloys have increasingly been subjected to more and more regulations or even banned [25]. However, piercing of other parts of the body has increased in the last years [26].

Today we find ourselves with a Ni "cocktail effect," in an allergic population, pierced, tattooed, and with orthodontic devices made of poor stainless steels. The presence of an orthopedic implant, especially a failed metal one, has been shown to predispose patients to dermal sensitivity when compared with the general population [27].

A part of the orthopedic implants is made of alloys which contain Cr, mainly used as an alloying element in steels, where it contributes to hardness, tempering, and resistance to oxidation. Such implants will release Cr ions. Because of the increasing number of arthroplasies in young patients with osteoarthritis, the exposure time to the released chromium may be over 50 years in these cases. Cr^{6+} has been labeled as a class 1 human carcinogen by the International Agency for Research on Cancer, signifying carcinogenesis as a potential long-term biological effect in patients with Cr alloy implants [15]. The subsequent chromium ion metabolism is complex. The Cr released during the degradation of the Co-Cr, Co-Cr-Mo, or NiCr alloys is Cr^{3+}, but it can be oxidized to Cr^{6+} at the cellular level. Cr^{6+} is mutagenic and carcinogenic; but its potential biological effects are controversial, as it is metabolized in the cytoplasm and cell's nucleus in Cr^{3+}, which is not involved in DNA and chromosomal damage. Effects as reduction in CD8 lymphocyte levels and possible hypersensitivity reactions (ALVAL) are controversial [15].

ALVAL may represent an immunological response to metal wear debris [28, 29] which may appear in tissues around metallic implants. Such infiltration was reported absent in case of implants without Co, Cr, and Ni [28] and in metal-on-polyethylene implants [30]. The toxic effects of the released metal ions and wear debris affect cells and tissues which are in the proximity and distant from the implant, as well. Histological studies carried out on tissues recovered from explanted metal prostheses revealed areas of necrotic tissue, with visible metal particles [15]. Elevated metal ion concentrations in serum [31, 32], erythrocytes [33], urine [34], whole blood [35, 36], tissue [37, 38], and organs [39] have all been reported in patients with implants [15]. Both Cr^{6+} and Cr^{3+} are described as allergens. According to ECHA 130,000 people allergic to Cr^{6+} are reported, and their number increases.

Titanium alloys are indicated for orthopedic implants because of the favorable combination of mechanical properties, low density, tissue tolerance, high strength-to-weight ratio, good resistance to corrosion by body fluids, biocompatibility, low density, nonmagnetic properties, and ability to join with the bone. Ti induces the formation of a fibrous tissue barrier when placed in contact with a healthy bone and facilitates subsequent bone growth. Contaminations of Ti alloys with elements like hydrogen and oxygen may occur during melting, thermic treatment, and surface hardening and must be avoided, due to their embrittling effect. Ti alloy corrosion resistance is superior to that of stainless steels [4]. This is the reason why we decided to present part of our research regarding titanium alloys.

2. Resistance evaluation of titanium alloys to cyclic fatigue by dynamic tests with crevice corrosion stimulation

The role of biomaterials is to aid or totally replace the functions of living tissues. In case of orthopedic implants, the loading response has to match the natural bone. The average load on a hip bone, estimated to be thrice the body weight, may increase to a value of 10 times the body weight during heavy exercise [15]. Therefore, the ideal orthopedic implant should manifest appropriate mechanical properties and be highly biocompatible with existing tissues [40]. In case of a metallic implant, the potential corrosion of the material in the body environment has to be considered.

Corrosion fatigue is defined as the process in which a metal fractures prematurely under conditions of simultaneous corrosion and repeated cyclic loading at lower stress levels or fewer cycles than would be required in the absence of the corrosion environment (ASTM G15-97).

The term cyclic dynamic test (fatigue) with crevice corrosion stimulation covers various phenomena, namely, crevice corrosion, fatigue, and tribocorrosion. The term stress corrosion (static) with crevice stimulation covers two entangled phenomena, namely, crevice corrosion and stress corrosion. Stress corrosion cracking is the result of a joint action between corrosion and a constraint of reaction or of static compression, applied or residual.

The aim of our research was to assess the mechanical properties of two titanium alloys currently used for orthopedic implants, namely, Ti6Al7Nb and Ti6Al4V, as reference.

The tested modular prostheses, type PL-06, consist of a distal and a proximal module, interlocked by a screw (**Figure 1**).

Two sample series were used for testing. The main characteristics are given in **Table 1**.

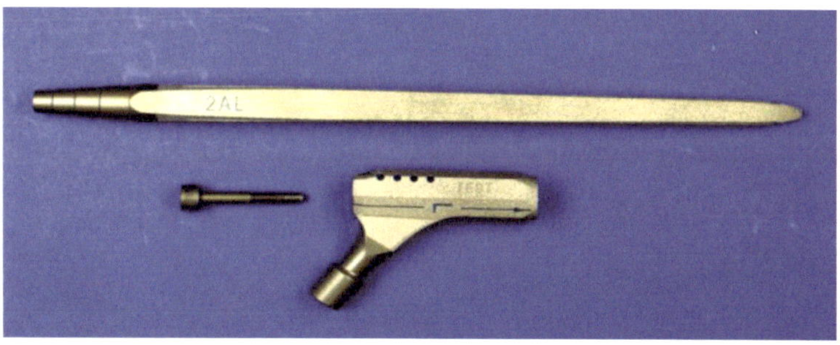

Figure 1.
The modular prostheses used for testing.

	First sample series	Second sample series
Composition	Ti6Al7Nb	Ti6Al4V
Surface	No specific treatment	type 2 ("Ti anodizing")
Tolerances	Uniforms	Adapted to each cone level

Table 1.
Main characteristics of the sample series.

For each series, four samples were evaluated: three in cyclic dynamic test (fatigue) with crevice stimulation and one constrained with crevice stimulation (static) (**Table 2**).

The organization and coding of the samples subjected to the tests are given in **Table 2**. Samples # 1 and # 2 were used for the cyclic adjustments of the fatigue test machine and verification of electrochemical corrosion programs.

For the fatigue tests, a Walter & Bai AG, Switzerland, LFV 10 KN series machine, adapted for fatigue testing (**Figure 2a**) and research of biomedical implants (hip implant prostheses according to ISO 7206-4 and 6), was used. This compact testing system has a hydraulic power pack integrated in its base. The crosshead features automatic adjustment with hydraulic unlocking and hydraulic moving through two long stroke actuators. The prosthesis, more precisely the distal module, was embedded in a non-conductive composite resin, held by a metal sample holder (**Figure 2b**). The embedding of the tapered shape was up to 1 cm from the boundary between the two modules (distal module/proximal module). The specimens were loaded with an average stress of 1.4 and amplitude of 1.1 mm at a frequency of 10 Hz.

Two types of mechanical tests were conducted, the first one under dynamic loading for 5 million fatigue cycles and the second one under a static force of 981 N, during the equivalent time corresponding to 5 million dynamic fatigue cycles, which correspond to approximately 5 years of walking for a person with a bodyweight of 100 kg. The parameters of sample placement followed the requirements of the ISO 7206-6:2013(E) standard.

The potentiostatic measurement technique (controlled-potential coulometry), adapted according to the ASTM F746-87 standard, consists in performing an excitation at a given potential for a very short period of time and then positioning itself on a fixed potential for a certain time. The composition of a measurement cycle is shown

Sample series no.	Code	Mechanical solicitation	Crevice electrochemical test
	#1 #2	Simulation, setting tests	Simulation, test, and control of the corrosion program
1	#3	Fatigue	Yes
1	#4	Fatigue	Yes
1	#5	Fatigue	Yes
1	#6	Constraint	Yes
2	#7	Fatigue	Yes
2	#8	Fatigue	Yes
2	#9	Constraint	Yes
2	#10	Fatigue	Yes

Table 2.
Organization and coding of the samples subjected to the tests.

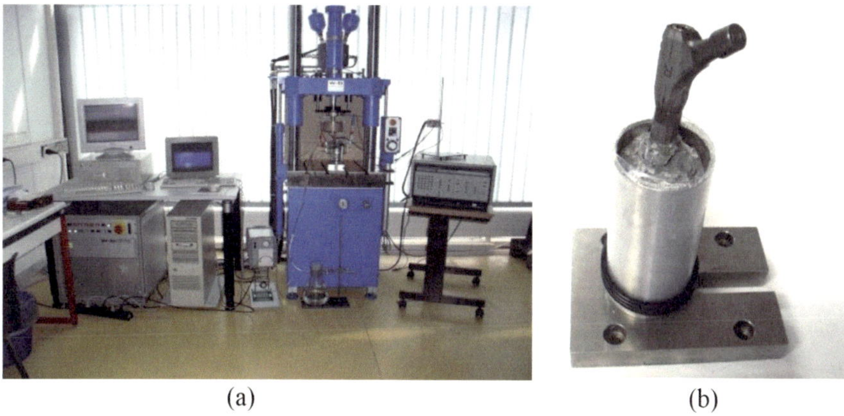

Figure 2.
The fatigue testing system. (a) The fatigue testing machine and (b) the metal sample holder, specifically adapted to the fatigue testing machine.

Potential type	Level (mV)	Time
Stimulated potential	800	60 s
Preselected potential	600	36 min
Stimulated potential	800	60 s
Preselected potential	650	36 min
Stimulated potential	800	60 s
Preselected potential	700	36 min
Stimulated potential	800	60 s
Preselected potential	750	36 min
Preselected potential	−500	20 min

Table 3.
Composition of an electrochemical measurement cycle.

in **Table 3**. Ten cycles correspond to 1 million cycles of mechanical fatigue. The test was carried out in increments of 1 million cycles for a total of 5 million cycles.

The choice of the measuring technique is motivated by the following considerations:

- The materials the components are made of do not pose any significant problems regarding the general corrosion resistance. On the other hand, electrolyte infiltration into the spaces of the distal/proximal module joint may lead to localized crevice corrosion, accompanied by a tribocorrosion process.
- The values of the electrical potential suggest that no decomposition process of the electrolyte into hydrogen and oxygen takes place.
- A cathode potential equalizer at −500 mV SCE was added in order to analyze the depassivation-repassivation capacity of the materials in the tested areas.
- Using this technique, the electrical charge used for the experiment may be easily measured. For analytical estimations the total electrical charge passed in the experiment is easily related to the concentration of electroactive species in the cell.

The potentiostat used is a model PAR 273A, EG&G (Princeton Applied Research). The electrochemical cell has been specially designed for these types of measurements. It is a cell with three electrodes: working electrode (green wire), platinum counter electrode (red wire), and the saturated calomel reference electrode on the right (**Figure 3a**). It is fitted on the head of the cyclic fatigue machine (**Figure 3b**). The test environment was a solution of NaCl at a concentration of 9 g/l (ASTM F746-1998) in ultrapure water (electrical resistivity 18 MΩ cm).

After the assembly of the two modules, there will always be a space which will allow the diffusion of the fluids (**Figure 4a**). Thus the presence of fluids in the interstice can generate crevice corrosion. After the fatigue corrosion test, it is possible to notice the corrosion by strong staining (**Figure 4b**).

Behavior to localized corrosion of samples #3, #4, and #5 (series 1) during 5 million mechanical fatigue cycles is shown in **Figure 5a**.

Behavior to localized corrosion of samples #7, #8, and #10 (series 2) during 5 million mechanical fatigue cycles is shown in **Figure 5b**. Sample #6 and #9 were evaluated for corrosion resistance without cyclic dynamic forces but under a load of 100 kg (**Figure 5**). The comparative behavior to localized corrosion for samples #6 and #9 is presented in **Figure 6**.

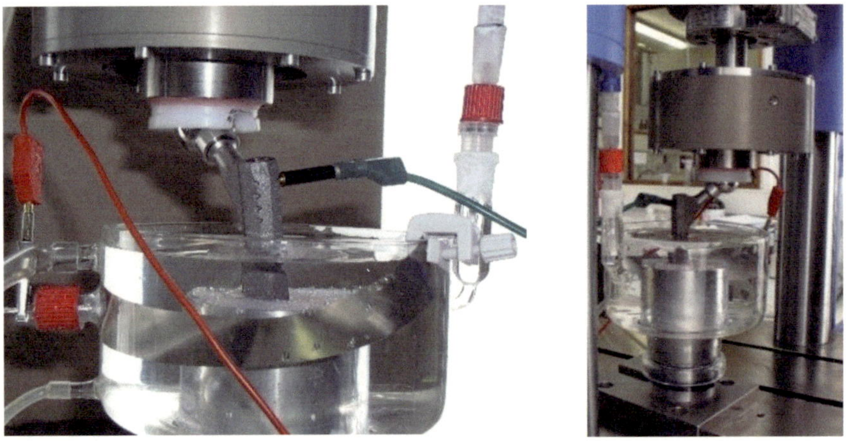

(a) (b)

Figure 3.
The electrochemical cell adapted on LFV 10 KN machine.

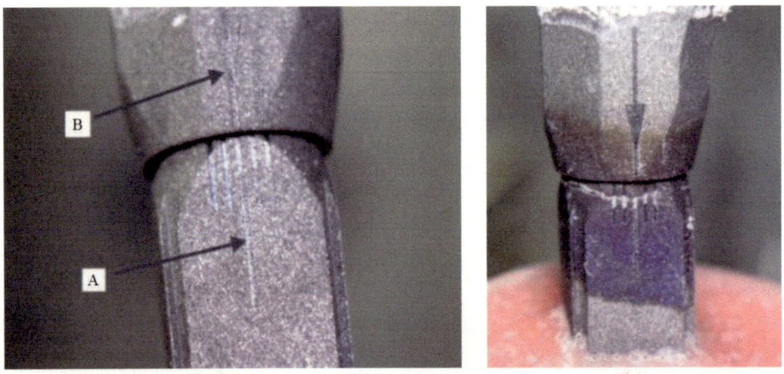

(a) (b)

Figure 4.
Sample #3 before and after testing. (a) (A) Distal module and (B) proximal module. Before testing. (b) Coloration of the two modules after the corrosion test.

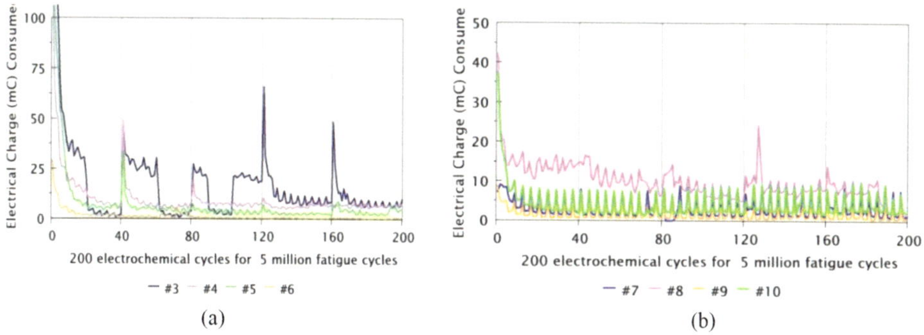

Figure 5.
Behavior to localized corrosion of tested samples. (a) Behavior to localized corrosion of samples #3, #4, #5, and #6 (series 1-cyclic fatigue dynamic test) and (b) behavior to localized corrosion of samples #7, #8 #9, and #10 (series 2-cyclic fatigue dynamic test).

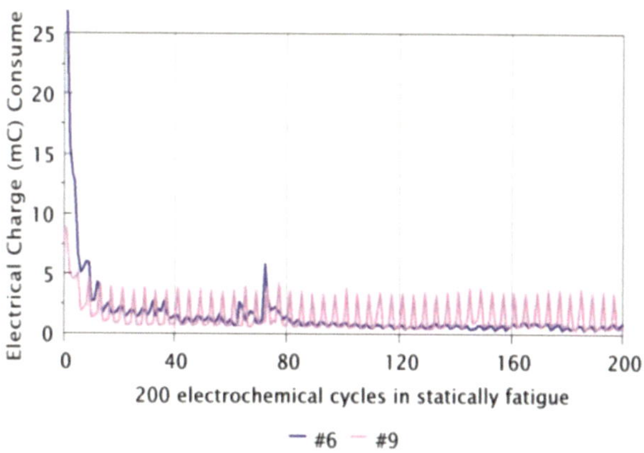

Figure 6.
Behavior to localized corrosion of samples #6 and #9 (static).

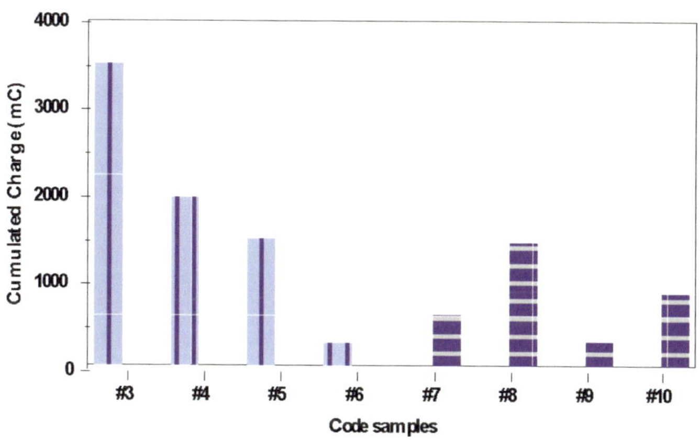

Figure 7.
Electrical charges accumulated during testing.

By analyzing the sum of the accumulated charges during the total duration of the test, we note a worse behavior of the series 1 samples compared to series 2 (**Figure 7**) (samples #6 and #9 were tested under a stress of 100 kg).

It should be noted that the curves presented in **Figure 5a** and **b** are very different from one series to another. The harmonic shape of the series 2 curves is probably due to the surface treatment.

Since the last stage of the measurement cycle (**Table 3**) is −500 mV vs. SCE; thus a depassivation cathodic plateau, the repassivation capacity of the layer designed for the first series samples shows an ability of slower repassivation. In the first stage (600 mV), one still finds higher amounts of electrical charges compared to the later stage (750 mV) vs. SCE (**Tables 4** and **5**). The curves of the series 2 samples have no correlation with the testing fatigue cyclic motion.

In case of series 1 samples, after about 3 million cycles of the dynamic tests, crystallized sodium chloride can be found in the area of the tightening screws of the modular prostheses (**Figure 8**). This means that there is an electrolyte pumping effect in the assembly space of the two parts, and therefore crevice corrosion is quite possible to occur. For series 2 samples, this process can be noticed earlier, starting from about 2.5 million cycles. In the case of electrochemical static measurements, no phenomena of electrolyte pumping were observed. The quantities of electrical charges consumed in the corrosion process (**Tables 4** and **5**) are greater in series 1 than in series 2 (**Figure 7**).

When interpreting the microscopic observations, it is essential to take into account that the various phenomena involved, namely, crevasse corrosion, fatigue (cyclic dynamic test), stress (static test), and tribocorrosion (cyclic dynamic test), are impossible to be considered separately. These processes are intimately intertwined by complex mechanisms; observation is limited to their combined effects. This is why the presented phenomenon is referred to as corrosion fatigue.

The distal module of sample #3, observed by scanning electron microscopy (SEM) at a 560× magnification, is presented in **Figure 9**.

In the analyzed area, cracks have developed (**Figure 9a**), which are probably due to the cyclic dynamic process (fatigue), accompanied by the corrosion processes (**Figure 9b**).

In case of samples #4 and #5, similar phenomena may be observed. In case of sample 4, the deposition is present at the level of the crevice as well as the interferential colorations (same as in **Figure 4b**). The EDX analysis of the area reveals only Al, Ti, and Nb.

Cycle no.	600 mV	650 mV	700 mV	750 mV	Total
1	108.2	69.7	54.1	50.3	282.3
2	30.9	26.3	25.9	25.3	108.4
3	23.0	18.4	18.9	20.3	80.6
4	19.3	15.4	16.0	16.7	67.4
5	15.7	13.2	13.3	14.8	57.0
6	10.8	10.5	10.2	10.5	42.0
7	10.7	8.3	7.9	8.6	35.5
8	8.7	6.8	7.3	7.8	30.6
9	8.1	6.4	6.4	6.8	27.7
10	7.8	5.7	5.9	6.1	25.5
Total	243.2	180.7	165.9	167.2	757.0

Table 4.
Quantity of electrical charge (mC) consumed for the first million fatigue cycles during testing of sample #4, series 1.

Cycle no.	600 mV	650 mV	700 mV	750 mV	Total
1	37.3	21.9	17.8	16.5	93.5
2	12.3	7.4	7.3	8.1	35.1
3	10.1	5.3	5.1	5.7	26.2
4	8.9	4.1	3.8	4.0	20.8
5	8.4	3.5	3.0	3.3	18.2
6	8.2	3.1	2.7	2.9	16.9
7	7.8	2.9	2.4	2.6	15.7
8	7.9	2.8	2.3	2.5	15.5
9	7.9	2.7	2.2	2.4	15.2
10	8.1	2.7	2.2	0.9	13.9
Total	116.9	56.4	48.8	48.9	271.0

Table 5.
Quantity of electrical charge (mC) consumed for the first million fatigue cycles during testing of sample #10, series 2.

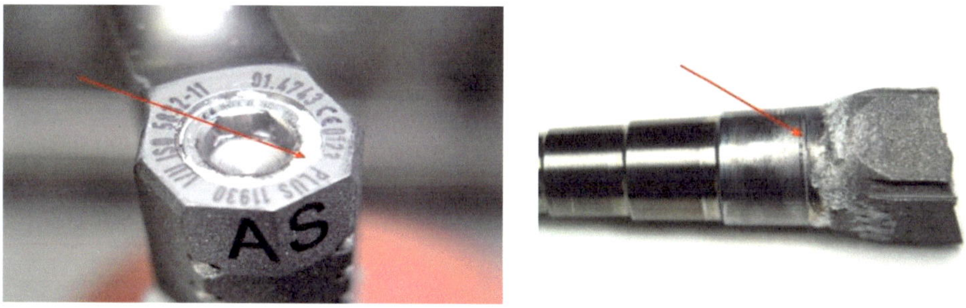

(a) (b)

Figure 8.
Sample #3. (a) Salt deposition in sample #3 and (b) crack of the distal module of sample #3.

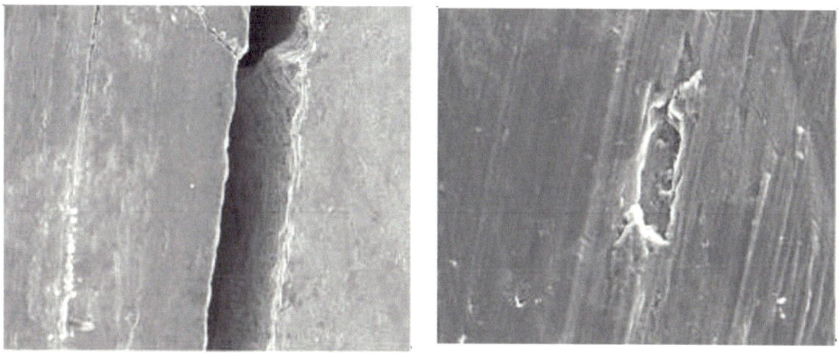

(a) (b)

Figure 9.
Sample #3. Distal module. (a) Sample #3. Distal module. SEM 560× and (b) sample #3. Distal module. SEM 560×.

Sample #5 shows gaps in the distal part (**Figure 10b**) at the level of the crevice as well as coloration at the same level on the proximal module (**Figure 10a**). The spectrum EDX analyses 1, 2, and 3 (**Figure 10c**) reveal depassivation (lack of oxygen in spectrum 1) of the gap interior; the gap's margins contain Na, Cl, Si, and K.

At the distal/proximal part interface of the distal module, deposits may be observed. The EDX analyses particularly show the presence of elements C, Na, Si, Cl, and K (**Figure 10d**).

Sample #7 reveals the same phenomenon of electrolyte penetration at the interface of the distal/proximal modules. The optical examination establishes the presence of wear and corrosion in this area, revealed by a rough appearance of the proximal module surface, as well as on the distal module but to a lesser extent. Examination of the distal module does not reveal the presence of cracks. The phenomenon is probably due to tribocorrosion and crevice corrosion.

Sample #8 also shows an increase in the roughness of the surface (**Figure 11a** and **b**). **Figure 11c** and **d** show the EDX spectrum analysis in the rough zone of the distal modules and reveal the presence of Na, Cl, and Si elements.

In case of sample #9, having undergone only a static test, the electrolyte does not enter the distal/proximal module interface, and no corrosion phenomenon is highlighted. These remarks are also valid for sample #6, which was subjected to the same static test. The EDX analysis does not reveal any traces of corrosion, the chemical composition being the same in the three zones subjected to evaluation.

Figure 12a shows a sectional view of the proximal module of sample #10; as in case of samples #7 and #8, salt deposits and a phenomenon of wear and/or corrosion in the crevice area are present. On the other hand, the distal module (**Figure 12b**) is much less marked by this phenomenon than the distal module of sample # 8. The EDX analysis in **Figure 12d** is measured on the distal module in the crevice area (**Figure 12c**).

According to the results obtained, the electrochemical quantities are examined, and the optical observations reveal a better corrosion behavior on the part of the series 2 samples (Ti6Al4V—anodized type 2) compared to the series 1 samples (Ti6Al7Nb).

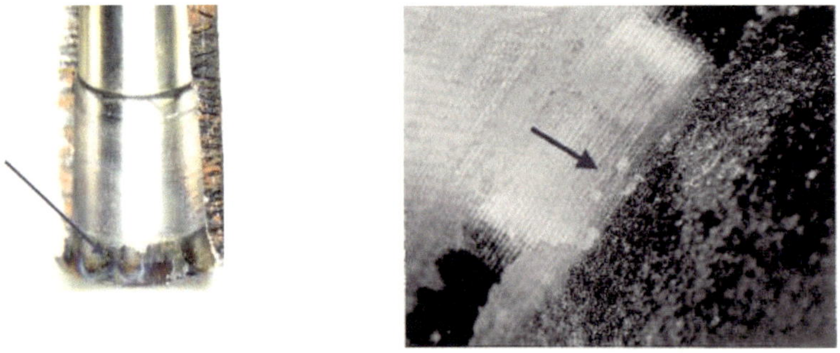

(a) (b)

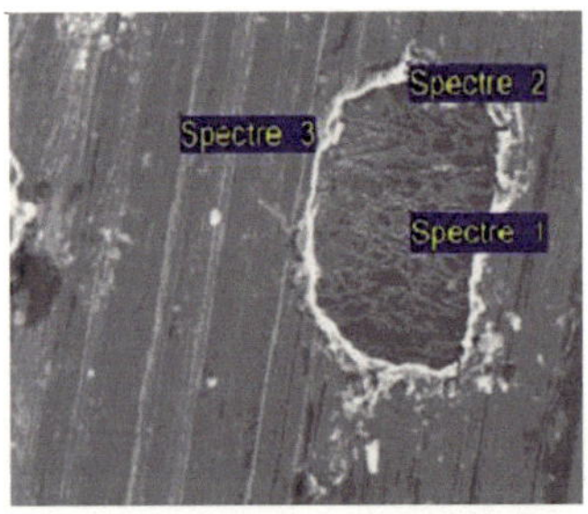

	Ti	Al	Nb	O	Na	Si	Cl	K
1	88.20	6.95	4.84	--	--	--	--	--
2	82.84	6.11	--	--	3.28	0.71	3.63	3.43
3	72.60	5.69	7.18	14.54	--	--	--	--

(c) (d)

Figure 10.
Sample #5. (a) Sample #5. Proximal module, (b) sample #5. Distal module, (c) sample #5. Distal module. SEM 230× and (d) spectrum EDX analysis.

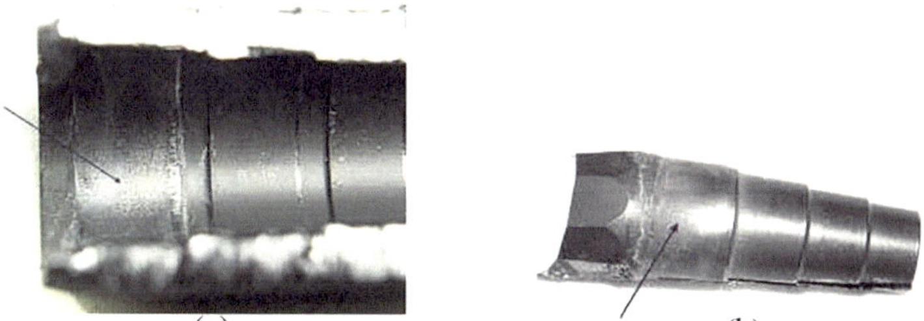

(a) (b)

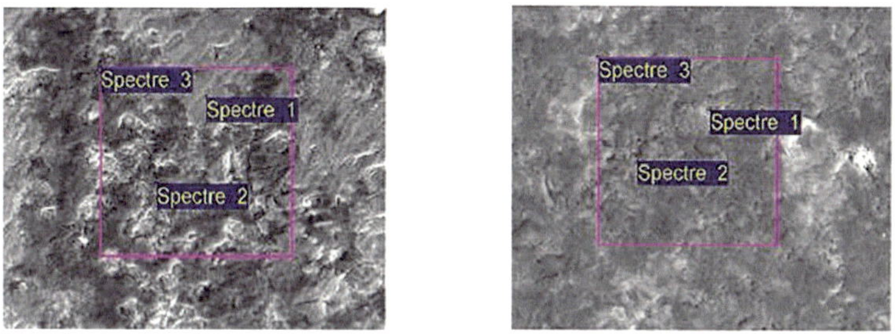

	Ti	Al	V	O	Si	Cl	Zr
1	92.64	2.47	3.94	--	0.95	--	--
2	70.76	1.96	3.68	20.11	2.77	--	0.72
3	84.98	2.19	3.96	--	8.67	0.20	--

(c)

	Ti	Al	V	Na	Si
1	71.12	3.19	2.50	0.56	22.62
2	78.10	3.52	3.71	--	14.67
3	75.51	3.67	3.08	0.76	16.98

(d)

Figure 11.
Sample #8. (a) Sample #8. Proximal module. Rough area, (b) sample #8. Distal module. Rough area, (c) sample #8. Distal module. SEM 560×. Spectrum EDX analysis and (d) sample #8. Distal module. SEM 560×. Spectrum EDX analysis.

(a) (b)

	Ti	Al	V	Na	Si	Cl
1	77.09	6.25	3.49	0.65	12.53	--
2	63.33	4.98	2.90	1.59	26.89	0.31
3	63.10	4.41	3.15	0.72	28.62	--

(c) (d)

Figure 12.
Sample #10. (a) Sample #10. Proximal module, (b) sample #10. Distal module, (c) sample #10. Distal module. Crevice area and (d) spectrum EDX analysis.

In cyclic dynamic tests with crevice stimulation, the electrolyte enters the interface between the distal and proximal modules, which is not the case during static tests (#6 and #9).

Samples #3 and #4 of series 1 reveal cracks in the distal module. Samples #3 and #5, also series 1, reveal holes in the crevice proximity. Metallic interferential staining of the distal/proximal module interfaces of the series 1 samples (#3, #4 and #5) is indicative of electrolyte reactions with the substrate and helps highlighting the corrosion process. This coloration does not appear in case of series 2 samples.

Series 2 samples (#7, #8, and #10) do not show cracks or holes as observed in case of series 1 samples. On the other hand, at crevice level, the surface of the proximal module and to a lesser extent the surface of the distal module present an increase of roughness after the cyclic dynamic corrosion test with crevice stimulation. This phenomenon is particularly visible on sample #8.

The observation of the samples only subjected to the static test does not reveal any sign of corrosion.

3. Components of Ti6Al7Nb and Ti6Al4V explanted modular prostheses (proximal module)

Two components of modular orthopedic prostheses—proximal module (**Figure 13**)—were explanted after 7 years. The proximal parts are made of Ti6Al7Nb (Lot 02-PL-06, Plus 662, PI 793—Gr. BS, art. 11,940, 1 series, lot 02.1424) and Ti6Al4V (Lot 03-PL-06, Plus 662, Gr. BL, art. 11,973—2 series, lot 03.463).

The aim was to observe and analyze the two components of the explanted modular prostheses (proximal module) and to compare them with the results of the cyclic fatigue dynamic tests with crevice corrosion stimulation (**Figure 14a** and **b**).

Figure 14a and **b** show the penetration of biological fluids at the interface of the proximal/distal modules and reveal significant similarities, in particular with regard to the phenomena of electrolyte penetration, product deposition, and corrosion.

Figure 15a reveals the presence of mechanical wear (SEM). The presence of a deposit originating from the biological fluids which have penetrated the interface is noted by the presence of C and Na (**Figure 15b**). Pumping effects of the electrolyte may be observed after 3 million cycles in laboratory tests. This type of effect is also observed on the explanted Ti6Al7Nb prosthesis.

The comparison between **Figure 16a** (explanted Ti6Al4V proximal module) and **Figure 16b** (Ti6Al4V proximal module, sample #8, after the cyclic fatigue

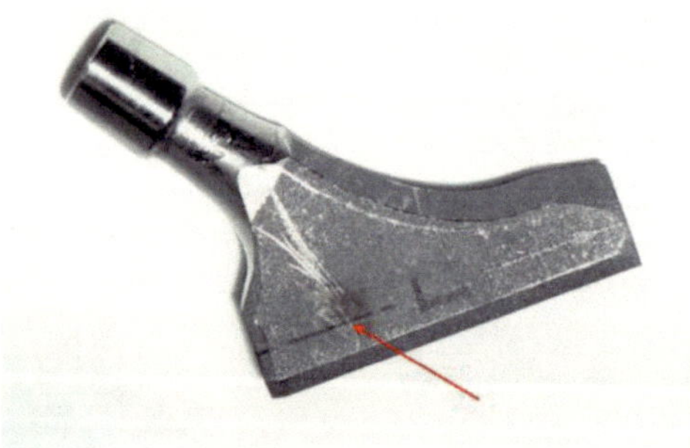

Figure 13.
Proximal module explanted.

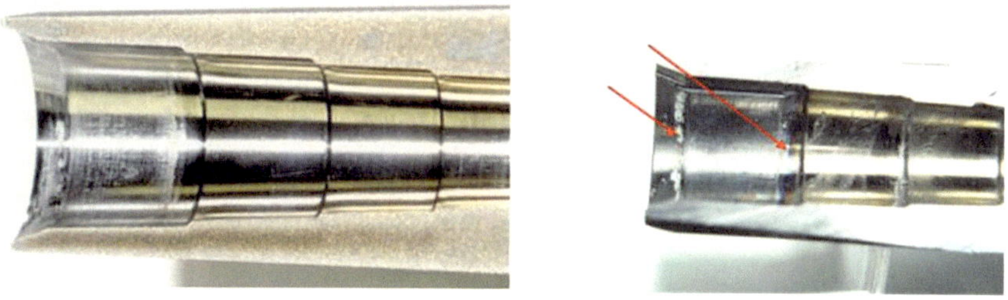

(a) (b)

Figure 14.
Ti6Al7Nb proximal modules. (a) Lot 02, Ti6Al7Nb proximal module explanted and (b) Ti6Al7Nb proximal module, sample #4, previously subjected to cyclic fatigue dynamic test with crevice corrosion stimulation.

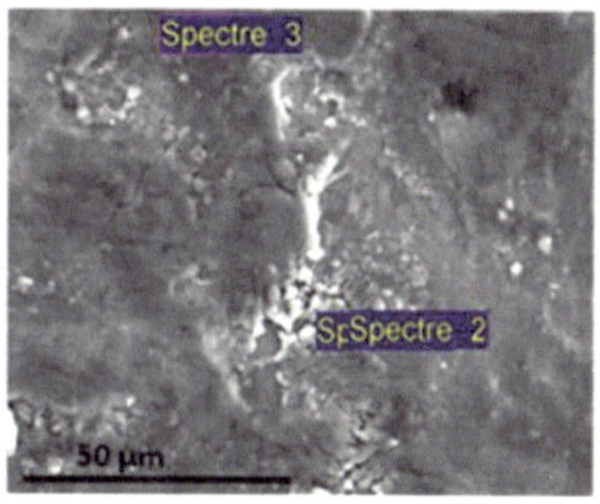

	C	O	Na	Al	Ti	Nb
1	4.7	41.6	0.4	1	48.6	3.8
2	2.8	37.2		3.2	53.1	3.7
3	4.5	43.1	0.4	1	46.6	4.4

(a) (b)

Figure 15.
Lot 02-Ti6Al7Nb proximal module explanted. (a) Lot 02-Ti6Al7Nb proximal module explanted (SEM) and (b) lot 02-Ti6Al7Nb proximal module explanted. Spectrum EDX analysis.

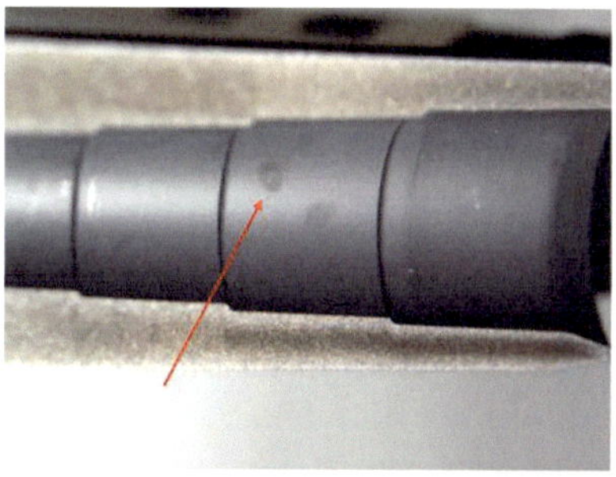

(a)

(b)

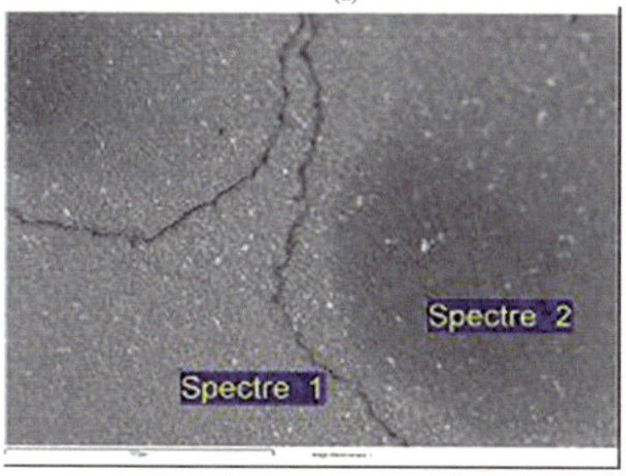

(c)

	C	O	Na	Al	Si	Ti	V
1	2.76	62.68	0.72	1.74	13.61	17.69	0.80
2	27.43	45.26	0.32	1.42	8.02	16.89	0.66

(d)

Figure 16.
Ti6Al4V proximal module. (a) Lot 03-Ti6Al4V proximal module explanted, (b) Ti6Al4V proximal module, sample #8, prior subjected to cyclic fatigue dynamic test with crevice corrosion stimulation, (c) lot 03-Ti6Al4V proximal module explanted (SEM) and (d) spectrum EDX analysis.

dynamic test with crevice corrosion stimulation) shows that the location of the visible spots on the proximal explanted module approximately corresponds to the electrolyte deposits observed during cyclic fatigue dynamic tests with crevice corrosion stimulation. SEM observation does not reveal obvious localized corrosion in the spots area (**Figure 16c**). In exchange, the EDX analysis (**Figure 16d**) reveals the presence of C and Na, which suggests that biological fluids have penetrated and diffused at the proximal/distal module interface. The evaluation of cyclic dynamic corrosion with crevice stimulation on Ti6Al4V modular prostheses shows a similar analogue phenomenon of electrolyte pumping at the interface of the proximal/distal modules.

The comparison of the explanted proximal parts with modular prostheses of the same type evaluated by cyclic fatigue dynamic tests with crevice corrosion stimulation reveals that there are significant similarities, in particular with regard to the electrolyte diffusion, deposition of products, and corrosion. Thus, these observations justify the use of cyclic fatigue dynamic tests with crevice corrosion stimulation in order to compare and evaluate different types of materials for the development of modular prostheses.

4. Cation extraction

Cation extraction tests were carried out only for Ti6Al7Nb prostheses that have undergone very particular surface treatments. **Table 6** presents the characteristics of the two series of prostheses used.

To minimize the volume/surface ratio, Pyrex glass reactors have been developed and adapted to the prosthesis shape (**Figure 17**). The orthopedic implant has a total area of 115.9 cm^2. An electrolyte solution of HCl 0.07 N (300 ml) prepared from Titrisol® 1.0 N (Merck) was used. For extraction tests the release solution volume (ml)/total sample surface (cm^2) ratio was equal to 3. The choice of electrolyte extraction was based on thermodynamic considerations [40] (solubility). Standards (EN-71-3) concerning bioavailability [41] and constraints of the analysis technique (simple matrix causes no perturbations) were considered.

For washing the glass reactors, concentrated nitric acid was used for 24 h. Afterwards they were thoroughly rinsed with deionized water (18 MΩ cm), in order to completely eliminate the acid and then, finally, dried. The extraction tests were conducted at 37 ± 2°C. The prosthesis were kept in the extraction solution for 168 h and then removed, rinsed, and dried. 50 ml of the extraction solution was used for the analysis using ICP-OES/MS method (PerkinElmer Elan DRC). A blanc solution was measured as a reference. The release of elements in the diluted solution of hydrochloric acid 0.07 N shows significant differences between the two types of

Series	Code	Type	Surface treatment
Series 1	#A	SL-Plus r NT Stem	Corundum blasted + mechanical-chemical cleaning
	#B		
	#C		
Series 2	#D	SL-Plus r Stem	Corundum blasted
	#E		
	#F		

Table 6.
Description of Ti6Al7Nb prosthesis used for cation extraction.

Figure 17.
Glass reactors for cation extraction tests.

Element	Blanc	Series 1			Series 2		
		#A	#B	#C	#D	#E	#F
As	<1	<1	1.4	<1	<1	1.7	2.6
Ba	0.24	0.51	0.24	0.48	0.63	0.8	0.77
Be	<0.1	<0.1	<0.1	<0.1	<0.1	<0.1	<0.1
Br	3.54	6.32	4.89	4.72	4.13	5.73	7
Cd	<0.02	0.18	0.02	<0.02	0.09	<0.02	<0.02
Co	0.02	0.78	0.76	0.79	0.12	0.22	0.16
Hg	0.81	0.62	0.18	0.13	0.1	0.13	<0.05
Li	0.11	0.16	0.16	0.15	0.17	<0.1	0.15
Mo	<0.02	<0.02	<0.02	<0.02	<0.02	<0.02	<0.02
Nb	<0.02	130	110	110	73	55	62
Pb	1.9	0.93	0.46	0.71	0.84	0.95	1.0
Sb	0.06	0.1	0.08	0.08	0.06	0.03	0.07
Sn	0.13	0.13	0.15	0.14	0.12	0.1	0.08
Zr	<0.02	0.03	<0.02	0.04	0.16	0.19	0.15
Al	18	193	143	135	194	180	199
Ca	0.0	0.1	0.0	0.1	0.0	<0.02	<0.02
Cr (total)	<0.5	0.9	0.9	0.9	2.1	1.8	2.2
Cu	<2	<2	<2	<2	<2	6.1	<2
Fe	<2	51.5	36	48.4	32.6	30.1	41.7
Ni	<2	<2	<2	<2	<2	<2	<2
P	<10	<10	<10	<10	<10	<10	<10
S	<0.02	<0.02	<0.02	<0.02	<0.02	<0.02	<0.02
Ti	0.7	1600	1590	1500	1510	1350	1500
V	<0.2	<0.02	<0.02	<0.02	<0.02	<0.02	<0.02
Zn	<2	15.9	4.2	6.9	5.5	5.9	7.7

Table 7.
Cations released in solution.

prostheses (**Table 7**). The following elements, which correspond to the detection limit (Be, Mo, Ni, P, S, and V) as well as those that show a released value identically to the blanc solution (Br, Hg, Pb and Sn), were not taken into account.

The presence of the alloying elements, Ti and Al, is comparable for both series of implants and confirms the literature data [42–45]. Spriano et al. [42] also reported an increase of metal ion concentration after a long time exposure, for the Ti6Al7Nb alloy in a SBF solution. On the other hand, the concentration of Nb cations for series 2 is significantly smaller than for series 1 (60–120 μg l^{-1}, respectively). The most important impurity is Fe (between 30 and 50 μg l^{-1}), almost identical for the two types of prostheses. The samples of series 1 released less chromium than series 2, respectively, 0.9 and 2 μg l^{-1}. The prostheses of sample #E released 6 μg l^{-1} Cu. Samples #B, #E and, #F released 1.4–2.6 μg l^{-1} As. The specimens from series 1 released less Ba than series 2. Series 1 released 0.8 μg l^{-1} Co, 0.03 μg l^{-1} Zr and series 2, 0.2 μg l^{-1} Co and respectively 0.17 μg l^{-1} Zr.

Part of the cations released in solution (Pb and Sn) probably originates from the glass reactor or the HCl composition (according to Merck information). As Cd and Cu are considered to be accidental impurities, their presence is not related to the affiliation of the tested sample to one or the other series of prostheses.

5. Conclusions

Various biomaterials have been used for orthopedic implant manufacturing. Polymeric materials, as a result of their mechanical weakness, have been considered unsuitable for the stress deformation requirements of orthopedic implant components, while ceramics have good biocompatibility but are brittle, and designs should take this into account. Alloys are known for their good mechanical properties, but poorer biocompatibility, due to the systemic release of ions [46]. An orthopedic implant is frequently made of a metallic or ceramic component articulating with a metal, ceramic, or polyethylene surface [19]. Different possible combinations are possible: metal (stainless steel or Co-Cr) on ultrahigh molecular weight polyethylene, metal on metal, ceramic on polyethylene, ceramic on ceramic, or ceramic on metal [47]. Coatings such as bioinert films, which have the main purpose of hindering corrosive processes of the underlying metal and bioactive films, which are capable of improving biological compatibility, avoiding inflammation or implant-associated infection processes, are used more and more often. The ideal coating is a system in which anticorrosion, anti-infection, and osseointegration can be obtained simultaneously [48]. Because of their favorable characteristics, Ti alloys are the first choice material in case of orthopedic implants. Even in case of Co-Cr-Mo alloys, Ti-vacuum-plasma-sprayed (VPS) coatings decrease the release of the substrate elements (Co, Cr, and Mo) considerably, but they do not suppress it completely [49].

Titanium remains the predominant material used for medical implants. Despite its high strength and good resistance to corrosion, multiple studies have demonstrated that degradation products of titanium alloys may be detected in neighboring tissues as well as in distant organs. Titanium particles are released from the implant's surfaces for many reasons, such as mechanical wear, contact with chemical agents, and bacteria embedded in adherent biofilm and inflammatory cells [16].

It is obvious that none of the orthopedic prosthetic materials are "inert". However the question regarding their toxicological behavior "Which are the long-term consequences for humans?" still stands.

The near future of multicomponent alloys for biomedical applications does not only belong to high-quality Co-Cr, Ti, Ta, or Zr alloys but also to customized orthopedic prostheses, manufactured by 3D printing techniques, based on a CT or MRI scan, which fit perfectly. One can also imagine the not-so-distant future, which seems to belong to the bioprinting techniques, in this case, bone-made orthopedic implants.

Conflict of interest

The authors declare no conflict of interest.

Author details

Lucien Reclaru[1], Lavinia Cosmina Ardelean[2]*, Alexandru Florian Grecu[3] and Catalin Adrian Miu[4]

1 Biomaterials and Medical Devices, Marin-Neuchâtel, Switzerland

2 Department of Technology of Materials and Devices in Dental Medicine, "Victor Babes" University of Medicine and Pharmacy from Timisoara, Timisoara, Romania

3 Orthopaedics and Traumatology, University of Medicine and Pharmacy Craiova, Craiova, Romania

4 3rd Department of Orthopaedics-Traumatology, "Victor Babes" University of Medicine and Pharmacy from Timisoara, Timisoara, Romania

*Address all correspondence to: lavinia_ardelean@umft.ro

References

[1] Ardelean L, Reclaru L, Bortun C, Rusu LC. Assessment of dental alloys by different methods. In: Aliofkhazraei M, editor. Superalloys. Rijeka: InTech; 2015. pp. 141-170. DOI: 10.5772/61115

[2] Reclaru L, Ardelean LC. Alternative processing techniques for CoCr dental alloys. In: Narayan R, editor. Encyclopedia of Biomedical Engineering. Vol. 1. Amsterdam: Elsevier; 2019. pp. 1-15. DOI: 10.1016/B978-0-12-801238-3.11100-6

[3] Wassell RW, Walls AWG, Steele JG. Crowns and extra-coronal restorations: Materials selection. British Dental Journal. 2002;**192**:199-211. DOI: 10.1038/sj.bdj.4801334a

[4] Choroszyński M, Choroszyński MR, Skrzypek SJ. Biomaterials for hip implants-important considerations relating to the choice of materials. Bio-Algorithms and Med-Systems. 2017;**13**:133-145. DOI: 10.1515/bams-2017-0017

[5] Jaishankar M, Tseten T, Anbalagan N, Blessy BM, Beeregowda KM. Toxicity, mechanism and health effects of some heavy metals. Interdisciplinary Toxicology. 2014;**2**:60-72. DOI: 10.2478/intox-2014-0009

[6] Hamann D, Hamann CR, Thyssenn JP. The impact of common metal allergens in daily devices. Current Opinion in Allergy and Clinical Immunology. 2013;**13**:525-530. DOI: 10.1097/ACI.0b013e32836457bf

[7] Ziegenhagen R, Reclaru L, Ardelean LC, Grecu AF. Corrosion resistance of stainless steels intended to come into direct or prolonged contact with the skin. Materials. 2019;**12**:987. DOI: 10.3390/ma12060987

[8] Ardelean LC, Reclaru L, Bortun CM, Ghiban B. Laser welding of different dental alloys. In: Cevik S, editor. Superalloys for Industry Applications. London: InTech; 2018. pp. 85-107. DOI: 10.5772/65570

[9] Reclaru L, Ardelean L, Rusu L. Toxic materials, allergens and mutagens and their impact on the dental field. Medicine in Evolution. 2008;**14**:98-102

[10] Regulation (EC) No 1907/2006 of the European Parliament and of the Council of 18 December 2006 concerning the Registration, Evaluation, Authorisation and Restriction of Chemicals (REACH), Establishing a European Chemicals Agency, Amending Directive 1999/45/EC and Repealing Council Regulation (EEC) No 793/93 and Commission Regulation (EC) No 1488/94 as Well as Council Directive 76/769/EEC and Commission Directives 91/155/EEC, 93/67/EEC, 93/105/EC and 2000/21/EC [Internet].2006. Available from: https://echa.europa.eu/regulations/reach/legislation [Accessed: 10 October 2018]

[11] ECHA (European Chemical Agency) [Internet]. Available from: https://echa.europa.eu/home [Accessed: 10 October 2018]

[12] ECHA-Legislation-REACH-Cooperation with authorities and stakeholders-PBT Expert Group [Internet]. Available from: https://echa.europa.eu/pbt-expert-group [Accessed: 11 March 2019]

[13] Diepgen TL, Ofenloch RF, Bruze M, Bertuccio P, Cazzaniga S, Coenraads PJ, et al. Prevalence of contact allergy in the general population in different European regions. British Journal of Dermatology. 2016;**174**:319-329. DOI: 10.1111/bjd.14167

[14] Reclaru L, Eschler PY, Lerf R, Blatter A. Electrochemical corrosion and metal ion release from Co-Cr-Mo

prosthesis with titanium plasma spray coating. Biomaterials. 2005;**26**:4747-4756. DOI: 10.1016/j.biomaterials.2005.01.004

[15] Afolaranmi GA, Tettey J, Meek RMD, Grant MH. Release of chromium from orthopaedic arthroplasties. The Open Orthopaedics Journal. 2008;**2**:10-18. DOI: 10.2174/1874325000802010010

[16] Noumbissi S, Scarano A, Gupta S. A literature review study on atomic ions dissolution of titanium and its alloys in implant dentistry. Materials. 2019;**12**:368. DOI: 10.3390/ma12030368

[17] Wataha JC. Biocompatibility of dental casting alloys: A review. Journal of Prosthetic Dentistry. 2000;**83**:223-234. DOI: 10.1016/S0022-3913(00)80016-5

[18] Ingham E, Fisher J. Biological reactions to wear debris in total joint replacement. Proceedings of the Institution of Mechanical Engineers Part H. Journal of Engineering in Medicine. 2000;**214**(H1):21-37. DOI: 10.1243/0954411001535219

[19] Sargeant A, Goswami T. Hip implants-paper VI-ion concentrations. Materials and Design. 2007;**28**:155-171

[20] MacDonald SJ, McCalden RW, Chess DG, Bourne RB, Rorabeck CH, Cleland D, et al. Metal-on-metal versus polyethylene in hip arthroplasty: A randomized clinical trial. Clinical Orthopaedics and Related Research. 2003;**406**:282-296. DOI: 10.1097/01.blo.0000043066.62337.9d

[21] Godrey A, Abdel-Moneim A, Sepúlveda MS. Acute mixture toxicity of halogenated chemicals and their next generation counterparts on zebrafish embryos. Chemosphere. 2017;**181**:710-771. DOI: 10.1016/j.chemosphere.2017.04.146

[22] Polychronis G, Al Jabbari YS, Eliades T, Zinelis S. Galvanic coupling of steel and gold alloy lingual brackets with orthodontic wires: Is corrosion a concern? The Angle Orthodontist. 2018;**88**:450-457. DOI: 10.2319/092917-655.1

[23] European Parliament and council directive 94/27/EC of June 30, 1994 [Internet]. Available from: https://publications.europa.eu/en/publication-detail/-/publication/0a5a20d2-063e-4892-b94d-d82c9b223de0/language-en [Accessed: 11 March 2019]

[24] Commission Directive 2004/96/EC of 27 September 2004 Amending council directive 76/769/EEC as regards restrictions on the marketing and use of nickel for piercing post assemblies. Official Journal of the European Union 2004, L301, 51

[25] Schnuch A, Wolter J, Geier J, Uter W. Nickel allergy is still frequent in young German females—Probably because of insufficient protection from nickel-releasing objects. Contact Dermatitis. 2011;**64**:142-150. DOI: 10.1111/j.1600-0536.2010.01841.x

[26] Gutsche P, Schmalz G, Landthaler M. Prevalence of piercing in a German population. European Journal of Dermatology. 2008;**18**:26-28. DOI: 10.1684/ejd.2008.0306

[27] Hallab N, Merritt K, Jacobs JJ. Metal sensitivity in patients with orthopaedic implants. The Journal of Bone and Joint Surgery. American Volume. 2001;**83A**:428-436

[28] Willert HG, Buchhorn GH, Fayyazi A, Flury R, Windler M, Köster G, et al. Metal-on-metal bearings and hypersensitivity in patients with artificial hip joints. A clinical and Histomorphological Study. The Journal of Bone and Joint Surgery. American Volume. 2005;**87**:28-36. DOI: 10.2106/JBJS.A.02039pp

[29] Roberts P, Grigoris P, Bosch H, Talwaker N. Resurfacing arthroplasty of the hip. Current Orthopaedics. 2005;**19**:263-279. DOI: 10.1016/j.cuor.2005.05.004

[30] Davies AP, Willert HG, Campbell PA, Learmonth ID, Case CP. An unusual lymphocytic perivascular infiltration in tissues around contemporary metal-on-metal joint replacements. The Journal of Bone and Joint Surgery. American Volume. 2005;**87**:18-27. DOI: 10.2106/JBJS.C.00949

[31] Clarke MT, Lee PTH, Arora A, Villar RN. Levels of metal ions after small- and large-diameter metal-on-metal hip arthroplasty. Journal of Bone and Joint Surgery. British Volume (London). 2003;**85B**:913-917

[32] Rasquinha VJ, Ranawat CS, Weiskopf J, Rodriguez JA, Skipor AK, Jacobs JJ. Serum metal levels and bearing surfaces in total hip arthroplasty. The Journal of Arthroplasty. 2006;**21**:47-52. DOI: 10.1016/j.arth.2006.05.005

[33] MacDonald SJ. Can a safe level for metal ions in patients with metal-on-metal total hip arthroplasties be determined? The Journal of Arthroplasty. 2004;**19**:71-77

[34] Iavicoli I, Gianluca F, Alessandrelli M, Cresti R, De Santis V, Salvatori S, et al. The release of metals from metal-on-metal surface arthroplasty of the hip. Journal of Trace Elements in Medicine and Biology. 2006;**20**:25-31. DOI: 10.1016/j.jtemb.2005.10.002

[35] Hart AJ, Hester T, Sinclair K, Powell JJ, Goodship AE, Pele L, et al. The association between metal ions from hip resurfacing and reduced T-cell counts. Journal of Bone and Joint Surgery. British Volume (London). 2006;**88B**:449-454. DOI: 10.1302/0301-620X.88B4.17216

[36] Lhotka C, Szekeres T, Steffan I, Zhuber K, Zweymuller K. Four-year study of cobalt and chromium blood levels in patients managed with two different metal-on-metal total hip replacements. Journal of Orthopaedic Research. 2003;**21**:189-195. DOI: 10.1016/S0736-0266(02)00152-3

[37] Hennig FF, Raithel HJ, Schaller KH, Dohler JR. Nickel-, chrom- and cobalt-concentrations in human tissue and body fluids of hip prosthesis patients. Journal of Trace Elements and Electrolytes in Health and Disease. 1992;**6**:239-243

[38] Dorr LD, Bloebaum R, Emmanual J, Meldrum R. Histologic biochemical, and ion analysis of tissue and fluids retrieved during total hip-arthroplasty. Clinical Orthopaedics and Related Research. 1990;**261**:82-95

[39] Urban RM, Jacobs JJ, Tomlinson MJ, Gavrilovic J, Black J, Peoc'h M. Dissemination of wear particles to the liver, spleen, and abdominal lymph nodes of patients with hip or knee replacement. The Journal of Bone and Joint Surgery. American Volume. 2000;**82A**:457-477

[40] Pourbaix M. Atlas of Electrochemical Equilibria in Aqueous Solutions, National Association of Corrosion Engineers. National Association of Corrosion Engineers: Houston, Texas; 1974. 644 p

[41] EN-71-3, Safety of toys, part 3: Migration of certain elements, directive 88/378/EEC (C 188, 2005-08-02). 1994. Available from: https://eur-lex.europa.eu/legal-content/ES/TXT/?uri=LEGISSUM%3Al21005 [Accessed: 11 March 2019]

[42] Spriano S, Bosetti M, Bronzoni M, Vern E, Mainac G, Bergo V, et al. Surface properties and cell response of low metal ion release Ti-6Al-7Nb alloy after multi-step chemical and

thermal treatments. Biomaterials. 2005;**26**:1219-1229

[43] Zaffe D, Bertoldi C, Consolo U. Element release from titanium devices used in oral and maxillofacial surgery. Biomaterials. 2003;**24**:1093-1099

[44] Martini D, Fini M, Franchi M, DePasquale V, Bacchelli B, Gamberini M, et al. Detachment of titanium and fluorohydroxyapatite particles in unloaded endosseous implants. Biomaterials. 2003;**24**:1309-1316. DOI: 10.1016/S0142-9612(02)00508-2

[45] Wang J, Wicklund B, Gustilo R, Tsukayama D. Titanium, chromium and cobalt ions modulate the release of bone-associated cytokines by human monocytes/macrophages in vitro. Biomaterials. 1996;**17**:2233-2240

[46] Katti KS. Biomaterials in total joint replacement. Colloids and Surfaces. B, Biointerfaces. 2004;**39**:133-142. DOI: 10.1016/j.colsurfb.2003.12.002

[47] Brown C, Williams S, Tipper JL, Fisher J, Ingham E. Characterisation of wear particles produced by metal on metal and ceramic on metal hip prostheses under standard and microseparation simulation. Journal of Materials Science. Materials in Medicine. 2007;**18**:819-827. DOI: 10.1007/s10856-006-0015-z

[48] Cometa S, Bonifacio MA, Mattioli-Belmonte M, Sabbatini L, De Giglio E. Electrochemical strategies for titanium implant polymeric coatings: The why and how. Coatings. 2019;**9**:268. DOI: 10.3390/coatings9040268

[49] Steinemann SG. Titanium—The material of choice. Periodontology 2000. 2007;**17**:7-21. DOI: 10.1111/j.1600-0757.1998.tb00119.x

Chapter 12

High-Entropy Alloys for Micro- and Nanojoining Applications

Ashutosh Sharma

Abstract

The aim of this chapter is to provide a basic understanding of the metal-ceramic joints and high-entropy alloy (HEA) research in microjoining applications. We will first overview the issues in metal-ceramic brazing and solutions to overcome those issues using various fillers. Various approaches are available for joining ceramic to metallic materials. One approach will be to look for brazing alloys with the so-called high-entropy characteristics which exhibit a solid solution phase. The conventional alloy design and arc melting, Bridgman solidification, and advanced powder metallurgy techniques will be studied, including high-energy ball milling (HEBM) for the mechanical alloying process, and hot-press and spark plasma sintering (SPS) techniques are utilized for improved densification and phase transformation. We also summarize the various thermodynamic relations to obtain the high-entropy phase and present future possibilities of high-entropy alloys in microjoining research at the later stage of this chapter.

Keywords: brazing, entropy, microjoining, welding, alloy, wetting, strength

1. Introduction

There are various types of technologies to bond metal-metal, metal-ceramic, and ceramic-ceramic couple. To achieve this we need a bonding agent known as filler or brazing alloy [1]. The most popular one is active metal brazing where the filler consists of a reactive element, like Ti, Zr, Nb, etc. Another approach is to coat the ceramic body with an active Ti layer. However, to utilize the maximum benefit of the active element, additional surface-active material is required to prevent the generation of residual stresses [2]. Also, the unwanted reaction products by the addition of a number of elements in the filler alloy further deteriorate the joint by creating additional intermetallic compounds (IMCs) as crack generation points. Thus, the formation of IMCs also needs to be controlled for better reliability of the joint [3].

At present, several researchers have used brazing of Al_2O_3, ZrO_2, Si_3N_4, SiC, etc. by using active filler brazing. Active metal brazing makes joining of high melting point materials feasible without melting the contact specimens [4, 5]. Various active filler metals, such as Ag-Cu, Ag-Cu-Ti, Ag-Cu-Sn-Ti, Ag-Cu-In-Ti, Ag-Cu-Ti, Ag-Ti, Ag-Zr, Ti-Zr-Ni-Cu, TiNi, etc., have been investigated for brazing of ceramics. The addition of several active elements can also contribute to various IMCs, Cu-Ti, Cu_3Ti, Ti_3SiC_2, etc. Due to these brittle IMCs, the joint cannot be operated at high temperatures [6–12]. Therefore, the key factor to braze ceramics to metals is to look for new fillers for brazing where active elements (Ti, Zr, etc.) distribute randomly in solid solution. In addition, the alloy system should be free of any additional undesirable IMCs. Recently discovered high-entropy alloys have already shown that with proper design strategy, they might replace traditional materials for a variety of applications [13].

2. Metal-ceramic brazing

In recent microjoining communities, ceramic-metal bonding is of prime importance due to its various applications in hybrid joints, such as ceramic packaging, ceramic sealants, thermoelectrics, and semiconductor transducers. The various types of microjoining processes are given in **Figure 1**.

However, due to the wide limitations of joining this couple owing to their wide gap in physical, mechanical, and thermodynamic behavior, they have always imposed a difficulty. Ceramics are lightweight, harder, and always inert toward the liquid metal and have a wide difference in thermal expansion behavior. Since metals are soft and malleable, for various functional applications, their joining is needed in semiconductors. There are various types of filler materials already available, such as Ag-Cu-Ti brazing system which is most suitable in metal-ceramic joining [7, 14].

The joining processes include various soldering, brazing, and welding processes depending upon the range of temperature of the filler. Soldering is a low-temperature joining but weaker for most of the ceramics due to high-temperature

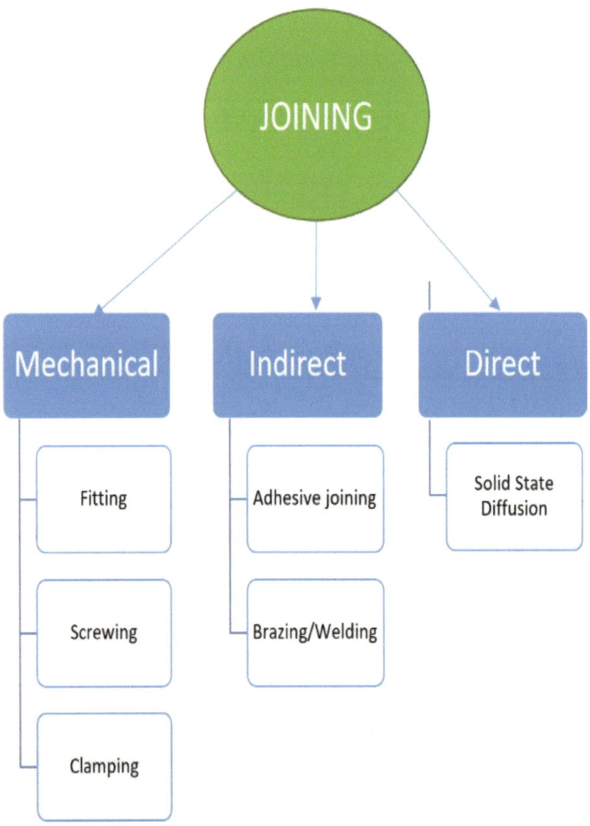

Figure 1.
Various types of metal-ceramic joining processes.

Figure 2.
Various types of welding processes according to the temperature of the application.

requirements. A solder works below 450°C and may not work for high-temperature superalloy joining, while welding is a high-temperature process mostly used for steels [15–18]. In welding, the metal parts need to be melted at high temperatures using fillers known as welding rods which solidify and bond to the workpiece. However, this may not be suitable for the bonding of different metal parts where thermal damage is of concern. A brief detail of the various types of welding processes is given in **Figure 2**.

3. Necessity of developing brazing fillers

Enormous amount of work has been done on the lightweight materials for automotive, defense, and aerospace applications in the last few decades, mainly on steel, ceramics, aluminum, and magnesium-based alloys. However, not all the materials have led to mass production for practical use. The studies reported are preliminary wherein most of the time, microjoining is ignored even if it is an essential part of the automobile, aerospace, consumer, and defense electronic industries.

Therefore, a new structural material, for example, for automobiles, cannot be designed without any foundation for microjoining that is an essential component to maintain the structural integrity of the part in service. There are several important problems in microjoining, such as the presence of brittle IMCs at the interface, segregations, and casting defects (shrinkage porosity, gaseous inclusions, pores, blowholes, etc.). The situation becomes more serious for ceramics (ceramic-metal, ceramic-superalloy, ceramic-steel, etc.) [19]. For example, in various complex assembly applications (ceramic-metal plugs for igniting car engines), ceramics are difficult to join with metals due to their inherent poor wettability and difference in coefficient of thermal expansion (CTE). As shown in **Figure 3**, these two problems have limited the application of ceramics in combination with metals for many years. In this chapter we mainly focus on the brazing which is the most popular for ceramic-metal joints. However, first, we discuss a bit more about brazing processes.

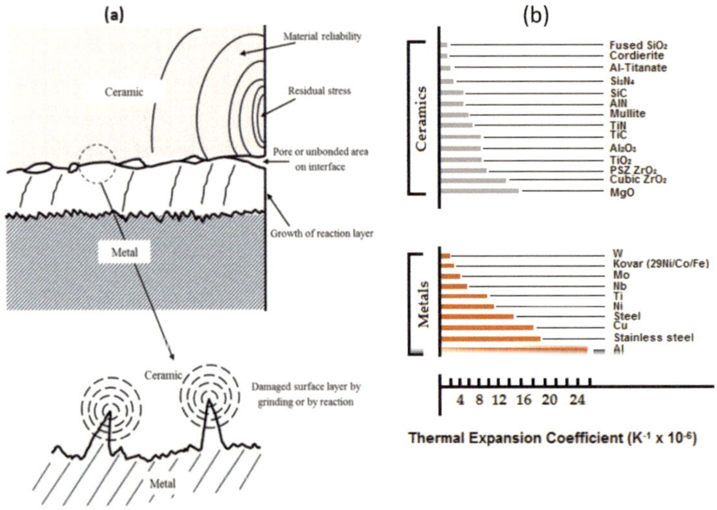

Figure 3.
Metal-ceramic bonding issues. (a) Schematic of factors influencing on the reliability of ceramic-metal joint and (b) comparison of thermal coefficients of metals and ceramics. © 2016 Uday et al. [19] under CC BY 3.0 license.

4. Active metal brazing

There are various brazing processes, such as diffusion bonding, transient liquid phase bonding, reactive bonding, etc., out of which active metal brazing is more popular due to its suitability to ceramics. Active metal brazing enhances the surface wetting of the contacts due to the presence of active elements in the fillers. Recently authors have bonded the ZrO_2 and Ti-6Al-4V using a composite filler of Ag-Cu-Ti to obtain a robust brazed joint [20]. Active metal brazing minimized the residual stresses and distortion in the joint poor strength and high-temperature stability are the issues in brazing industries. To understand brazing, we should have an idea of the wetting of the two contact surfaces. We describe the wetting theory in the following sections.

4.1 Theory of wetting and spreading

According to the wetting theory formulated by Thomas Young in 1805, the interfacial surface energies at the solid–liquid–vapor interface are related by an equation:

$$\gamma_{sv} - \gamma_{sl} = \gamma_{lv} \cos\theta \tag{1}$$

Here, γ_{sv}, γ_{sl}, and γ_{lv} correspond to the interfacial energies of solid-vapor, solid–liquid and liquid–vapor interface (**Figure 4(a)**). According to the spreading parameter,

$$S = \gamma_{sv} - \gamma_{sl} - \gamma_{lv} \tag{2}$$

$\gamma_{sv} - \gamma_{sl} - \gamma_{lv} = 0$ for minimum surface energy (high spreading) and $S < 0$ for poor wetting. **Figure 4(b)** gives the wetting angles of different metals at various temperatures [7].

We can see that the temperature has a great role in the wetting of metal fluid on the substrate. The increase of Ti and Sn in optimal proportion leads to an increment

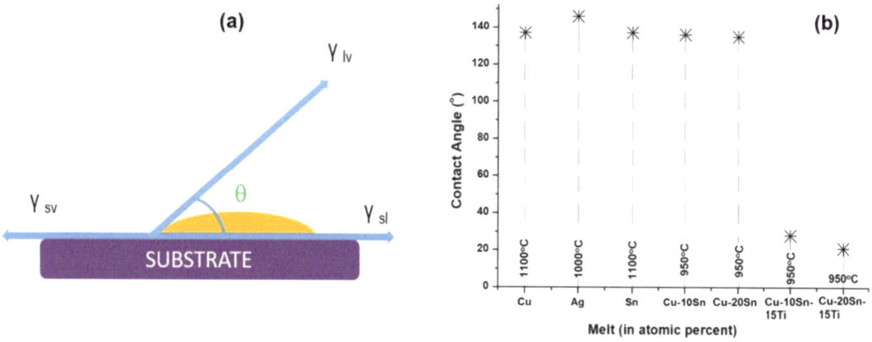

Figure 4.
(a) Schematic of filler wetting at the substrate and (b) contact angle of metals on cBN substrate. [the data in (b) is adapted from [7]].

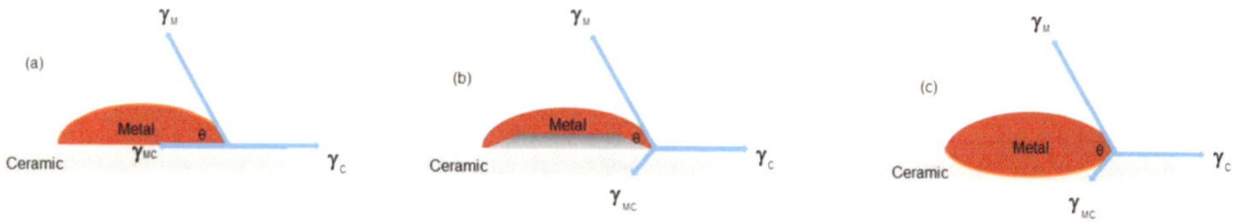

Figure 5.
Liquid metal drop shape depending on the contact time: (a) initial contact or rigid solid surface, (b) formation of a ridge (vertical scale exaggerated), and (c) final equilibrium configuration on deformable solid. © 2016 Uday et al. [19] under CC BY 3.0 license.

in the wettability of the contact surface. The various situations of wetting are given in **Figure 5** depending on the contact angles between the metal and ceramic substrate.

4.2 Factors influencing brazed joint

4.2.1 Issues in metal-ceramic wetting

The wetting is influenced by the surface energy, reactivity of the contact materials to brazing filler, and type of filler used. The reaction between the contact surfaces and brazing filler is key to improve the wetting. In the active metal brazing, a higher wetting can be achieved using Ti, Sc, Cr, or Zr active metals that minimize the surface energy and enhance wetting [21].

4.2.2 Ag-Cu-Ti filler system

Various active fillers are developed in the past. The authors will focus mainly on Ag-Cu-Ti among conventional fillers. This alloy system has various features such as solid solution Ag(Cu) and intermetallic phase Ti_2Cu and Ti_2Ag compounds. The melting range is around 800–900°C and, therefore, is useful to bond ceramics-metals for most of the systems. Ag promotes the activity of Ti and reaction proceeds at a faster rate. There is no ternary intermetallic in Ag-Cu-Ti system [22].

4.2.3 Surface condition

Sharp edges and corners may raise the stress concentration and completely fail the ceramic-metal joint. There should be no surface cracks as they propagate to form a bigger crack which ultimately leads to joint failure. The damage can be repaired through re-sintering the part [19].

4.2.4 Thermal characteristics

High-temperature gradients may cause the generation of thermal stresses and may not be desirable from the brazing point of view. In ceramic-metal bonding, if the ceramic has less CTE than metal, edge cracks will be formed, while core cracks are formed in the opposite case, as shown in **Figure 6** [19, 23, 24].

The remedy for minimizing these cracks is given by various researchers. Zhou [25] suggested the application of metal layer as an interlayer to release the stress by elastoplastic deformation, using composite layers, low-temperature joining, and proper heat treatment and joint design.

4.3 Interface characteristics

The presence of chemical bonding also affects the brazing performance. For example, a thick interfacial layer causes more mismatch in thermal expansion and weakens the joint compared to a thin layer [26]. The operating temperature and time are key parameters to control the interfacial layer. The presence of pores and edges should be avoided before brazing to control the contact angle of the filler with the substrate.

4.4 Type of filler

Filler type also affects the brazing operation. Researchers have also used nanotechnology to develop various brazing fillers, known as composite fillers [27–31]. These nanocomposite fillers refine the various IMCs present in the matrix and joint. However, the dispersion of nanoparticles in the composite matrix is cumbersome. For example, the active metal filler has been reinforced with metallic or ceramic additives to match the CTE and promote the brazing [20]. In the last few decades, the research community has found the solution to overcome the formation of IMCs in an alloy. Such a novel type of alloy design is known as high-entropy alloys as we describe in the following section.

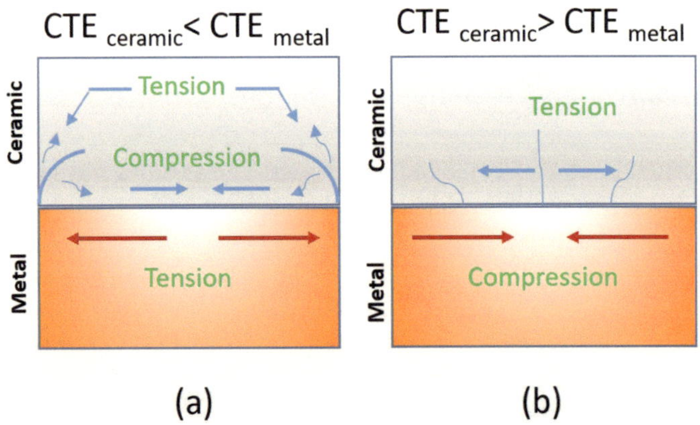

Figure 6.
Types of cracks in metal-ceramic joints. © 2016 Uday et al. [19] under CC BY 3.0 license. (a) Edge cracks in ceramic. (b) Core cracks in ceramic.

5. Advanced brazing fillers must be of high-entropy characteristics

Conventional alloy design suggests the formation of various intermetallic compounds (IMCs) or complex phases with multiple alloying elements. These IMCs act as stress raiser points in alloys and composites which are strictly not desirable. Specially in microjoining applications, these IMCs can cause poor joint properties and catastrophic failure of the entire device. Yeh et al. broke this paradigm by suggesting high-entropy alloys (HEAs), composed of five or more elements in an equiatomic or near-equiatomic fraction varying from 5 to 35 at.% [32, 33]. Therefore, these HEAs contain usually simple solid solution phases rather than IMCs. Many HEAs with high strength, thermal stability, excellent corrosion and wear resistance have been reported. However, they are rarely explored for microjoining applications. Application of these novel HEAs in microjoining can provide several benefits. High entropy filler alloy realizes the active interfacial reaction with ceramic, and the solid solution forms the brazing seam during brazing. The interaction of additional IMCs in the brazing seam is prevented (unlike traditional brazing fillers). This further provides additional benefits like a reduced requirement of ceramic metallization partly with active metal (Ti, Zr), prevents stresses and distortion in joints, improves joint strength, depresses the brazing temperature, and saves energy.

6. Fillers processed by powder technology

Most of the brazing fillers so far available for ceramic-metal joining are silver-based alloys mostly produced by casting approach, e.g., Ag-Cu, Ag-Cu-Ti, Ag-Cu-Sn-Ti, Ag-Cu-In-Ti, etc. [19–26]. Several elements are added to enhance the wetting of the contact surfaces which further produce unwanted IMCs. Moreover, conventional casting or melting techniques are not suitable for industrialization as they produce a number of defects (shrinkage porosity, gaseous inclusions, pores, segregation of constituents, impurities), and there are limitations on the shape/size of the final product. In contrast, the advanced solid-state powder technology has rarely been used for HEA synthesis, and there is a lot more to explore. It offers a low-cost superior alternative over other expensive techniques (sputter deposition, evaporation, induction melting, arc and laser melting, etc.) Secondly, it is an environmentally benign and safe technique as compared to chemical routes (solution precipitation, plating, chemical vapor deposition) where toxic waste disposal and drainage costs are added up [17–19]. In addition, high-energy ball milling is a simple and effective way to produce novel nanostructured materials, homogeneous chemical distribution, and extension of solid solubility and widens the scope of HEA [34–40]. Further densification of milled powders by heat treatment, hot-pressing, cold isostatic pressing, or spark plasma sintering (SPS) technique gives the stable bulk HEA.

7. Research and development on HEA in microjoining

HEAs are the novel and advanced alloys which are composed of 5–35 at.% where all the elements serve as principal elements. They are unique because of their attractive properties compared to their conventional alloys such as strength and ductility trade-off, high thermal stability, high wear, and corrosion resistance. HEAs are first discovered in 1995 by Yeh et al. and coined as multicomponent alloys in 2004 by Cantor et al. [32–33]. Previous studies have shown that most of the HEAs are composed of simple face-centered cubic (FCC), body-centered cubic (BCC), or FCC + BCC solid solutions owing to their high configurational entropy. The composition of these HEAs can be tailored to obtain promising properties such as high hardness and ductility as well as high resistance to wear, oxidation, and corrosion.

7.1 Thermodynamic calculations for fabrication of HEA

In accordance with the Hume-Rothery rules, the factors affecting the binary solid solutions, i.e., atomic size difference, valence electron concentration (VEC), the crystal structure of the solute and solvent atoms, and the difference in electronegativity, will be used to design the HEAs. The criteria adopted for the solid solution phase formation is given by various parameters according to the laws of thermodynamics. According to Zhang et al., the various parameters for HEA formation are given as follows [41, 42]:

$$\Delta S_{mix} = -R\sum_{i=1}^{n} C_i \ln C_i \tag{3}$$

$$\Delta H_{mix} = 4\sum_{i=1, i\neq j}^{n} C_i C_j \Delta H_{mix}^{AB} \tag{4}$$

$$\Omega = |T\Delta S_{mix}/\Delta H_{mix}| \tag{5}$$

$$r = \sum_{i=1}^{n} C_i r_i \tag{6}$$

$$\delta^2 = \sum_{i=1}^{n} C_i \left[1 - \frac{r_i}{\sum_{i=1}^{n} C_i r_i}\right]^2 \tag{7}$$

δ is the atomic size difference; C_i is the atomic fraction of ith element with radius r_i. T denotes the temperature. The average radius of all elements is r. The enthalpy change before and after mixing of A and B elements is ΔH_{mix}^{AB}, and ΔH_{mix} and ΔS_{mix} are the enthalpy change and mixing entropy of the HEA. The parameter Ω is the interaction parameter of the HEA. For HEA formation, $\delta \leq 6\%$ and $\Omega \geq 1.1$.

7.2 Applications of HEA

These novel HEAs find applications in various fields such as aerospace, automobiles, submarines, and nuclear and power plant industries [43–45]. Therefore, it is of utmost importance to develop a reliable and feasible technique of microjoining using these HEAs as fillers. The important conventional brazing methods cause cracking issues in joining metal-ceramic components due to the presence of various interfacial compounds, in addition to the inherent residual stress in fusion welding approaches. In contrast, brazing avoids the cracking issues and distortion, but still, the creation of numerous IMCs at the interface is a serious concern. HEAs with an optimal balance of strength and ductility and mainly solid solution phases may overcome these interfacial compounds and can be an excellent tool for brazing industries. There is not enough work done in this area related to high-entropy brazing. Few studies talk about the laser brazing of Ni superalloy. Gao and his coworkers used FeCoNiMnCu non-equiatomic HEA to braze the Ni superalloy [46]. In this study, they also evacuated the effect of brazing time and foil thickness on the shear strength and found a remarkable improvement.

7.3 Advantages of HEA as filler

There are various reasons for using HEA as fillers:

1. High ductility of HEAs will lead to a robust tough joint. It can be fabricated easily into wires and rods.

2. HEAs will minimize the formation of IMCs in the joint and ensure a safe joint.

3. The cracking and distortion are minimized due to the IMCs and reaction compounds compared to the fusion bonding processes.

4. A single solid solution of HEAs will have a narrow or even zero solidus-liquidus range which is beneficial for brazing to avoid distortion and thermal damage.

8. Future and current status of HEA fillers

These fillers are generally prepared by using various methods such as arc melting, powder metallurgy, sputtering, laser cladding, and electrodeposition. Most of the HEAs are produced initially by arc melting starting from Cantor alloy and its derivatives.

8.1 Arc melting

Arc melting consists of melting the pure elements by an arc in a vacuum or argon environment followed by a cooling approach [46]. However, some vaporizable elements are difficult to use in this method and lead to porosity. The schematic diagram is shown in **Figure 7**.

8.2 Bridgman solidification

There is another variation of the solidification method for obtaining single crystals of HEA, known as Bridgman solidification method [46]. In this case the polycrystalline HEA is melted and cooled from the liquid state (usually an eutectic

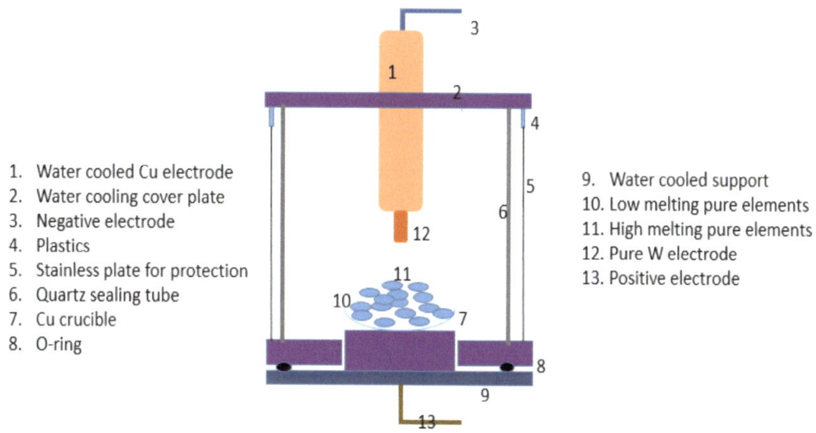

Figure 7.
Schematic for arc melting.

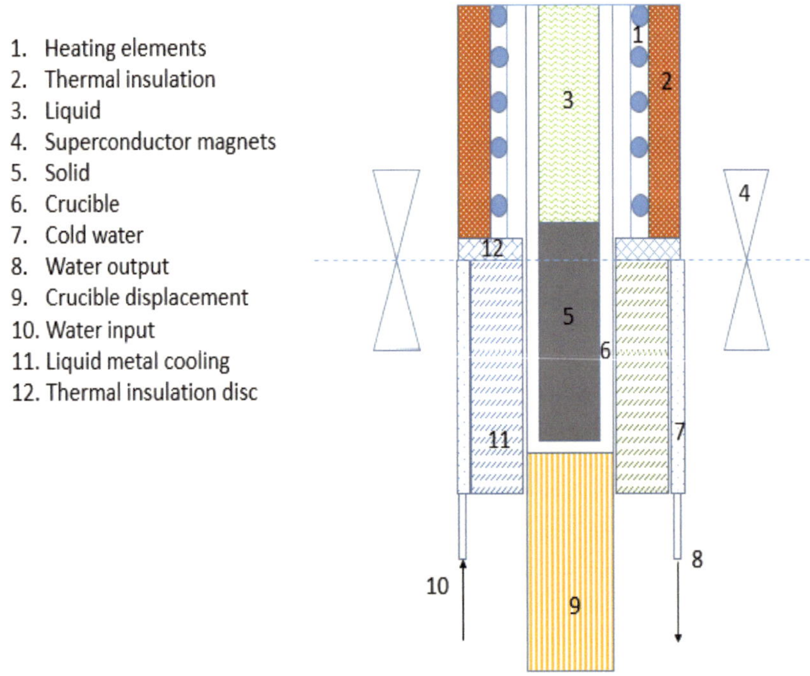

Figure 8.
Bridgman solidification.

Ga-In alloy) and stays in a liquid state at room temperature from the end of the container with a seed layer (**Figure 8**).

8.3 Powder metallurgy

Another important method is powder metallurgy, i.e., mechanical mixing of constituent powders followed by densification by heat treatment such as hot-pressing, cold isostatic pressing, or SPS treatment (**Figure 9**). The powder metallurgy method is beneficial in improving the homogeneity and distribution of powders.

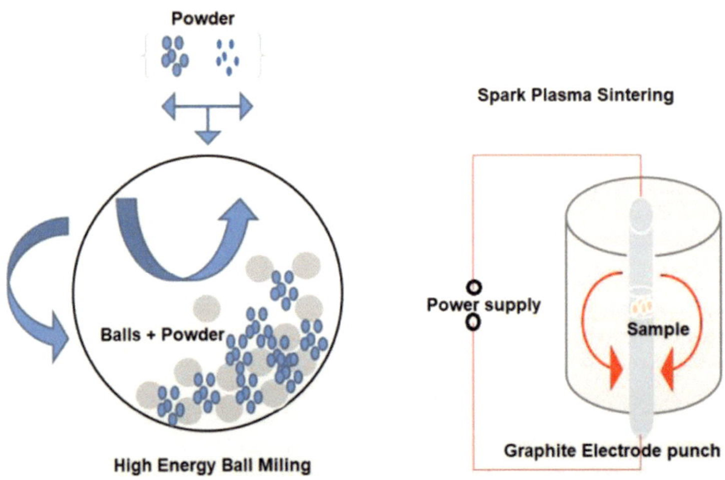

Figure 9.
Mechanical alloying and sintering diagram.

Heat treatment is used to prepare the bulk specimen which also relieves the stress in the powders to get the stable HEAs (**Figure 9**).

Varalakshmi and her co-workers produced nanocrystalline CuNiCoZnAlTi HEA with a BCC structure by powder metallurgy and showed a remarkable increment in the hardness and compressive strength of 8.79 GPa and 2.76 GPa, respectively, higher than the current hard-facing alloys [47]. Other researchers have also produced HEAs with sputtering, laser cladding, and electrodeposition in a few studies. However, until now, a wide exploration of these HEAs for microjoining is very limited.

9. Summary and guidelines

This HEA filler research has the ability to develop a new microjoining material which can join dissimilar materials (ceramic-metal, ceramic-superalloy, ceramic-steel, etc.) and will contribute to the integration of various metallic and ceramic components toward achieving high joint strength, protecting devices, and saving energy. This invention will largely improve the performance of the conventional IMC containing brazing fillers. The development of IMC free fillers for joining is the long-standing demand for microjoining industries. This high quality and the innovative project can be further expanded and commercialized to Korean companies looking for cheap and IMC-free Al alloys for joining for mass production. The combination of HEA filler with those of industry will provide the scale and credibility needed to bring high-entropy alloy technology fully into the market. Newly developed alloys will be patented and commercialized to global companies to realize a capital sum and raise the world economy.

Author details

Ashutosh Sharma[1,2]

1 Department of Energy Systems Research, Ajou University, Suwon, Republic of Korea

2 Department of Materials Science and Engineering, Ajou University, Suwon, Republic of Korea

*Address all correspondence to: stannum.ashu@gmail.com

References

[1] Jacobson DM, Humpston G. Principles of Brazing. Ohio, USA: ASM International; 2005

[2] Hanson WB, Ironside KI, Fernie JA. Active metal brazing of zirconia. Acta Materialia. 2000;**48**:4673-4676

[3] Kee SH, Park SY, Jung JP, Kim WJ. Active metal brazing applied to joining of ZrO_2-Ti alloy. Journal of Welding and Joining. 2012;**30**(3):236-241

[4] Nascimento RM, Martinelli AE, Buschinelli AJA. Review article: Recent advances in metal-ceramic brazing. Cerâmica. 2003;**49**:178-198

[5] Passerone A, Muoio ML. Joining technology in metal-ceramic systems. Materials and Manufacturing Processes. 2000;**15**(5):631-648

[6] Sharma A, Kee SH, Jung F, Heo Y, Jung JP. Compressive strength evaluation in brazed ZrO_2/Ti_6Al_4V joints using finite element analysis. Journal of Materials Engineering and Performance. 2016;**25**:1722-1728

[7] Mishra S, Sharma A, Jung DH, Jung JP. Recent advances in active metal brazing of ceramics and process. Metals and Materials International. 2019. DOI: 10.1007/s12540-019-00536-4

[8] Lin CC, Chen RB, Shiue RK. A wettability study of Cu/Sn/Ti active braze alloys on alumina. Journal of Materials Science. 2001;**36**:2145-2150

[9] Wang Y, Feng JC, Zhang LX, He P, Zhang JH. Microstructure of alumina ceramic/Ag-Cu-Ti brazing alloy/Kovar alloy joint. Materials Science and Technology. 2007;**23**:320-323

[10] Chang CT, Shiue RK. Infrared brazing Ti–6Al–4V and Mo using the Ti–15Cu–15Ni braze alloy. International Journal of Refractory Metals and Hard Materials. 2005;**23**:161-170

[11] He YM, Zhang J, Liu CF, SunY. Microstructure and mechanical properties of Si_3N_4/Si_3N_4 joint brazed with Ag–Cu–Ti + SiCp composite filler. Materials Science Engineering A 2010;**527**:2819-2825

[12] Miao Q, Ding W, Zhu Y, Chen Z, Xu J, Yang C. Joining interface and compressive strength of brazed cubic boron nitride grains with Ag-Cu-Ti/TiX composite fillers. Ceramics International. 2016;**42**:13723-13737

[13] George EP, Raabe D, Ritchie RO. High entropy alloys. Nature Review Materials. 2019;**4**:515-534

[14] Srivastava AK, Sharma A. Advances in joining and welding technologies for automotive and electronic applications. American Journal of Materials Engineering and Technology. 2017;**5**(1):7-13

[15] Fernie JA, Threadgill PL, Watson MN. Progress in joining of advanced materials. Welding and Material Fabrication. 1991;**59**(4):179-184

[16] Kah P, Martikainen J. Current trends in welding processes and materials: Improve in effectiveness. Reviews on Advanced Materials Science. 2012;**30**:189-200

[17] Sharma A, Das S, Das K. Pulse electrodeposition of lead-free tin-based composites for microelectronic packaging. In: Mohamed AMA, editor. Electrodeposition of Composite Materials. Croatia: IntechOpen; 2016. Available from: http://intechopen.com

[18] Sharma A, Das S, Das K. In: Aliofkhazraei M, editor. Pulse Electroplating of Ultrafine Grained

Tin Coating. In: Electroplating of Nanostructures. Croatia: IntechOpen; 2015. Available from: http://intechopen.com

[19] Uday MB, Ahmad-Fauzi MN, Noor AM, Rajoo S. Current issues and problems in the joining of ceramic to metal. In: Ishak M, editor. Joining Technologies. Croatia: IntechOpen; 2016. DOI: 10.5772/64524

[20] Sharma A, Ahn B. Brazeability, microstructure and joint characteristics of ZrO2/Ti-6Al-4V brazed by Ag-CuTi filler reinforced with cerium oxide nanoparticles. Advances in Materials Science and Engineering. 2019;**2019**:1-11. Article ID 8602632. DOI: 10.1155/2019/8602632

[21] Naidich YV, Zhuravlev VS, Gab II, Kostyuk BD, Krasovskyy VP, Adamovskyy AA, et al. Liquid metal wettability and advanced ceramic brazing. Journal of European Ceramic Society. 2008;**28**:717-728

[22] Dezellus O, Arroyave R, Fries SG. Thermodynamic modeling of the Ag–Cu–Ti ternary system. International Journal of Materials Research. 2011;**102**(3):286-297

[23] Mizuhara H, Huebel E, Oyama T. High-reliability joining of ceramic to metal. American Ceramic Society Bulletin. 1989;**68**(9):1591-1599

[24] Olofsson J. Friction and wear mechanisms of ceramic surfaces: With applications to micro motors and hip joint replacements [PhD thesis]. Sweden: Uppsala University; 2011

[25] Zhou Y. Microjoining and Nanojoining. 1st ed. Amsterdam: Elsevier; 2008

[26] Eustathopoulos N, Nicholas MG, Drevet B. Wettability at High Temperatures. 1st ed. Oxford: Pergamon; 1999

[27] Shin J, Sharma A, Jung DH, Jung JP. Effect of Sn content on filler and bonding characteristics of active metal brazed Cu/Al_2O_3 joint. Korean Journal of Metals and Materials. 2018;**56**(5):366-374

[28] Sharma A, Roh MH, Jung DH, Jung JP. Effect of ZrO_2 nanoparticles on the microstructure of Al-Si-Cu filler for low-temperature Al brazing applications. Metallurgical and Materials Transactions A. 2016;**47A**:510-521

[29] Sharma A, Lim D, Jung JP. Microstructure and brazeability of SiC nanoparticles reinforced Al-9Si-20Cu produced by induction melting. Materials Science and Technology. 2016;**32**(8):773-779

[30] Sharma A, Roh MH, Jung JP. Effect of La_2O_3 nanoparticles on the brazeability, microstructure, and mechanical properties of Al-11Si-20Cu alloy. Journal of Materials Engineering and Performance. 2016;**25**(8):3538-3545

[31] Sharma A, Di X, Jung JP. Effect of different nanoparticles on microstructure, wetting and joint strength of Al-12Si-20Cu braze filler. Materials Research Express. 2019;**6**(5):056526

[32] Yeh JW, Chen SK, Lin SJ, et al. Nanostructured high-entropy alloys with multiple principal elements: Novel alloy design concepts and outcomes. Advanced Engineering Materials. 2004;**6**(5):299-303

[33] Cantor B. High-entropy alloys. In: Buschow KHJ, Cahn RW, Flemings MC, Ilschner B, Kramer EJ, Mahajan S, et al. editors. Encyclopedia of Materials: Science and Technology. ISBN 978-0-08043152-9

[34] Bhattacharya S, Sharma A, Das S, Das K. Synthesis and properties of

pulse electrodeposited lead-free tin-based Sn/$ZrSiO_4$ nanocomposite coatings. Metallurgical and Materials Transactions A. 2016;**47**(3):1292-1312

[35] Sharma A. Effect of synthesis routes on microstructure of nanocrystalline cerium oxide powder. Materials Science and Applications. 2013;**4**:504-508

[36] Sharma A, Baek BG, Jung JP. Influence of La_2O_3 nanoparticle additions on microstructure, wetting, and tensile characteristics of Sn-Ag-Cu alloy. Materials and Design. 2015;**87**:370-379

[37] Sharma A, Yu H, Cho IS, Seo H, Ahn BA. ZrO_2 nanoparticle embedded low silver lead free solder alloy for modern electronic devices. Electronic Materials Letters. 2019;**15**:27-35

[38] Sharma A, Srivastava AK, Lee K, Ahn B. Impact of non-reactive ceria nanoparticles on the wettability and reaction kinetics between lead-free Sn-58Bi and Cu pad. Metals and Materials International. 2019;**25**:1027-1038

[39] Sharma A, Srivastava AK, Ahn B. Microstructure, mechanical properties, and drop reliability of CeO_2 reinforced Sn-9Zn composite for low temperature soldering. Materials Research Express. 2019;**6**:056520

[40] Sharma A, Sohn HR, Jung JP. Effect of graphene nanoplatelets on wetting, microstructure, and tensile characteristics of Sn-3.0Ag-0.5 Cu (SAC) alloy. Metallurgical and Materials Transactions A. 2016;**47**(1):494-503

[41] Yeh JW, Chen YL, Lin SJ, et al. High-entropy alloys—A new era of exploitation. Materials Science Forum. 2007;**560**:1-9

[42] Zhang Y, Zhou YJ, Lin JP, et al. Solid-solution phase formation rules for multi-component alloys. Advanced Engineering Materials. 2008;**10**(6):534-538

[43] Zhou YJ, Zhang Y, Wang YL, et al. Solid solution alloys of AlCoCrFeNiTix with excellent room temperature mechanical properties. Applied Physics Letters. 2007;**90**(18):1904

[44] Senkov ON, Wilks GB, Scott JM, Miracle DB. Mechanical properties of Nb25Mo25Ta25W25 and V20Nb20Mo20Ta20W20 refractory high entropy alloys. Intermetallics. 2011;**19**:698-706

[45] Lin CM, Tsai HL. Evolution of microstructure, hardness, and corrosion properties of high-entropy Al0.5CoCrFeNi alloy. Intermetallics. 2011;**19**(3):288-294

[46] Gao M. Development of new high entropy alloys for brazing of Ni-base superalloys [MS thesis]. Colorado School of Mines; 2017

[47] Varalakshmi S, Kamaraj M, Murty BS. Formation and stability of equiatomic and nonequiatomic nanocrystalline CuNiCoZnAlTi high-entropy alloys by mechanical alloying. Metallurgical and Materials Transactions. 2010;**41**(10):2703-2709

Section 5

Advances in Surface Engineering

Chapter 13

Laser Surface Treatment

Anas Ahmad Siddiqui and Avanish Kumar Dubey

Abstract

Laser surface treatment (LST) utilizes intense thermal energy of laser beam for modification, alloying, and cladding surface of substrate materials. In LST very high cooling rates of 104–106°C/s can be achieved. Such high cooling rate arrests the possibilities of segregation in the case of multicomponent systems. Moreover, very narrow heat-affected zone (HAZ) and easy automation make it suitable for large-scale industrial production. LST depends on many process parameters such as laser power, scan speed, focal length, spot size, substrate temperature, and type of material. Selection of proper range of process parameters for good surface quality is essential. Pores and cracks may arise due to improper selection of parameters. Multilayered, high-entropy, thermal barrier coatings using LST with good bonding with substrates have been developed.

Keywords: laser surface modification, laser surface alloying, laser cladding

1. Introduction

Metals and their alloys are one of the means to fulfill our imagination. With change in necessity, their utility is also changing. Today industries require materials that can meet the demands of challenging markets. In this age of miniaturization, we require materials that can form the framework for new technologies. Advanced biomaterials for bones and dentures with critical surface properties have been developed [1, 2]. These materials have shown to perform better than the available materials. Artificial bones of Mg and their alloys have been researched [3]. These may restrict the requirement of recursive surgeries in case of implants. In the nuclear industry, materials which can restrict the harmful radiations while themselves remaining neutral have been proposed [4]. Such materials may be able to improve the working conditions of nuclear industry workers and may restrict the radiation leakages in prolonged use. Chromium has been used for decades as a surface hardening and corrosion-resisting agent. In recent years there are some articles that discuss the effect of chromium on health [5, 6]. It has been found that few states of chromium may be the probable cause of cancer. Hence, alternate materials possessing properties similar to chromium coatings have been reported [7]. In an attempt to improve the efficiency of power plants, turbine blades which can handle high stresses have been tested [8]. These materials may help boost the limits of power plants, aircrafts, and other propulsion systems. The thrust of ever-expanding horizons of knowledge development of materials and surface properties has become essential.

Most of the advanced applications require superior surface properties such as high hardness, strength, wear resistance, corrosion resistance, high temperature oxidation resistance, and improved magnetic and chemical behavior. All these properties can be incorporated and developed by modifying the surface of the

Attributes	LSA	Electrodeposition	Thermal spray	CVD	PVD
Dilution	High	Nil	High	Nil	Nil
Bonding strength	High	Low	Moderate	Low	Low
Heat-affected zone	Low	Nil	High	Low	Low
Coating thickness	Moderate	Moderate	High	Low	Low
Repeatability	High	Low	Moderate	High	High
Controllability	High	Low	Moderate	High	High

Table 1.
Characteristics of different techniques.

components. There are wide varieties of surface modification techniques available. Some of these techniques are thermochemical coatings (nitriding, carburizing, cyaniding, etc.), electrodeposition, electroless deposition, spray coatings (flame spray, thermal spray coating, plasma spray coating, etc.), physical vapor deposition (PVD), chemical vapor deposition (CVD), laser surface modification (LSM), etc. These diverse techniques mutually form a branch termed as surface engineering. All these surface modification techniques have certain advantages and disadvantages. **Table 1** lists some of the desirable attributes and corresponding behavior observed with different processes. For precision coatings of thermally sensitive and multicomponent materials, usually laser material processing is employed. Due to its localized heating and rapid solidification rates, thermal distortion and segregation possibilities are diminished. Also, high energy density leads to melting of almost any metal [9]. High-energy-density laser beam produces high dilution and good bonding strength, and very low heat-affected zone can be developed. Other techniques usually suffer in one or the other reasons. Also, high repeatability and controllability makes it a suitable technique for industrial standards.

With the development in the automation sector, lasers having high accuracy and precision are available. Thus, in the last decade, a large number of literature dealing with application of lasers in various fields are available. These lasers may also be used to develop layer by layer lamina to develop a desired 3D structure. Laser-based techniques employed in 3D printing are selective laser melting and sintering. A part program of the 2D structure to be manufactured is developed. These 2D structures of the same or varying sections are developed above one another. These adjacent layers join together and form a required 3D structure. Hence, laser printing is very similar to surface treatment processes. This chapter in particular presents the ongoing trends of laser surface treatments in melt regime, i.e., it discusses techniques such as laser surface alloying, laser cladding (LC), selective laser melting, and laser glazing. Although the basis of these techniques is same, these techniques differ from one another in the desirability of final surface properties achieved. Numerical simulation and application of these techniques have been discussed.

2. Selection of laser

Many aspects are considered during selection of laser for LST processes. Some of the desirable characteristics are presented in the flow chart shown in **Figure 1**. The material to be processed is one of the factors for selection of laser type. Heat-sensitive materials and refractory materials are generally processed in pulsed mode [10–15]. It is also observed that the solidified structure of developed materials may be different in the case of continuous and pulse laser modes [16]. **Figure 2** shows the relationship between reflectivity of a material and wavelength of radiation [17]. Materials such as aluminum which have very low absorptivity are usually processed with low wavelength pulses [18]. Besides, these desirable properties also significantly affect the selection of laser for a particular application [19].

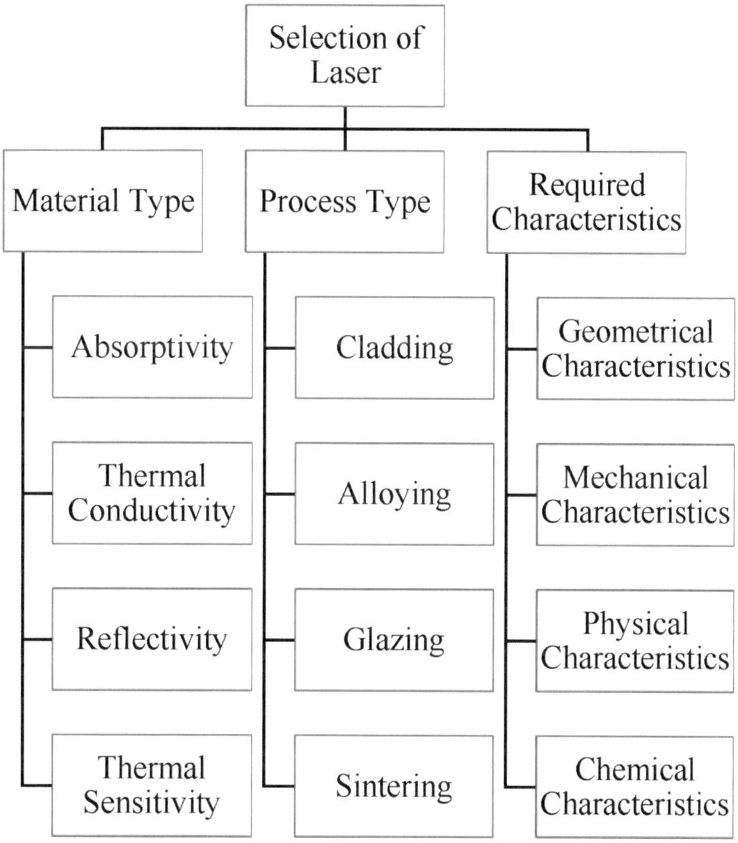

Figure 1.
Constraints considered in selection of laser source type.

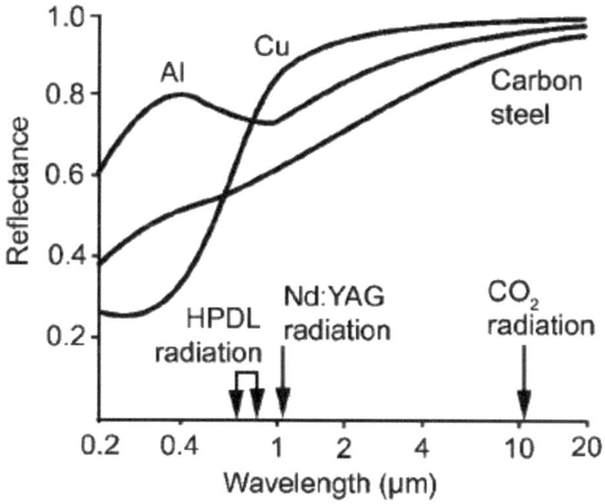

Figure 2.
Reflectance of substrate vs wavelength of radiation for some materials [20].

The type of process also affects the selection of laser. Generally surface alloying requires a large amount of heat to melt considerable amount of substrate surface. Thus high-energy lasers are required. Glazing and sintering usually employ low-energy lasers, whereas cladding uses intermediate-energy lasers.

3. Laser surface treatment processes

Extensive laser surface treatment techniques are available. Laser surface treatment usually modifies the topography, phase composition, and microstructure of a substrate material to improve its surface properties. When a laser is incident on a substrate material, laser radiations are absorbed by conduction electrons near the surface region (in nm range). These excited electrons collide with lattice ions and rapidly produce heat. The heat produced in this thin layer is conducted to the bulk substrate. This causes swift heating of a layer of material, having a thickness greater than the characteristic radiation absorption depth. As soon as the laser irradiation is stopped, the substrate material cools due to heat transfer. **Figure 3** presents a block diagram of laser interaction with substrate.

These thermal cycles may possibly cause phase transformations, topography, and microstructural variations. The extent of these changes depends on the behavior and type of material irradiated, the maximum temperature attained, and heating and cooling rates experienced. All the above said factors depend on the laser power density and interaction time between laser and substrate material. Laser surface treatment techniques are differentiated on the basis of temperature observed at the surface due to irradiation. If surface temperature attained due to laser irradiance is less than the melting temperature of a material, solid state transformations can be observed. Such a system is observed in hardening, shock peening, and engraving. When the surface temperature obtained is greater than the melting temperature but lower than the vaporization temperature of the substrate material, melting of substrate surface takes place. This is the most widely used regime for surface modification. Techniques such as laser cladding, laser alloying, laser glazing, and selective laser melting fall under this regime. If the surface temperature is greater than the vaporization temperature of the substrate, vaporization of substrate surface takes place. This regime is used in laser machining techniques such as laser drilling, cutting, and contouring. **Figure 4** presents a block diagram of this classification.

3.1 Laser cladding

Laser cladding technique is employed to produce coatings with enhanced surface properties or to repair surface defects of different components. LC employs high energy density of laser beams to melt and alloy the surface of substrate materials. Due to high energy density, most of the metals can be melted and alloyed. Usually when the dilution percent is less than 10%, LC is meaningful because low concentration of substrate is desired in LC. Thick to moderate layers of almost any material can be developed. **Figure 8(a)** shows a cross-sectional image of clad bead.

The material to be deposited on a substrate may be supplied using two techniques: preplaced powder deposition [21–25] or codeposition method [26–30].

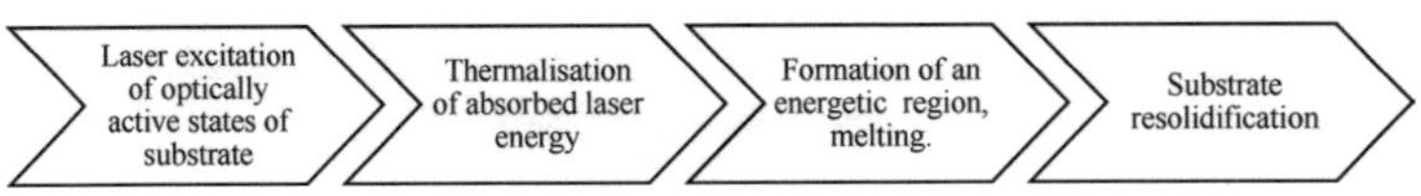

Figure 3.
Laser material interactions.

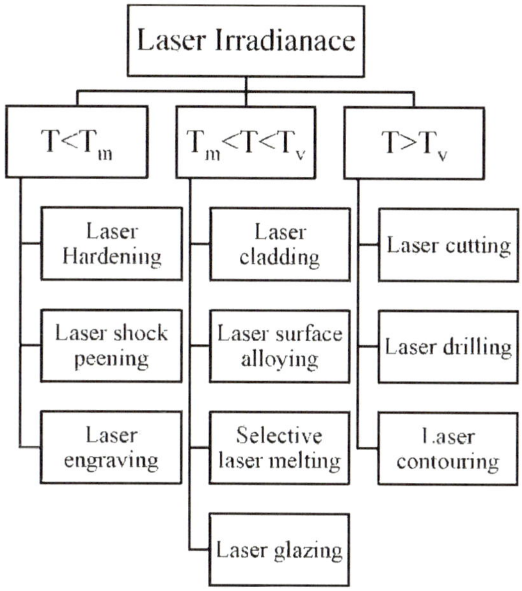

Figure 4.
Classification of laser material processing techniques.

These methods differ in the supply of clad material. In the first method as shown in **Figure 5(a)**, the powder to be clad is first mixed with certain adhesives (polyvinyl alcohol) to form slurry. This slurry is placed above the substrate as a uniform coat and allowed to dry and harden. This is done so that it can withstand the pressure of the shielding gas and particle nature of laser.

In the second method as shown in **Figure 5(b)**, the powder to be clad is fed through the powder feeder nozzle onto the laser beam and subsequently on the molten pool. This powder feeding can be done at various angles through the laser beam. When the angle of feeding is zero degrees, it forms a coaxial feeding system. Some authors have also studied LMP using different types of nozzles [31, 32]. Generally, off axis, four stream and coaxial nozzles have been employed. **Figure 6** shows the gas and particle flow patterns for various nozzles. Thus, cladding of substrate is possible using both the techniques.

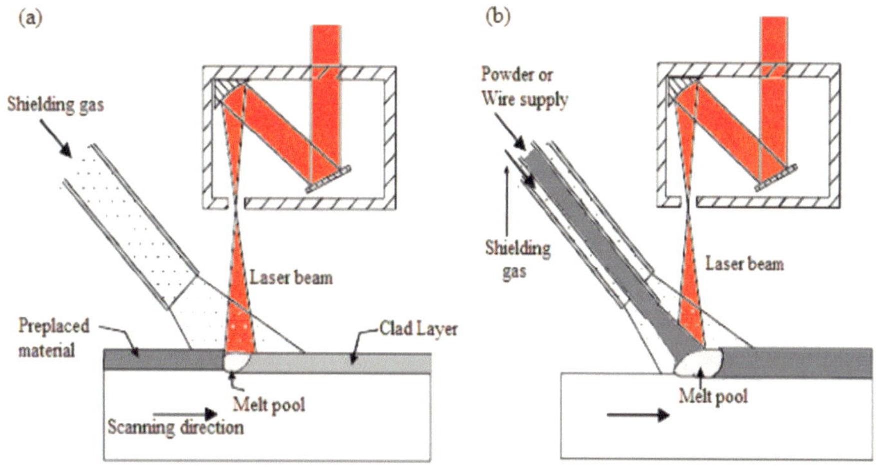

Figure 5.
(a) Preplaced technique and (b) codeposition technique [33].

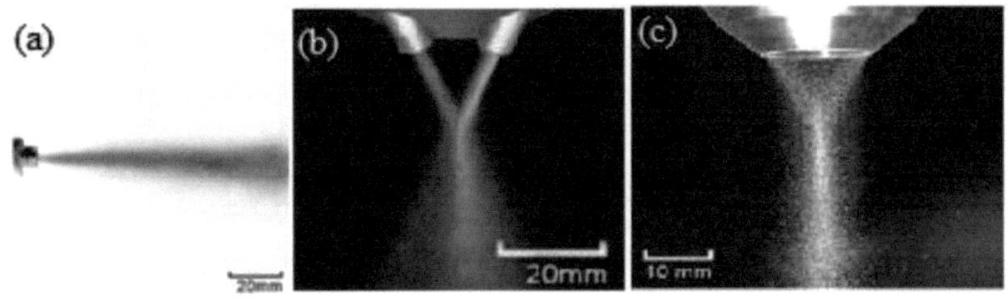

Figure 6.
Gas and particle flow for (a) off focus, (b) four stream, and (c) coaxial nozzles [32].

Through extensive review of articles, it was observed that beads produced using powder preplaced method are prone to more defects. This may be attributed to the presence of a bonding agent which vaporizes during laser beam interaction. Also, the height of clad can be manipulated in a codeposition method which is difficult in preplaced technique [34].

The type of microstructure formed depends on the temperature gradient (G) and solidification rate of crystal (R) [35–37]. High G/R ratio leads to planar structure, with decrease in G and increase in R columnar structure being achieved, and low G/R ratio leads to equiaxed dendritic structure. In LST high cooling rates (10^3–10^8 K/s) can be achieved [38]. Hence in LST generally dendritic structure is visible. Change in structure can also be realized with change in mode of laser. It is observed that in continuous laser mode, columnar dendritic structure was formed which was oriented towards the center of clad bead, whereas in pulsed laser mode, stacks of dendrite were randomly oriented. This was due to cyclic melting and resolidification phases, leading to progressive change in the molten pool. **Figure 7** shows the LC using continuous (a) and pulsed (b) mode.

3.2 Laser surface alloying

Laser surface alloying (LSA) is a similar process to LC but using high energy density. LSA sample is shown in **Figure 8(b)** [40]. It is observed that dilution percent is greater than 10%. Hence, no clear distinction up to a certain depth can be observed, and alloy bead has some proportions of substrate material. Usually, LC is employed in applications requiring entirely different properties at the surface and core, whereas LSA is employed in applications requiring change in properties for greater depth.

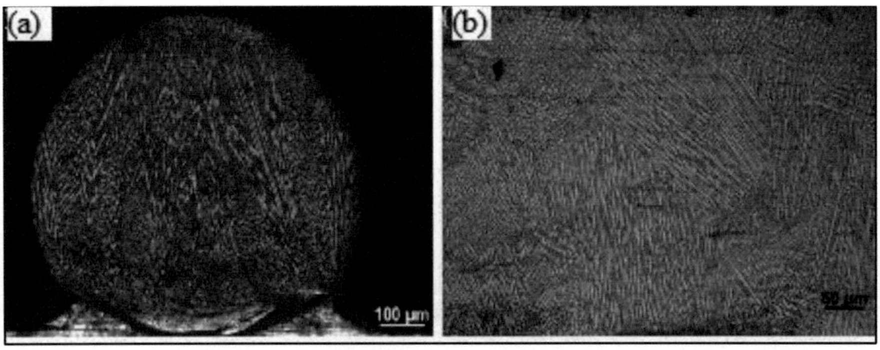

Figure 7.
LC using (a) continuous mode and (b) pulsed mode [39].

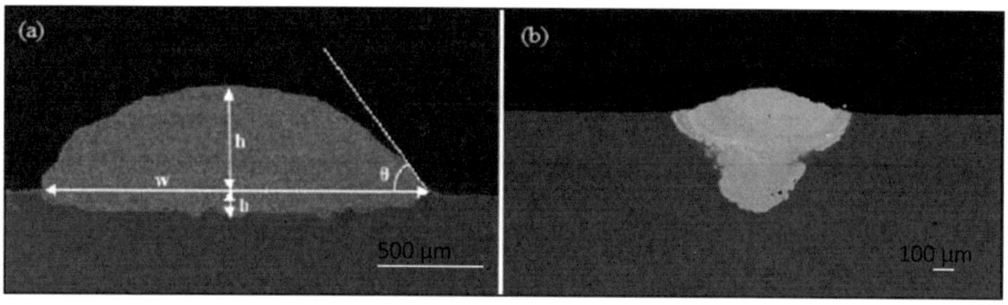

Figure 8.
Cross-sectional images of (a) laser clad [30] and (b) laser surface alloyed bead [40].

3.3 Selective laser melting

Selective laser melting (SLM) utilizes laser energy to create 3D parts using a 3D CAD sketch of the part geometry to be produced. The 3D model is then broken to a 2D stack of layers which form the required geometry. These 2D layers are created by laser scanning over the cross-sectional area. This scanning of laser melts and bonds particles together to form a thin layer. Repeating this process, a subsequent layer may be produced and altogether bonded to previously produced layers. These formed stacks of 2D layers represent the final 3D required geometry. Selective laser sintering (SLS) is a similar process to SLM, but in SLS complete melting of powder does not take place. SLS uses low-power lasers for fabrication of 3D parts compared to SLM. Thus, the final products formed using SLS usually have high porosity and require impregnation of different materials. **Figure 9** presents the steps in SLM.

Literatures suggests that SLM is successfully applied to aluminum and its alloys, high-speed steels, nickel, and copper alloys. The main problems with SLM are porosity, cracking, oxide inclusion, and loss of alloying elements. Porosity may be reduced by proper selection of laser energy density for specific material. Cracking can be reduced by decreasing the cooling and solidification rate.

3.4 Laser glazing

Laser glazing (LG) is a surface melting method using a continuous high-energy laser beam which traverses the surface of a substrate, generating a thin layer of melted material. After the solidification of this thin melted layer, the material's surface appears glassy; therefore this method is termed as laser glazing. Researchers have done LG to improve surface properties [42–44]. It employs low peak power;

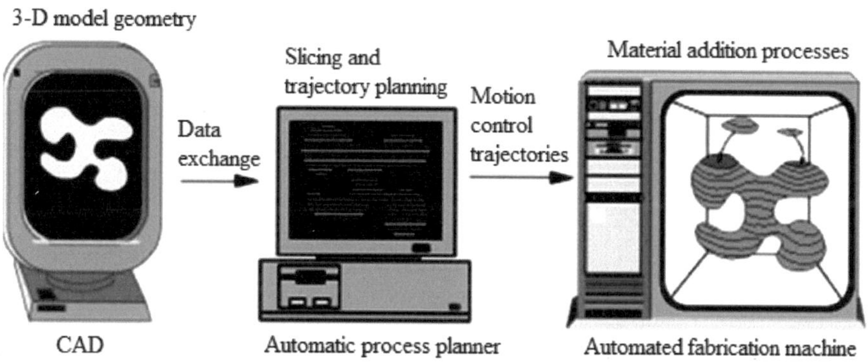

Figure 9.
Steps in SLM technique [41].

hence vaporization of surface materials is restricted. LG is used to level the surface and develop a dense re-melted layer. Due to such a system microhardness, erosion resistance and thermal shock behavior of the substrate may be improved.

4. Numerical modeling of laser surface treatment processes

Since LST processes are highly nonlinear processes, hence for clear understanding of the process, numerical modeling is essential [45]. Besides laser beam interaction with the substrate, LC and LSA impose additional constraints of clad and alloying materials that may be supplied in the form of preplaced powder or through powder feeder. In case powder is supplied through powder feeder, then the study of powder flow dynamics becomes important. Hence, LC and LSA processes can be broken in three stages: powder stream dynamics, melt pool region, and quality variables. Experimental study of laser surface treatments is not sufficient to understand complex phenomena such as powder stream pattern, laser and substrate interaction, heat transfer mode, and melt pool behavior. Hence, analytical models and numerical simulations have been developed. Some of the brief advancements and understandings in these areas are presented.

4.1 Powder stream dynamics

Powder stream dynamics is a significant area in the blown powder technique because we can predict powder stream characteristics such as motion, flow profile, powder with laser system interaction, etc. These parameters may be further used in optimization of parameters and calculate powder efficiency. The behavior of powder flow is governed by the type of nozzle employed. The turbulent flow of carrier gas can be assumed to be a continuum; hence Navier-Stokes equations can be used. A dispersed powder in carrier gas is generally solved using the Lagrangian equation [46]. During powder–laser interaction, attenuation of laser energy takes place due to absorption, reflection, radiation, beam scattering, and ricocheting of powder particles. Models based on particle attenuation [47], ray tracing [48], resolution [49], and light propulsion force model [50] have been reported to predict the behavior of attenuated laser energy fairly. Finally, powder substrate interaction depends on the standoff distance. For high convergence the substrate should lie at the focus of the nozzle [51].

4.2 Melt pool

The substrate melts due to absorption of remaining incident laser energy. The focuses of interest in this region are heat transfer, geometry of melt pool, fluid flow velocity, cooling rate, and solidification rate. These variables have been calculated using kinetic approach [52], volume method [53], and finite element method [54, 55]. Usually commercial multi-physics platforms such as ANSYS, COMSOL, FLUENT, and ABAQUS are employed for the problem.

4.3 Quality variables

The simulation of desirable properties is usually done in combination with developed thermal model. Phase transformation models along with thermal models provide base for the measurement of quality variables. Diffusion and diffusion-less phase transformations may occur in different material systems. Diffusion phase transformations have been modeled using the Johnson-Mehl-Avrami equation, while diffusion-less phase transformations are modeled using the Koinstinen Marburguer equation [56, 57]. Hardness measurement of a treated surface has been predicted by coupling thermo-kinetic relation and thermal model [58, 59]. Residual stresses develop in laser-treated parts; these lead to crack and distortion. The finite element method has been used to solve coupled elastic, plastic, and thermal strain equations with phase transformation equations [60, 61].

5. LST applications

LST has a wide application in aerospace, automobile, medical, nuclear, oil recovery, and refinery industries [62–66]. Aluminum and its alloys are widely used in aerospace industry; they have been efficiently clad with other novel metals to improve their surface properties [67]. Stainless steel is used in automobile and household applications [68, 69]. Titanium and its alloys are used in the medical sector [70]. LC on Ti6Al4V has been studied frequently to improve its surface properties [71].

Materials	Application	Improvement	Author, year
CMSX-4 (Ni-based super alloy)	Repair of turbine blades	This method helped to develop monocrystalline CMSX-4	Rottwinkel, 2016 [73]
Stellite-6/WC on B27 boron steel	Repair of tools for soil cultivation	Formation of intermetallic compounds improved the wear resistance	Bartkowski, 2016 [74]
NI40 and NI60 on C60 steel	Improvement of barrel-screw system in plastic injection molding	Ni-Cr alloy clad improved the microhardness	Zarini, 2014 [75]
CPM9V steel on H13 tool steel	Repair of molds and dies used in hot and cold working	Presence of compressive stress due to formation of martensite phase	Paul, 2017 [76]
Grade C wheel U75V rail with 316L, 420, 410	Repair of damaged railway wheels	The wear rates decrease with increased hardness of the clad materials	Zhu, 2019 [77]
Titanium hydroxylapatite on Nitinol	Coating on Nitinol implants to restrict nickel release	Modulus of elasticity of coated samples falls in the range of 6–30 GPa which is similar to the natural bone	Chakraborty, 2019 [78]
Mg-Zn-Dy alloy casted and laser melted	Restrict in vitro degradation and improve tissue integration	Improvement in in vitro degradation due to formation of insoluble protective layer	Rakesh, 2019 [79]
Powdered Co29Cr9W3Cu alloy	SLM is used to develop Co29Cr9W3Cu alloy joint prostheses	Initiation of crack is arrested due to plastic deformation caused by strain-induced martensitic transformation	Lu, 2019 [80]
Ti powder on Ti6Al4V substrate	Improve in vitro biocompatibility capacity of the titanium deposits to be used as medical implants	In vitro test of samples in Hank's solution shows that the leaching was within the desired values	Nyoni, 2016 [81]

Table 2.
Some applications of LST.

LST applications can be classified in two categories. First is remanufacturing or refurbishing products to restore their properties and dimensions [72]. Second is development of new materials with improved properties. **Table 2** presents some critical applications of LST.

6. Conclusion

Laser surface treatment may be employed to provide advanced surface properties to a substrate material. The melt pool regime in laser irradiance is utilized to completely modify the surface properties. The interaction of laser with alloy/clad powder, laser and substrate material, powder particles, and substrate materials is important for clear understanding of the problem. Numerical-based techniques have provided a way to optimize, standardize the processes, and reduce wastage during actual processing. These techniques have a vast application horizon, i.e., from medical implants to turbine blades, all can be modified using these techniques.

Author details

Anas Ahmad Siddiqui* and Avanish Kumar Dubey
Department of Mechanical Engineering, Motilal Nehru National Institute of Technology, Allahabad, Prayagraj, India

*Address all correspondence to: anas091991@gmail.com

References

[1] Rajendran A, Pattanayak DK. Nanoporous, bioactive and cytocompatible TiO_2 encapsulated Ti particles as bone augmentation material. Advanced Powder Technology. 2020;**31**:695-701. DOI: 10.1016/j.apt.2019.11.024

[2] Wu D, Spanou A, Diez-escudero A, Persson C. 3D-printed PLA/HA composite structures as synthetic trabecular bone: A feasibility study using fused deposition modelling. Journal of the Mechanical Behavior of Biomedical Materials. 2020;**103**:103608. DOI: 10.1016/j.jmbbm.2019.103608

[3] Lin Z, Zhao Y, Chu PK, Wang L, Pan H, Zheng Y. Biomaterials A functionalized TiO_2/ Mg_2TiO_4 nano layer on biodegradable magnesium implant enables superior bone implant integration and bacterial disinfection. Biomaterials. 2019;**219**(March):119372. DOI: 10.1016/j.biomaterials.2019.119372

[4] Was GS, Petti D, Ukai S, Zinkle S. Materials for future nuclear energy systems. Journal of Nuclear Materials. 2019;**527**:151837. DOI: 10.1016/j.jnucmat.2019.151837

[5] Yin J, Wang L, Chen Y, Zhang D, Hegazy AM, Zhang X. A comparison of accumulation and depuration effect of dissolved hexavalent chromium (Cr^{6+}) in head and muscle of bighead carp (*Aristichthys nobilis*) and assessment of the potential health risk for consumers. Food Chemistry. 2019;**286**(February):388-394. DOI: 10.1016/j.foodchem.2019.01.186

[6] Yang D, Liu J, Wang Q, Hong H, Zhao W, Chen S. Chemosphere Geochemical and probabilistic human health risk of chromium in mangrove sediments: A case study in Fujian, China. Chemosphere. 2019;**233**:503-511. DOI: 10.1016/j.chemosphere.2019.05.245

[7] Forsich C, Dipolt C, Heim D, Mueller T, Gebeshuber A, Holecek R. Potential of thick a-C:H:Si films as substitute for chromium plating. Surface and Coating Technology. 2014;**241**:86-92. DOI: 10.1016/j.surfcoat.2013.11.011

[8] Sexton L, Lavin S, Byrne G, Kennedy A. Laser cladding of aerospace materials. Journal of Materials Processing Technology. 2002;**122**:63-68

[9] Brown MS, Arnold CB. Laser Precision Microfabrication. Vol. 135. Springer; 2010. pp. 91-120. DOI: 10.1007/978-3-642-10523-4

[10] Darkwa KM, Gupta RK, Kumar D. Fabrication and characterization of hydroxyapatite-magnesium composite thin films on magnesium plates for implant application. ASME IMECE. 2012;**86918**:1-6

[11] Ganeev RA. Laser-Surface Interactions. Dordrecht Heidelberg London New York: Springer; 2014. DOI: 10.1007/978-94-007-7341-7

[12] Gladush GG, Smurov I. Physics of Laser Materials Processing. Springer Series in Material Science; 2011;**146**. DOI: 10.1007/978-3-642-19831-1

[13] Ansari M, Soltani R, Sohi MH, Valefi Z. Microstructural and hardness study of pulsed Nd:YAG laser surface alloyed aluminum with iron. Metallurgical and Materials Transactions A: Physical Metallurgy and Materials Science. 2016;**47**:1698-1704. DOI: 10.1007/s11661-015-3320-7

[14] Gorunov AI. Laser alloying of surface of Ti-5.5Al-2Zr-1Mo-1V titanium near α-alloy prepared via melted by pulsed laser radiation TiC particles. Lasers in Manufacturing and Materials Processing. 2019;**6**(1):26-40. DOI: 10.1007/s40516-018-0076-0

[15] Ansari M, Ramezani H, Yari S, Soltani R. Wear and corrosion resistance pulsed Nd:YAG laser surface alloying of AZ31 magnesium with nickel for improved wear and corrosion resistance. Journal of Laser Applications. 2016;**28**:012013. DOI: 10.2351/1.4937429

[16] Dadoo A, Mohammad S, Boutorabi A, Kheirandish S. Effect of titanium carbide concentration on the morphology of MC carbides in pulsed laser surface alloyed AISI H13 tool steel. Optics and Laser Technology. 2019;**112**:236-244. DOI: 10.1016/j.optlastec.2018.11.001

[17] Hamoudi WK, Ismail RA. Micro and nano laser pulses for melting and surface alloying of aluminum with copper. Lasers in Manufacturing and Materials Processing. 2017;**4**:24-35. DOI: 10.1007/s40516-017-0034-2

[18] Roy NK, Dibua OJ, Jou W, He F, Jeong J, Wang Y, et al. A comprehensive study of the sintering of copper nanoparticles using femtosecond, nanosecond, and continuous wave lasers. ASMEJ Micro Nano-Manufacturing. 2017;**6**:1-21

[19] Wang D, Hu Q, Zheng Y, Xie Y, Zeng X. Study on deposition rate and laser energy efficiency of laser-induction hybrid cladding. Optics and Laser Technology. 2016;**77**:16-22

[20] Sun RL, Yang DZ, Guo LX, Dong SL. Laser cladding of Ti-6Al-4V alloy with TiC and TiC_qNiCrBSi powders. Surface and Coatings Technology. 2001;**135**:307-312

[21] Chen E, Zhang K, Zou J. Laser cladding of a Mg based Mg-Gd-Y-Zr alloy with Al-Si powders. Applied Surface Science. 2016;**367**:11-18

[22] Liu YH, Guo ZX, Yang Y, Wang HY, Hu JD, Li YX, et al. Laser (a pulsed Nd:YAG) cladding of AZ91D magnesium alloy with Al and Al_2O_3 powders. Applied Surface Science. 2006;**253**:1722-1728

[23] Ma Q, Gao X, Li J. Microstructure performance and formation mechanism of laser alloying rare earth oxides modified nanocrystalline layer on TA7. Physica E: Low-dimensional Systems and Nanostructures. 2016;**77**:29-33

[24] Zhang H, Pan Y, He Y, Jiao H. Microstructure and properties of 6FeNiCoSiCrAlTi high-entropy alloy coating prepared by laser cladding. Applied Surface Science. 2011;**257**(6):2259-2263

[25] Zeng X, Tao Z, Zhu B, Zhou E, Cui K. Investigation of laser cladding ceramic-metal composite coatings: Processing modes and mechanisms. Surface and Coatings Technology. 1996;**79**(1-3):209-217

[26] Li Y, Zhang P, Bai P, Wu L, Liu B, Zhao Z. Microstructure and properties of Ti/TiBCN coating on 7075 aluminium alloy by laser cladding. Surface and Coating Technology. 2018;**334**:142-149

[27] Qin R, Zhang X, Guo S, Sun B, Tang S, Li W. Laser cladding of high Co-Ni secondary hardening steel on18Cr2Ni4WA steel. Surface and Coating Technology. 2016;**285**:242-248

[28] Yue TM, Xie H, Lin X, Yang HO, Meng GH. Solidification behaviour in laser cladding of AlCoCrCuFeNi high-entropy alloy on magnesium substrates. Journal of Alloys and Compounds. 2014;**587**:588-593

[29] Siddiqui AA, Dubey AK, Paul CP. A study of metallurgy and erosion in laser surface alloying of $Al_xCu_{0.5}$FeNiTi high entropy alloy. Surface and Coatings Technology. 2019;**361**:27-34. DOI: 10.1016/j.surfcoat.2019.01.042

[30] Siddiqui AA, Dubey AK, Paul CP. Geometrical characteristics

in laser surface alloying of a high-entropy alloy. Lasers in Engineering. 2019;**43**:237-259

[31] Nagulin KY, Iskhakov FR, Shpilev AI, Gilmutdinov AK. Optical diagnostics and optimization of the gas-powder flow in the nozzles for laser cladding. Optics and Laser Technology. 2018;**108**:310-320. DOI: 10.1016/j.optlastec.2018.07.001

[32] Karmakar DP, Gopinath M, Harmalkar S, Nath AK. Investigation of wear and corrosion characteristics of Stellite-6 and Stellite-21 layers deposited by co-axial laser cladding. Proceedings of the ASME MSEC. 2017;**2841**:1-6

[33] Schneider M. Laser cladding with powder [Ph.D. thesis]. Enschede, The Netherlands: University of Twente; 1998. ISBN: 90-365-1098-8

[34] Abioye TE, Farayibi PK, Clare AT. A comparative study of Inconel 625 laser cladding by wire and powder feedstock. Materials and Manufacturing Processes. 2017;**32**:1653-1659. DOI: 10.1080/10426914.2017.1317787

[35] Alam MK, Edrisy A, Urbanic J. Microstructural analysis of the laser-cladded AISI 420 martensitic stainless steel. Metallurgical and Materials Transactions A: Physical Metallurgy and Materials Science. 2019;**50**(5):2495-2506. DOI: 10.1007/s11661-019-05156-6

[36] Liu Z, Li T, Ning F, Cong W, Kim H, Jiang Q, et al. Effects of deposition variables on molten pool temperature during laser engineered net shaping of Inconel 718 superalloy. International Journal of Advanced Manufacturing Technology. 2019;**102**(1-4):969-976

[37] Cai Y, Luo Z, Chen Y, Ao S. Influence of CeO_2 on tribological behaviour of TiC/Fe-based composite coating. Surface Engineering. 2017;**34**:936-943. DOI: 10.1080/02670844.2017.1309742

[38] Yan H, Wang A, Xu K, Wang W, Huang Z. Microstructure and interfacial evaluation of Co-based alloy coating on copper by pulsed Nd:YAG multilayer laser cladding. Journal of Alloys and Compounds. 2010;**505**(2):645-653

[39] Muvvala G, Karmakar DP, Nath AK. Online monitoring of thermo-cycles and its correlation with microstructure in laser cladding of nickel based super alloy. Optics and Lasers in Engineering. 2017;**88**:139-152

[40] Fogagnolo JB, Rodrigues AV, Leva ES, Lima MSF, Caram R. Surface stiffness gradient in Ti parts obtained by laser surface alloying with Cu and Nb. Surface and Coatings Technology. 2016;**297**:34-42

[41] Olakanmi EO, Cochrane RF, Dalgarno KW. A review on selective laser sintering/melting (SLS/SLM) of aluminium alloy powders: Processing, microstructure, and properties. Progress in Materials Science. 2015;**74**:401-477. DOI: 10.1016/j.pmatsci.2015.03.002

[42] Kumar SRD, Duraiselvam M, Natarajan S, Panwar SS, Jana T, Khan SS. Enhanced ablation resistance through laser glazing of plasma sprayed $LaTi_2Al_9O_{19}$ based functionally graded thermal barrier coating. Ceramics International. 2016;**42**(8):10184-10190. DOI: 10.1016/j.ceramint.2016.03.134

[43] Salim MS, Ibrahim TK, Fouad S. Novel improvement glazing depth measurement model for concrete surface treated by CO_2 laser. Optik. 2017;**142**:412-420. DOI: 10.1016/j.ijleo.2017.06.019

[44] Zhang Y, Wang Y, Ophelia M, Zhong X, Li Q, Cao X. Laser glazing of lanthanum magnesium hexaaluminate. Optics and Lasers in Engineering. 2008;**46**:601-603. DOI: 10.1016/j.optlaseng.2008.04.001

[45] Tamanna N, Crouch R, Naher S. Progress in numerical simulation of the

laser cladding process. Optics and Lasers in Engineering. 2019;**122**(May):151-163. DOI: 10.1016/j.optlaseng.2019.05.026

[46] Zhu G, Li D, Zhang A, Tang Y. Numerical simulation of metallic powder flow in a coaxial nozzle in laser direct metal deposition. Optics and Laser Technology. 2011;**43**:106-113. DOI: 10.1016/j.optlastec.2010.05.012

[47] Tabernero I, Lamikiz A, Martínez S, Ukar E, López De Lacalle LN. Modelling of energy attenuation due to powder flow-laser beam interaction during laser cladding process. Journal of Materials Processing Technology. 2012;**212**:516-522. DOI: 10.1016/j.jmatprotec.2011.10.019

[48] Devesse W, De Baere D, Guillaume P, Brussel VU. Modeling of laser beam and powder flow interaction in laser cladding using ray-tracing. Journal of Laser Applications. 2015;**27**:0-8. DOI: 10.2351/1.4906394

[49] Liu J, Li L, Zhang Y, Xie X. Attenuation of laser power of a focused Gaussian beam during interaction between a laser and powder in coaxial laser cladding. Journal of Physics D: Applied Physics. 2005;**38**:1546-1550. DOI: 10.1088/0022-3727/38/10/008

[50] Kovaleva IO, Kovalev OB. Simulation of the acceleration mechanism by light propulsion for the powder particles at laser direct material deposition. Optics and Laser Technology. 2012;**44**:714-725. DOI: 10.1016/j.optlastec.2011.09.016

[51] Lin J. Laser attenuation of the focused powder streams in coaxial laser cladding. Journal of Laser Applications. 2000;**12**:28. DOI: 10.2351/1.521910

[52] Mirzade FK, Niziev VG, Panchenko VY, Khomenko MD, Grishaev RV, Pityana S. Kinetic approach in numerical modeling of melting and crystallization at laser cladding with powder injection. Physica B: Condensed Matter. 2013;**423**:69-76. DOI: 10.1016/j.physb.2013.04.053

[53] Kumar A, Roy S. Effect of three-dimensional melt pool convection on process characteristics during laser cladding. Computational Materials Science. 2009;**46**:495-506. DOI: 10.1016/j.commatsci.2009.04.002

[54] Hofman JT, De Lange DF, Pathiraj B, Meijer J. FEM modeling and experimental verification for dilution control in laser cladding. Journal of Materials Processing Technology. 2011;**211**:187-196. DOI: 10.1016/j.jmatprotec.2010.09.007

[55] El Cheikh H, Courant B, Branchu S, Hascoet JY, Guillen R. Analysis and prediction of single laser tracks geometrical characteristics in coaxial laser cladding process. Optics and Lasers in Engineering. 2012;**50**:413-422. DOI: 10.1016/j.optlaseng.2011.10.014

[56] Fan Y, Cheng P, Yao YL, Yang Z, Egland K, Fan Y. Effect of phase transformations on laser forming of Ti6Al4V alloy. Journal of Applied Physics. 2005;**98**:013518. DOI: 10.1063/1.1944202

[57] Suarez A, Tobar MJ, Yanez A, Perez I, Sampedro J, Amigo V. Modeling of phase transformations of Ti6Al4V during laser metal deposition. Physics Procedia. 2011;**12**:666-673

[58] Farahmand P, Balu P, Kong F, Kovacevic R. Investigation of thermal cycle and hardness distribution in the laser cladding of AISI H13 tool steel produced by a high power direct diode laser. In: Proceedings of the ASME 2013 International Mechanical Engineering Congress and Exposition. 2015. pp. 1-12

[59] Santhanakrishnan S, Kong F, Kovacevic R. An experimentally based

thermo-kinetic hardening model for high power direct diode laser cladding. Journal of Materials Processing Technology. 2011;**211**:1247-1259. DOI: 10.1016/j.jmatprotec. 2011.02.006

[60] Bailey NS, Katinas C, Shin YC. Laser direct deposition of AISI H13 tool steel powder with numerical modeling of solid phase transformation, hardness, and residual stresses. Journal of Materials Processing Technology. 2017;**247**:223-233. DOI: 10.1016/j.jmatprotec.2017.04.020

[61] Suarez A, Amado JM, Tobar MJ, Yanez A, Fraga E, Peel MJ. Study of residual stresses generated inside laser cladded plates using FEM and diffraction of synchrotron radiation. Surface and Coating Technology. 2010;**204**:1983-1988. DOI: 10.1016/ j.surfcoat.2009.11.037

[62] Liu Z, Jiang Q, Li T, Dong S, Yan S, Zhang H, et al. Environmental benefits of remanufacturing: A case study of cylinder heads remanufactured through laser cladding. Journal of Cleaner Production. 2016;**133**:1027-1033

[63] Orlovskii VP, Komlev VS, Barinov VS. Hydroxyapatite and hydroxyapatite based ceramics. Inorganic Materials. 2002;**38**:973-984

[64] Cheng B, Kim YJ, Chou P. Improving accident tolerance of nuclear fuel with coated Mo-alloy cladding. Nuclear Engineering and Technology. 2016;**48**:16-25

[65] Kim WJ, Kim D, Park JY. Fabrication and material issues for the application of SiC composites to LWR fuel cladding. Nuclear Engineering and Technology. 2013;**45**:565-572

[66] Yan H, Zhang P, Gao Q, Qin Y, Li R. Laser cladding Ni-based alloy/nano-Ni encapsulated h-BN self-lubricating composite coatings. Surface and Coatings Technology. 2017;**332**(March):422-427

[67] Riquelme A, Rodrigo P, Rodriguez MDE, Rams J. Analysis and optimization of process parameters in Al-SiC_p laser cladding. Optics and Lasers in Engineering. 2016;**78**:165-173

[68] Nazemi N, Urbanic J. An experimental and simulation study for powder injection multitrack laser cladding of P420 stainless steel on AISI 1018 steel for selected mechanical properties. Journal of Manufacturing Science & Engineering—ASME. 2017;**140**(1):1-12

[69] Saqib SM, Urbanic RJ. Investigation of the transient characteristics for laser cladding beads using 420 stainless steel powder. Journal of Manufacturing Science & Engineering—ASME. 2017;**139**(8):1-12

[70] Ghaith ES, Hodgson S, Sharp M. Laser surface alloying of 316L stainless steel coated with a bioactive hydroxyapatite titanium oxide composite. Journal of Materials Science. Materials in Medicine. 2015;**26**:1-8. DOI: 10.1007/s10856-015-5399-1

[71] Tian YS, Chen CZ, Li ST, Huo QH. Research progress on laser surface modification of titanium alloys. Applied Surface Science. 2005;**242**(1-2):177-184

[72] Yao JH, Zhang QL, Kong FZ. Laser Remanufacturing to Improve the Erosion and Corrosion Resistance of Metal Components. Series in Metals and Surface Engineering. Woodhead Publishing Limited. 2012. pp. 320-354. DOI: 10.1533/9780857095831.2.320

[73] Rottwinkel B, Nölke C, Kaierle S, Wesling V. Laser cladding for crack repair of CMSX-4 single-crystalline turbine parts. Lasers in Manufacturing and Materials Processing. 2017;**4**:13-23. DOI: 10.1007/s40516-016-0033-8

[74] Bartkowski D, Bartkowska A. Wear resistance in the soil of Stellite-6/WC

coatings produced using laser cladding method. International Journal of Refractory Metals and Hard Materials. 2017;**64**:20-26. DOI: 10.1016/j.ijrmhm.2016.12.013

[75] Zarini S, Vedani M, Previtali B, Rovatti L. Cracks susceptibility elimination in fiber laser cladding of Ni based alloy with addition of Tungsten Carbides. Conference on Engineering Systems Design and Analysis. 2014;**20623**:1-7

[76] Paul S, Thool K, Singh R, Samajdar I, Yan W. Experimental characterization of clad microstructure and its correlation with residual stresses. Procedia Manufacturing. 2017;**10**:804-818. DOI: 10.1016/j.promfg.2017.07.081

[77] Zhu Y, Yang Y, Mu X, Wang W, Yao Z, Yang H. Study on wear and RCF performance of repaired damage railway wheels: Assessing laser cladding to repair local defects on wheels. Wear. 2019;**431**:126-136. DOI: 10.1016/j.wear.2019.04.028

[78] Chakraborty R, Raza MS, Datta S, Saha P. Synthesis and characterization of nickel free titanium-hydroxyapatite composite coating over Nitinol surface through in-situ laser cladding and alloying. Surface and Coatings Technology. 2019;**358**:539-550. DOI: 10.1016/j.surfcoat.2018.11.036

[79] Rakesh KR, Bontha S, Ramesh MR, Das M, Krishna V. Laser surface melting of Mg-Zn-Dy alloy for better wettability and corrosion resistance for biodegradable implant applications. Applied Surface Science. 2019;**480**:70-82. DOI: 10.1016/j.apsusc.2019.02.167

[80] Lu Y, Yang C, Liu Y, Yang K, Lin J. Characterization of lattice defects and tensile deformation of biomedical Co29Cr9W3Cu alloy produced by selective laser melting. Additive Manufacturing. 2019;**30**:100908. DOI: 10.1016/j.addma.2019.100908

[81] Nyoni E, Akinlabi ET. Process parameter interaction effect on the evolving properties of laser metal deposited titanium for biomedical applications. Thin Solid Films. 2016;**620**:94-102. DOI: 10.1016/j.tsf.2016.09.060

Chapter 14

Building and Architecture Paints and Coatings

Valentina Loganina and Yerkebulan Mazhitov

Abstract

Information is given on the strength of the coatings of cement concrete for the exterior walls of buildings. It was found that the strength of the coating depends on the quality of its appearance. A strength model is proposed depending on the surface roughness of the coating. The influence of the scale factor on the change in the strength of coatings is established. To assess the long-term strength of the coatings, we studied the temperature-time dependence of strength. The values of the activation energy of the destruction process of some coatings are experimentally determined. The dependence of the long-term strength of the coatings on tensions is given. The kinetics of changes in the short-term strength of coatings during aging is considered from the perspective of the kinetic concept of the strength of solids. The condition for coating cracking is obtained. Taking into account the influence of the scale factor and the conditions of brittle fracture of coatings, a method for choosing the optimal coating thickness is proposed.

Keywords: coatings, structure, properties, coating strength, coating appearance quality, mathematical model of strength

1. Introduction

Construction and maintenance of buildings and structures require a large number of paints. The share of building paints and varnishes, including repair materials, accounts for up to 5–55% of the total volume of manufactured paints and varnishes, of which 46–48% are paints and 5–8% are varnishes. The main market share of coatings is architectural and decorative coatings.

Currently, for the decoration of building facades, compositions based on polymer binders are widely used: water-dispersion, perchlorovinyl, organosilicon, polymer-cement, silicate paint, sol silicate, and polymer silicate paint [1–3]. The proportion of organosoluble products in the total consumer market of paints and varnishes has now stabilized at about 15%, and the share of water-dispersion varnishes and paints is more than 60%.

The problem of reliability and durability of protective and decorative coatings of the exterior walls of buildings is one of the urgent scientific and technical problems in the field of materials science [4, 5]. It is known that the durability of coatings depends on the type of binder, the technology of applying the paint composition, operating conditions, etc. [6–8].

Crack resistance is the main characteristic that characterizes the durability of finishing coatings. The main reasons for the occurrence of cracks are significant shrinkage deformations that occur during the hardening of finishing coatings as well as during operation. The property values of the coats on a cement backing are known to be variable and depend on a number of factors (the roughness and porosity of the backing, technological factors, etc.) [9, 10]. However, at present, many issues of strength and durability of coatings of cement concrete are not considered.

2. Strength of protective and decorative coatings

One of the most common types of destruction of coatings is cracking and peeling [11]. Coating cracking occurs when internal tensile stresses reach the cohesive strength of the coating material, i.e.,

$$\sigma = \text{R kog} \tag{1}$$

Thus, to assess the resistance of coatings to cracking, it is necessary to identify patterns of change in the strength of coatings.

The presence of defects on the surface of coatings will undoubtedly affect the physicomechanical properties of coatings.

The probability of destruction of protective and decorative coatings depending on the presence of defects on their surface can be defined by the formula:

$$P = 1 - e^{-\rho S} \tag{2}$$

where p is the defect concentration and S is the surface area.

As can be seen from formula (2), for the same surface area S, the probability of coating failure increases with increasing defect concentration [8].

In the study the following paint is used: alkyd enamel PF-115 grade, polystyrene paint brands PS-160, silicone enamel KO-168, polyvinyl acetate cement (PVAC) paint, silicate paint, polymer-lime paint, lime paint, and perchlorovinyl XB-161.

The surface roughness of the paintwork was evaluated which is determined by profilograph TR-100 [12].

Assessment deformation of coating was carried out with the help of a tensile machine IR 5057–50 with the samples after 28 days of curing. The method is based on the sample stretching until it ruptures (deformation speed of 1 mm/minute).

The 1 × 1 × 5 cm samples were fixed in the clips of the tensile machine so that their longitudinal axis was in the direction of stretching and the force was applied equally all over the sample section. The tests were carried out at the temperature of 20°C and relative air humidity of 60%. The ultimate tensile strength estimation was carried out for not less than four samples of each compound. The ultimate tensile strength R_{kog} for each sample is derived from the formula:

$$R_{\text{kog}} = \frac{F_{Pi}}{S_{Oi}} \tag{3}$$

where F_{Pi} is the stretching loading at the time of a rupture, N, and S_{Oi} is the initial cross-sectional area of a sample, mm^2.

The modulus of elasticity was calculated according to the chart "tension deformation" on an inclination tangent of angle to abscissa axis of the tangent, which was drawn to an initial straight section of the chart.

The modulus of elasticity for each sample (E_{upr}) in MPa is derived from the formula:

$$E_{upr} = \frac{R'_{kogi}}{\varepsilon'_i} \cdot 100 \tag{4}$$

where R'_{kogi} is the ultimate tensile strength at the time of the tangent separation from the chart "tension deformation," MPa, and ε'_i is the relative lengthening at the time of rupture, %.

The presence of defects on the surface of the coatings will undoubtedly affect the physicomechanical properties of paint coatings. It is revealed that the elastoplastic character of the destruction is characteristic for the coatings studied.

Regardless of the type of paint composition, the strength and relative deformations are reduced, the plastic deformation is increased, and the elastic surfaces are reduced with increasing roughness (**Figures 1–4**) [13]. Thus, when the surface of coating roughness based on paint PS-160 is Ra = 0.74 μm, the tensile strength R_p is 69.4 kgf/cm^2, the relative strain ε = 3.1%, and with a roughness Ra = 0.86 μm - R_p =

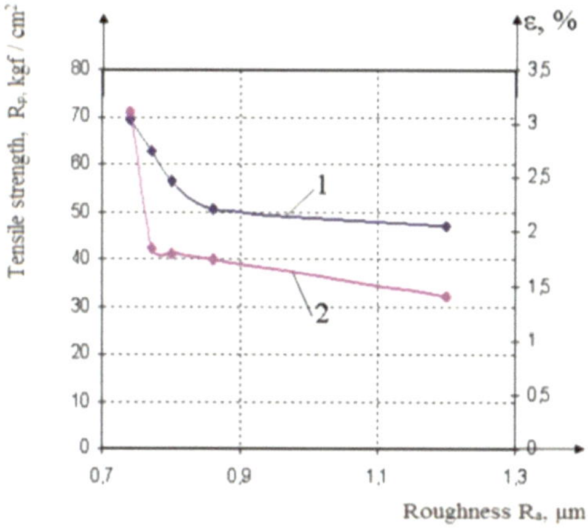

Figure 1.
Dependence of the tensile strength (1) and the relative elongation (2) on the roughness of the film surface based on paint PS-160.

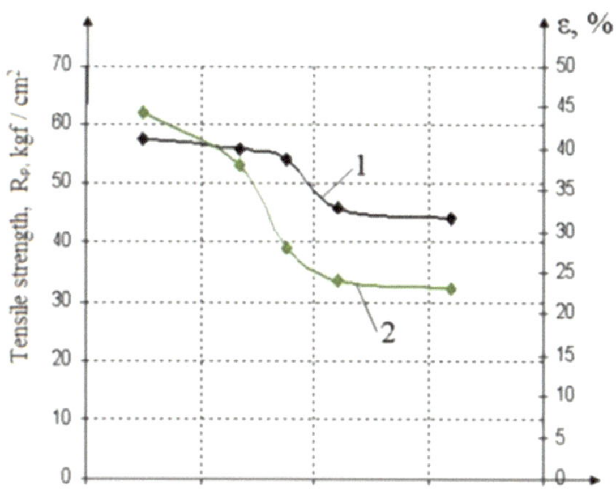

Figure 2.
Dependence of the tensile strength (1) and the relative elongation (2) on the roughness of the film surface on based of paint PF-115.

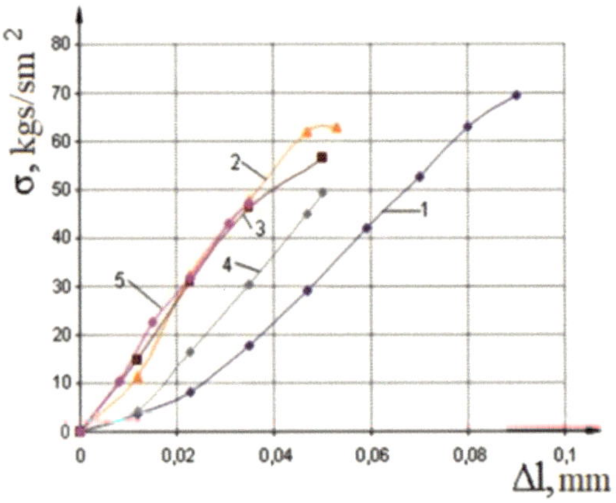

Figure 3.
Tensile diagrams of films based on PS paint, 160: (1) roughness of 0.74 µm; (2) roughness of 0.77 microns; (3) roughness of 0.8 microns; (4) roughness of 0.86 microns; (5) roughness 1.2 µm.

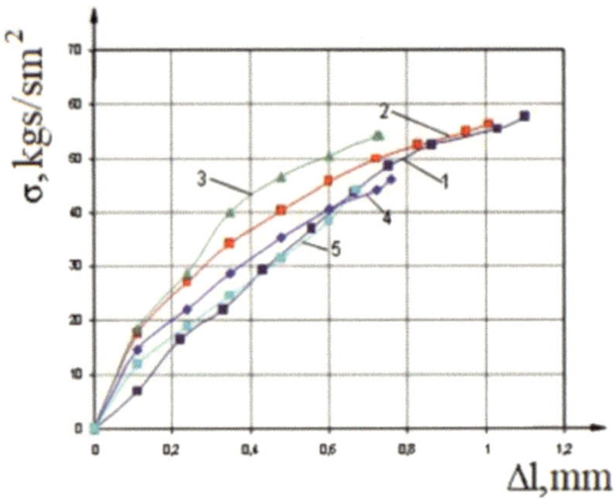

Figure 4.
Tension diagrams of films based on PF paint, 115: (1) roughness 1.2 µm; (2) roughness of 1.37 µm; (3) roughness of 1.45 µm; (4) roughness of 1.54 microns; (5) roughness of 1.74 µm.

50.5 kgf/cm^2 and 1.75%, respectively. At a roughness of the film based on the paint PF-115 Ra = 0.74 µm, the tensile strength R_p is 57.7 kgf/cm^2, and the relative strain ε = 44.3%, at a roughness Ra = 1.74 µm - R_p = 44.1 kgf/cm^2 and 23%, respectively.

For films on the basis of the investigated paint, the flowing character of the curve "tensile strength-roughness" with a sharp decline in strength to a certain value is observed, constituting 45–55 kg/cm^2 at the roughness of the film, respectively, on the basis of the paint PF-115 (1.4–1.6 micrometers) and on the basis of the paint PS-160 (0.8–1 microns). With further increase in the roughness of the film surface, at least a significant reduction in tensile strength is observed.

Analysis of the data shown in **Figures 1** and **2** shows that the dependence of the tensile strength on the roughness of the surface of the films can be approximated by an expression of the form

$$R_p = a \cdot e^{b \cdot R_a} \quad (5)$$

where R_a is the surface roughness, µm; b is the coefficient that takes into account the degree of reduction in strength from roughness, µm^{-1}; and a is the coefficient that characterizes the value of tensile strength, at R_a = 0 (ideal model).

For films based on PS-160 paint, model (5) has the form

$$R_p = 110.5 \cdot e^{-0.761 \cdot R_a} \tag{6}$$

For films based on PF-115 paint,

$$R_p = 114.5 \cdot e^{-0.548 \cdot R_a} \tag{7}$$

The presented models make it possible to evaluate the expected tensile strength depending on the surface roughness of the coatings.

The influence of the scale factor on the tensile stress of the films is revealed. **Figures 5** and **6** show the results of a study of the effect of film thickness on a change in tensile strength of PVAC and silicate of the coating.

An analysis of the data shows in **Figure 5** that an increase in the thickness of liquid glass-based films from 0.16 to 0.52 mm leads to a decrease in tensile strength from 9.44 to 5.56 MPa, respectively. Similar patterns are also characteristic of films based on a mixture of water glass and styrene acrylic dispersion.

An analysis of the data obtained (**Figures 5** and **6**) shows that the dependence of the tensile strength on the film thickness can be approximated by the expression

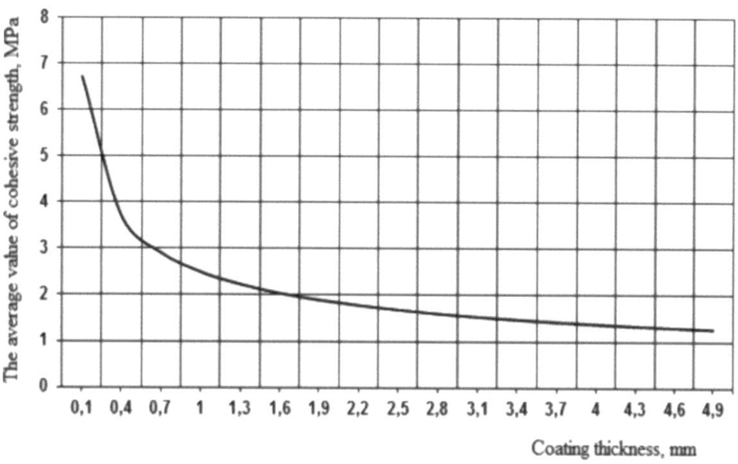

Figure 5.
Dependence of cohesive strength of PVAC coatings on their thickness.

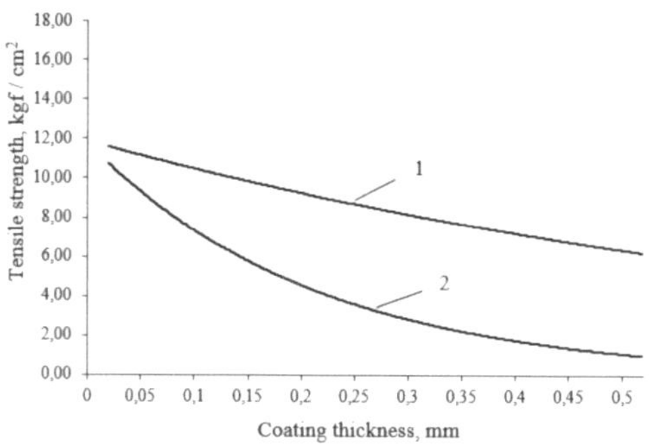

Figure 6.
Dependence of cohesive strength on film thickness: (1) based on silicate paint; (2) based on polymer silicate paint.

$$R_p = Ah^b \tag{8}$$

where h is the film thickness.

Mathematical data processing (curves 1 and 2, **Figure 6**) showed that they are well described by the expression:

For silicate-based films

$$R_p = 5.030h^{-0.37} \tag{9}$$

For films based on polymer silicate paint

$$R_p = 0.886h^{-1.3} \tag{10}$$

The correctness of the equations was checked by Fisher's criterion.

3. Development of sol silicate paint

Sol silicate paint were used a filler microcalcite MK-2 (TU 09 5743-001-91892010-2011) and talc MT-GShM (GOST 19284-79) and a pigment titanium dioxide 230 rutile form (TU 2321-001-1754-7702-2014), ocher (GOST 8019-71), iron red oxide (GOST 8135-74), ultramarine UM-1 (OST 6-10 - 404 - 77), and chromium oxide OHP-1 (GOST 2912-79). To determine the content of the pigment (filler), the viscosity was measured using a viscometer VZ-4. To obtain different shades, titanium dioxide is mixed with an appropriate pigment [14].

Figure 7 shows the dependence of viscosity of paint on the content of pigment and filler. As can be seen from the obtained data, when filling in the range of about $0 < \varphi < 0.12$, the viscosity increase is insignificant, and the polymer matrix only partially passes into the film state. With a low concentration of pigment (filler), the boundary layers of distant particles do not constitute an independent phase in the bulk of the material that can influence its properties. With further filling ($\varphi > 0.12$), there is a significant change in the ratio of bulk and film phases of the matrix, and a sharp increase in the viscosity of the composition is observed.

In **Figure 8** the dependence of viscosity on the volume fraction of pigment in the coordinates lg□ - C (where C is the concentration of pigment and filler in the system) is shown. This dependence consists of two intersecting straight lines.

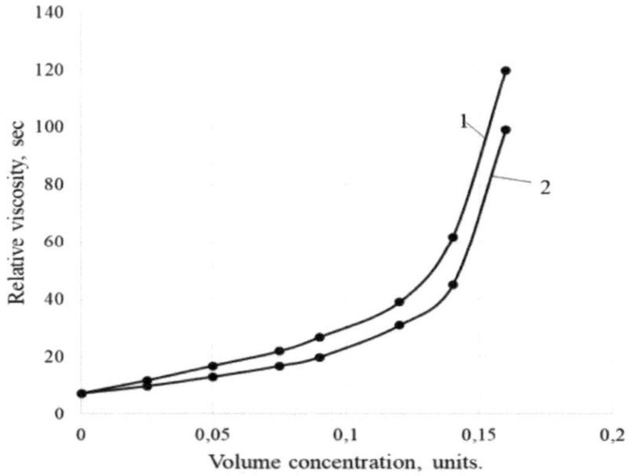

Figure 7.
The dependence of the viscosity of the sol of silicate paint on the content of pigment and filler: (1) sol silicate paint; (2) sol silicate paint with glycerin.

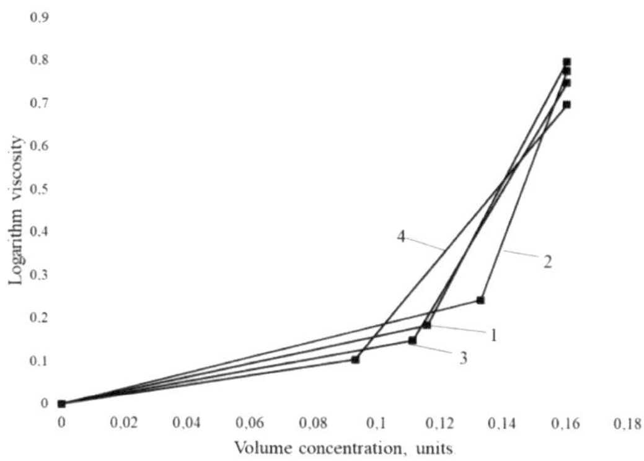

Figure 8.
The dependence of viscosity on the volume fraction of pigment in the coordinates lgη- C*: (1) red iron oxide; (2) ocher; (3) ultramarine; (4) chromium oxide.*

The point of intersection projected on the x-axis will be the critical volume concentration of the pigment (CVCP) [4, 15].

As the results in **Figure** 7 show, the viscosity of the paint increases during filling, while its change at low degrees of filling can be described by the Einstein equation:

$$\varphi = \varphi_O(1 + 2.5\varphi) \tag{11}$$

where η_0 is the viscosity of the unfilled system and φ is the volume fraction of the pigment (filler).

With increasing concentration of the dispersed phase (more than 0.08), the interaction between particles increases, and deviations from the Einstein equation have been detected. These deviations are apparently due to the interaction of the particles and the formation of a structure in which the particles of the dispersed phase are oriented relative to each other in a certain way (structuring systems). The results of the calculations show that the model changes the viscosity of the paint from the volume concentration of the pigment (filler) which can be described by a polynomial

$$\eta = \eta_o\left(a + b\varphi + c\varphi^2 + d\varphi^3\right) \tag{12}$$

where η_0 is the viscosity of the unfilled system and φ is the volume fraction of the pigment (filler).

The following equations were obtained:

$$\eta = \eta_o\left(0.334 + 119\varphi - 2110.47\varphi^2 + 12217.83\varphi^3\right) \text{–for sol silicate paint} \tag{13}$$

Testing the adequacy of the model showed that Eq. (2) is valid in the filling region up to φ = 0.16.

In view of the foregoing, the cohesive properties and the ability of the polysilicate binder to wet the surface of the pigment (filler) were also investigated. In the work, polysilicate solutions were obtained by the interaction of stabilized solutions of colloidal silica (sols) with aqueous solutions of alkaline silicates (liquid glasses). A sodium liquid glass with a modulus M = 2.78 was used, and a potassium liquid glass with a modulus M = 3.29.

We was calculated the work of adhesion of liquid glass and polysilicate solution to the pigment (filler). The contact angle of wetting was determined on the KRUSS DSA-30 [13].

To determine the wetting contact angle, tablets are mixed from the mixture of pigment and filler with the automatic hydraulic press Vaneox-40 t, with a pressure of 18 tons for 11 seconds. The powder compressed in a dry state, without further processing. The surface tension of the solutions is determined by the stalagmometric method. The stalagmometric method based on measuring the number of droplets formed when a liquid flows out of a vertical tube of a small radius.

Analysis of the data (**Table 1**) shows that for the potassium polysilicate solution, a large work of adhesion to the filler (pigment) is characteristic. Thus, the work of adhesion of the potassium polysilicate solution to the filler (pigment) is 103.85 mN/m, while the work of adhesion of potassium liquid glass is 87.74 mN/m. Similar regularities are observed when using sodium liquid glass and sodium polysilicate solution. A potassium polysilicate solution is also characterized by a large wetting work of 39.786 mN/m.

In determining the wetting contact angle, it was found that sodium glass droplets on the surface of the sample formed an angle much larger than that of the potassium and for 5 minutes remained unchanged on the surface, while the drops from the potassium for half a second remained in shape and then blurred. Drops based on the sodium polysilicate solution were quickly absorbed into the material, forming a pyramidal shape. Drops based on the potassium polysilicate solution are more stable and retained on the sample for up to 2 minutes.

The availability of more complete wetting of the surface of the filler and the pigment in the case of the use of a potassium polysilicate solution promotes the formation of a denser coating structure and an increase in the physicomechanical properties. This is evidenced by data on the change in the tensile strength of films based on colorful compositions.

It was found that the cohesive strength of membranes based on sol silicate paint is 2.65 MPa and based on silicate paint 1.8 MPa (**Figure 9**). An increase in the relative deformations is observed, which is 0.06 mm/mm for membranes based on sol silicate paint and 0.033 mm/mm based on silicate paint.

In continuation of further research, frost resistance tests were carried out by alternate freezing and thawing painted mortar samples. The samples were painted with silicate and sol with silicate paint with intermediate drying for 20 minutes. After the coatings were cured, frost resistance tests were carried out. Evaluation of the appearance of the coatings was carried out according to GOST 6992-68 "Paint coatings. Test method for resistance to atmospheric conditions". The status of coating, assessed as AD3 and AZ4, was taken as a "failure" [7]. The energy of interaction between particles of the coating was estimated by value of the Hamaker constant [16]. It was established that the condition of the coating based on silicate

Name of film-forming	Surface tension, mN/m	°Angle of wetting, °	Adhesion work, mJ/m^2	Wetting operation, mN/m
Water	72.8	46.2	123.18	50.38
Binder				
Potassium liquid glass	55.22	53.9	87.74	32.52
Potassium polysilicate solution (15% Nanosil 20)	64.064	51.6	103.85	39.786
Sodium liquid glass	51.66	74.7	65.3	13.64
Sodium polysilicate solution (15% Nanosil 20)	55.22	62.5	80.73	25.51

Table 1.
The work of adhesion of a polysilicate binder to a filler.

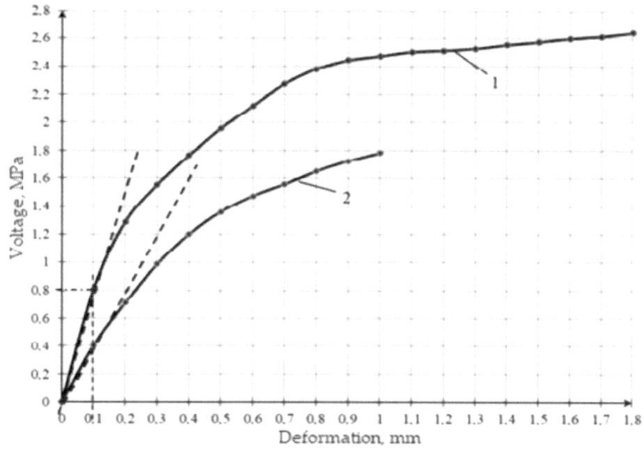

Figure 9.
Change in the relative deformations when tensile samples based on: (1) sol silicate paint; (2) silicate paint.

paint after 35 cycles is estimated as AD1 and AZ1, which corresponds to the state of coating with no discoloration, chalking, and dirt retention. Silicate-based coatings are more susceptible to degradation. The condition of coating based on silicate paint is estimated as AD3 and AZ3. The test results showed that the "failure" of coating based on silicate paint occurred after 40 freeze-thaw cycles, while the state of the coating based on polysilicate solution was evaluated as AD2 and AZ2. The "failure" of the coating based on the polysilicate solution occurred after 50 test cycles. Adhesion strength after 50 test cycles in accordance with GOST 31149 "Paint and varnish materials. Determination of adhesion by the lattice cut method" for silicate coatings and for coatings based on polysilicate solution was by 1 point.

The surface energy of the coatings was calculated using the critical surface tension of the liquid at the boundary with a solid (method of Zisman). The dispersion contribution to the intermolecular interaction between the particles of the coatings was estimated, for which the value of the complex Hamaker constant A* was additionally determined, which takes into account the complex action of the two components—the interparticle interaction between homogeneous particles and the interfacial interaction at the solid-solution interface.

The measurement data of the contact angle showed that for all samples studied, a linear dependence $\cos\theta = f(\sigma_{lig})$ is observed (**Figure 9**). By extrapolating the dependence $\cos\theta = f(\sigma_{lig})$ by $\cos\theta = 1$, we obtained value of the critical surface tension of a solid surface (coating). The energy of interaction between particles of the coating was estimated by value of the Hamaker constant calculated by the equation

$$\cos\theta - 1 = \frac{A^*}{12 h_{\min} \sigma_{ж}} \quad (14)$$

where h_{min} is the smallest membrane thickness, which corresponds to the van der Waals distance (0.24 nm); $\sigma_{ж}$ is the surface tension of the liquid; and A^* is the complex constant of Hamaker in interaction of a liquid with a solid at the boundary with air.

To calculate the complex Hamaker constant, functional dependences $\cos\theta - 1 = f(1/\sigma_{lig})$ were built. **Figures 10–12** present results for coatings based on silicate and sol silicate paints [17].

Table 2 presents the calculated values of the surface tension of coatings and the Hamaker constant A^*.

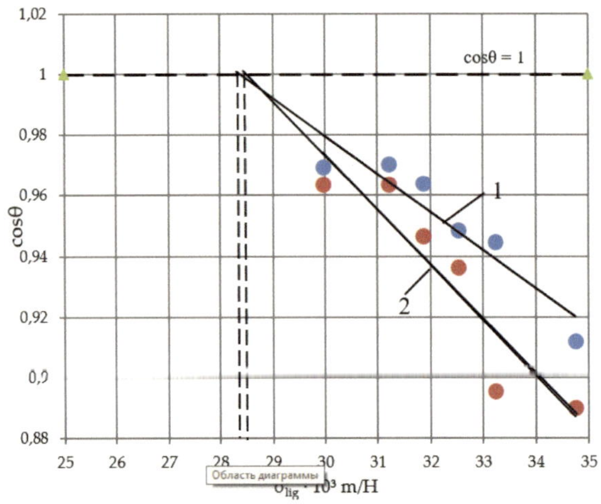

Figure 10.
The dependence $cos\theta = f(\sigma_{lig})$: (1) on coating based on silicate paint; (2) on coating based on sol silicate.

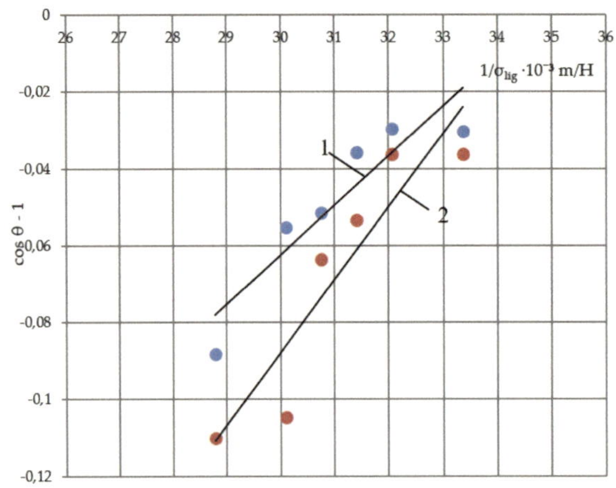

Figure 11.
Functional type dependency $cos\theta\text{-}1 = f(1/\sigma_{lig})$ before testing coatings based on: (1) silicate paint; (2) sol of silicate paint.

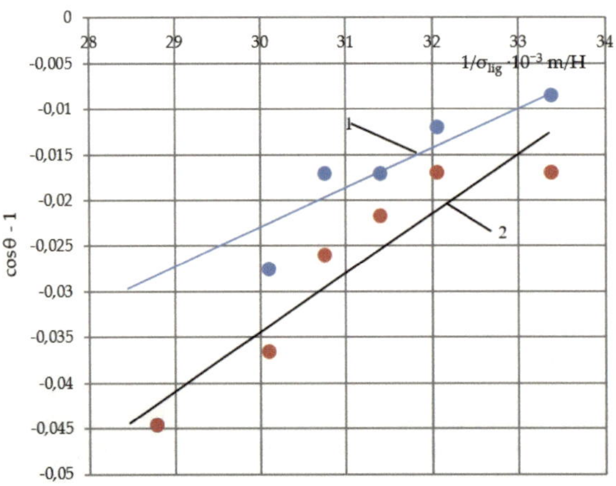

Figure 12.
Functional dependence of the form $cos\theta\text{-}1 = f(1/\sigma_{lig})$ after 50 test cycles of coatings based on: (1) silicate paint; (2) sol of silicate paint.

Type of paint coating	Value critical surface tension of the coating [mN/m]	Constant value Hamaker $A^{*}10^{-20}$ [Дж*]
Based on potassium liquid glass	28.4	2.85 1048
Based on potassium polysilicate solution	28.7	4.09 1524

Above the line is the value of the Hamaker constant for coatings prior to frost resistance test, below the line—after the test.

Table 2.
Surface tension values of silicate coatings.

Name indicators	Values
Portability	Good
Class of quality of appearance of coatings	IV
Viscosity to B3-4 [s]	17–20
Shrinkage, the presence of cracks	No
Viability [day]	More 90
Drying time [minute], to degree 5	15–25
Adhesion [points]	1
Adhesion [MPa]	1.1–1.3
Coefficient of vapor permeability [mg/m × hPa]	0.00878
Relative hardness	0.47
Impact strength [kgcm]	50
Washability [g/m^2]	No more 2
Water resistance (appearance after 24 hours in water)	Absence of white matte spots, flaking, rashes, bubbles
Frost resistance, brand	F35

Table 3.
Properties of the paint composition and coatings based on it.

Analysis of experimental data shown in **Figures 10–12** indicates that the critical surface tension of the coatings is almost the same, which is apparently explained by the almost identical nature components of coating. The value of the Hamaker constant for coatings based on polysilicate solution, amounting to 4.09×10^{-20} J, is higher than the coatings based on silicate paints. This is confirmed by data on the higher strength of coatings based on polysilicate solutions. A higher value of the Hamaker constant for coatings based on polysilicate solution after frost resistance testing, equal to 1524×10^{-20} J, indicates a greater preservation of interparticle interaction in the coating.

Table 3 shows the values of the properties of coatings based on sol silicate paint.

4. Long-lasting durability of paint coatings

The temperature-time dependence of the strength of paint and varnish materials can be described by the Zhurkov Equation [18, 19]:

$$\tau = \tau_o \exp\left[(U_o - \gamma\sigma)/RT\right] \qquad (15)$$

where γ is the structural-sensitive factor characterizing overstrain of bonds in the structure of the material; U_o is the activation energy of the process of destruction; R is the universal gas constant; and T is the absolute temperature.

The values of the activation energy of the fracture process and the structure-sensitive factor for polyvinyl acetate cement PVAC, organosilicon KO-168, and silicate coatings were calculated.

Figures 13–17 in semilogarithmic coordinates show the experimental dependence of the long-term cohesive strength of coatings from the value of tensions and temperatures for the coatings under study. The values of the structure-sensitive factor and the activation energy of the process of destruction of coatings are given in **Table 4.**

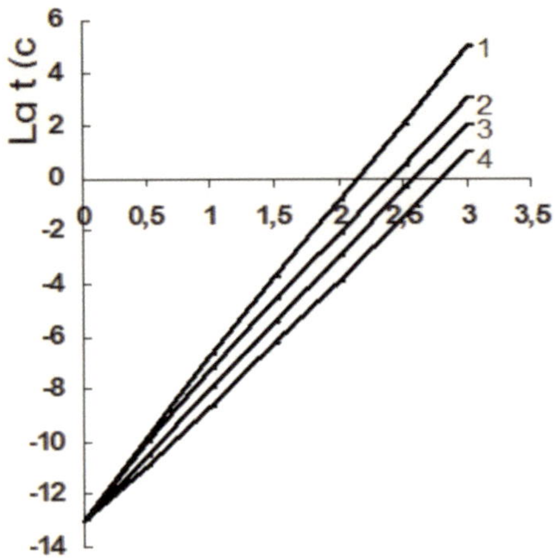

Figure 13.
The temperature dependence of lgτ *(σ): (1) coating of PVAC, σ = 0.166 MPa; (2) coating of PVAC, σ = 0.124 MPa; (3) coating of KO-168, σ = 0.144 MPa; (4) coating of KO-168, σ = 0.17 MPa.*

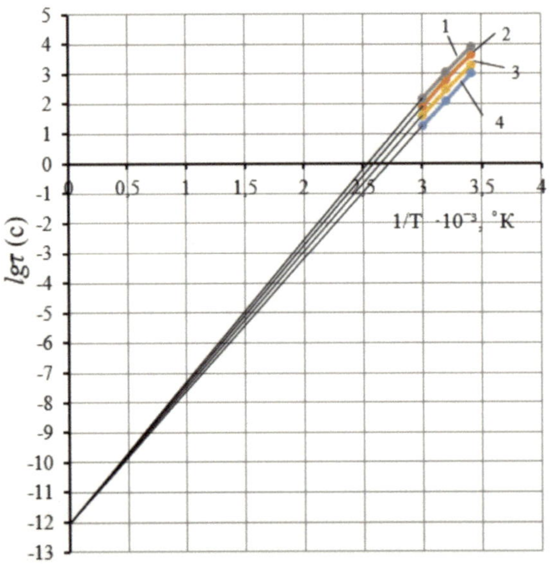

Figure 14.
The temperature dependence of coating based on silicate paint lgτ(σ): (1) stress σ = 0.14 MPa; (2) stress σ = 0.28 MPa; (3) stress σ = 0.42 MPa; (4) stress σ = 0.56 MPa.

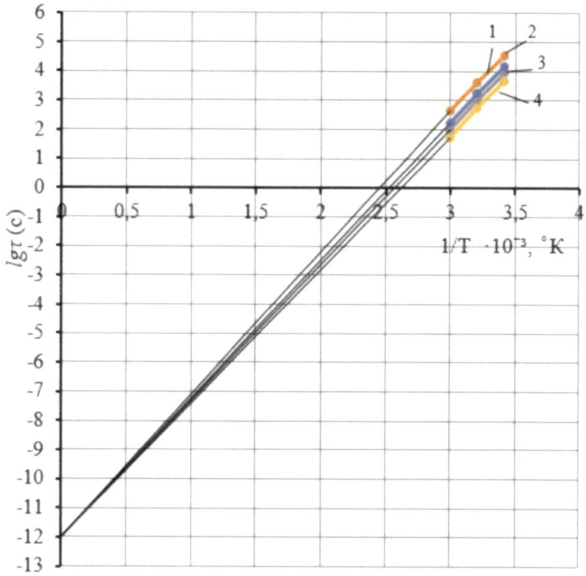

Figure 15.
The temperature dependence of coating based on sol silicate paint lgτ (σ): (1) stress σ = 0.14 MPa; (2) stress σ = 0.28 MPa; (3) stress σ = 0.42 MPa; (4) stress σ = 0.56 MPa.

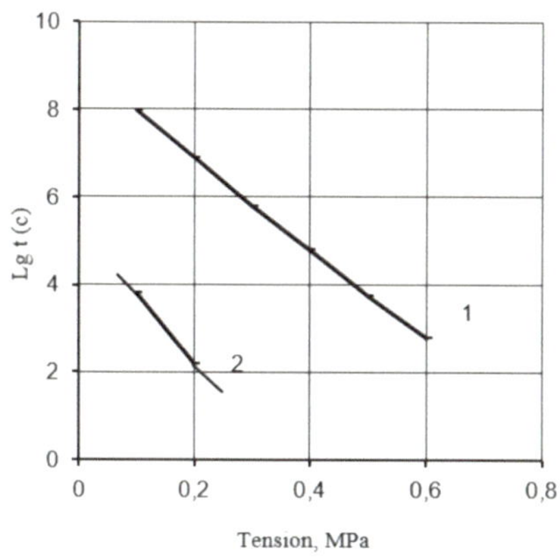

Figure 16.
Dependence of the long-term strength of coatings from tensions: (1) coating of PVAC; (2) coating of KO-168.

Analysis of the data in **Table 4** shows that the activation energy of cohesive destruction of coatings decreases with increasing stresses acting on the coatings. A higher value of the activation energy and a lower value of the structurally sensitive factor indicate high strength of polyvinyl acetate cements compared with organosilicon coatings.

The duration of maintaining the cohesive strength of coatings during operation is also determined by the resistance to periodic exposure to environmental factors: wetting, drying, freezing, thawing, etc. It was determined the change in cohesive strength as a function of wetting time. To this end, stretched film samples were sprinkled.

The duration of preservation of cohesive strength of coatings during operation is also determined by the resistance to periodic effects of environmental factors: wetting-drying, freezing-thawing, etc. In this connection, the influence of humidification on the change in the duration of cohesive strength was assessed. To this

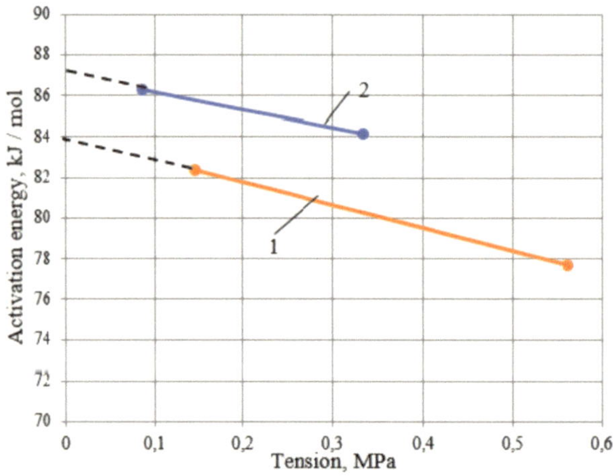

Figure 17.
Values of the activation energy of the process of destruction of coatings depending on tensions: (1) coating based on silicate paint; (2) coatings based on sol silicate paint.

Type of coating	U_o	γ
Polyvinyl acetate cement	122.81	27.98
Silicone KO-168	102.53	42.86
Silicate	84	11.14
Sol silicate	87	8.55
Note: For coatings $\iota_o = 10^{-13}$ s.		

Table 4.
The values of U and γ of coatings.

end, stretched coating samples were subjected to sprinkling. The results of the tests are shown in **Figure 18**.

Consider the condition of brittle cracking of polymer coatings under investigation under the action of internal tensions. In the case of brittle failure, the cracking condition has the form

$$\sigma \geq 0.5R \tag{16}$$

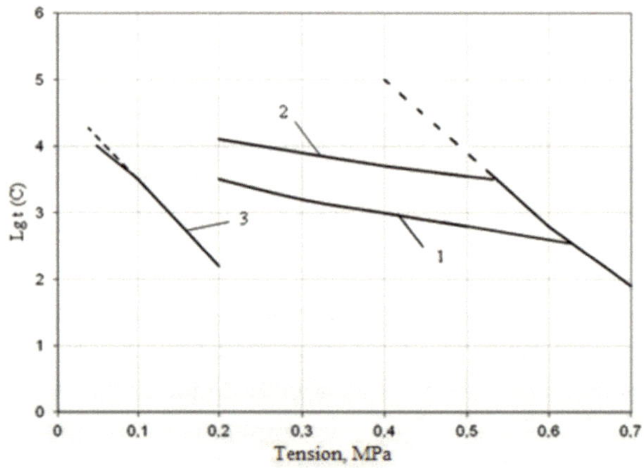

Figure 18.
Dependence of the long-term strength of coatings in the process of humidification: (1) coating of PVAC; (2) coating of PVAC with surface hydrophobization; (3) coating of KO-168.

Name of coatings	Internal stresses, MPa	Cohesive at strength of coatings, MPa	σ/R	Time of occurrence cracks, day
PVAC	0.08	0.38	0.21	Not observed
Polymer-lime	0.062	0.13	0.47	27
Lime	0.075	0.09	0.9	30

Table 5.
Physicomechanical properties of coatings after moistening.

Consider the change in internal stresses and the strength of coatings in the aging process on the example of humidification. It was found that at the first instant, humidification, a sharp increase in internal stresses, and then their slow decrease, i.e., an effect of the time factor, are observed. For example, for PVAC coatings, stabilization of internal stresses on level σ = 0.08 MPa is observed after 20 days of moistening, and stabilization of internal stresses in polymer-lime coatings is observed after 18 days of humidification on level σ = 0.062 MPa. The value of short-term strength after 60 days of moistening is given in **Table 5.**

The polyvinyl acetate cement coating after 2 months of moistening did not have cracks; the ratio σ/R for such coatings does not exceed 0.21.

Consider the kinetics of changes in the short-term strength of coatings during aging from the perspective of the kinetic concept of the strength of solids. It is known that the number of structural bonds N determines the strength of a material

$$R = f(N) \tag{17}$$

During operation, there is a change in the number of structural bonds.

$$n_1 u_o = \gamma\sigma \tag{18}$$

where u_0 is the unit bond breaking energy and γ is the structurally sensitive coefficient, characterizing the overvoltage of bonds.

The state of the material structure at a time t_1 is characterized by the value of the activation energy of bond breaking:

$$U_t = U_o - \gamma\sigma = U_o - n_1 u_o \tag{19}$$

The increase in the number of severed bonds at time due to the increase dU will be equal to

$$dn_1/N_o = \alpha dU \tag{20}$$

Integration of Eq. (20) leads to the form

$$n_1 = N_o \exp(-\alpha N_1) \tag{21}$$

In view of Eq. (18),

$$n_1 = N_o \exp[-\alpha(U_o - \gamma\sigma)] \tag{22}$$

During long-term operation, short-term strength is reduced in proportion to the number of broken bonds n_1:

$$R_t/R_o = [(N_o - n_1)/N_o] : [(\tau - t)/\tau] \tag{23}$$

where τ is the durability of the coating and t is the operating time.

After the transformation, Eq. (23) has the form

$$R_t = \{R_o \tau \exp(U/RT)[1 - \exp(-\alpha U)]\}/(\tau_o \exp(U/RT)) - t \quad (24)$$

Thus, the cracking condition has the form

$$\sigma_{\max} = 0.5\{R_o \tau \exp(U/RT)[1 - \exp(-\alpha U)]\}/(\tau_o \exp(U/RT)) - t \quad (25)$$

The obtained cracking condition (25) was used in the analysis of coating cracking due to wind load.

Consider the work of the coating in the pore zone, unfilled with a colorful composition. Suppose that the distance with pores that are not filled with a colorful composition is larger than the size of the pores themselves. Such a coating can be considered as a thin round plate pivotally supported along the contour, while the load is evenly distributed over the area. The coating thickness does not exceed 1/5 of the smallest pore size. The calculations established that the deflection boom does not exceed 1/5 of the coating thickness; therefore, such a plate can be considered rigid.

In accordance with the theory of plates, the magnitude of the stresses arising in the coatings can be determined by the formula

$$\sigma = \left(1.5 - 0.262\alpha^2 - 1.95\ln\alpha\right)q(d/h)^2 \quad (26)$$

Name cities	Floor	Tension, $\sigma \cdot 10^{-1}$ MPa					
		Pore diameter, mm					
		0.1	0.2	0.5	1.0	2.0	3.0
Moscow	1–2	0.01	0.03	0.18	0.71	2.85	6.41
	3	0.01	0.01	0.11	0.44	1.77	4.00
	6	0.01	0.04	0.23	0.93	3.73	8.33
	12	0.01	0.02	0.14	0.57	2.31	5.20
		0.01	0.05	0.30	1.21	4.84	10.89
		0.01	0.03	0.18	0.75	3.00	6.80
		0.02	0.06	0.39	1.57	6.26	14.09
		0.01	0.03	0.24	0.96	3.86	8.70
Penza	1–2	0.01	0.04	0.23	0.93	3.71	8.36
	3	0.01	0.02	0.14	0.58	2.32	5.22
	6	0.01	0.05	0.30	1.21	4.83	10.86
	12	0.01	0.03	0.18	0.75	3.01	6.78
		0.01	0.06	0.39	1.58	6.31	14.21
		0.01	0.04	0.24	0.99	3.94	8.87
		0.02	0.08	0.51	2.04	8.17	18.38
		0.01	0.05	0.39	1.27	5.10	11.49
Vladivostok	1–2	0.02	0.07	0.46	1.86	7.43	16.76
	3	0.01	0.04	0.29	1.16	4.64	10.40
	6	0.02	0.1	0.6	2.41	9.66	21.73
	12	0.01	0.07	0.45	1.83	7.34	16.50
		0.03	0.13	0.79	3.16	12.63	28.41
		0.01	0.08	0.49	1.97	7.89	17.70
		0.04	0.16	1.02	4.09	16.34	36.77
		0.01	0.1	0.63	2.55		22.90

Note: Above the line are voltage values for PVAC coatings with a thickness of 200 μm and below the line for coating XB-161 with a thickness of 80 μm.

Table 6.
Tensions in coatings due to wind load.

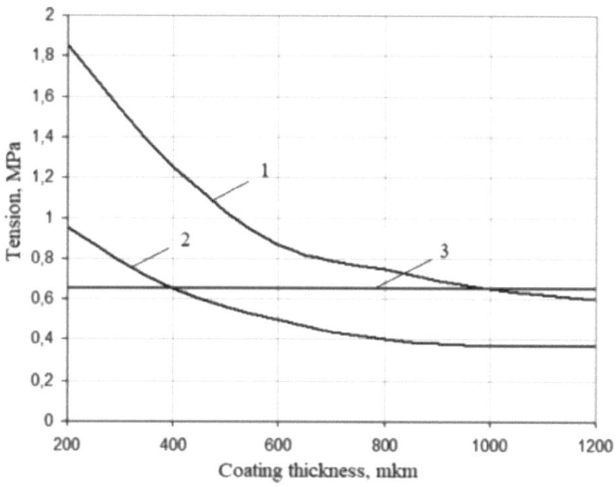

Figure 19.
Choosing the optimal coating thickness taking into account the duration of the voltage: (1) current voltage; (2) short-term strength; (3) long-lasting strength.

where h is the coating thickness; d is the pore diameter; and q is the load.

The external wall self-supporting panel of the building experiences the effect of horizontal wind load. In accordance with SNiP 2.01.07 "loads and impacts," the normative value of the average component of the wind load was selected. The results of calculating the stresses arising in thin-coat paint coatings during operation from the action of the wind load are given in **Table 6**.

Given the influence of the scale factor and the condition of brittle fracture of the coatings, the optimum coating thickness was selected (**Figure 19**).

Line 1 characterizes the numerical values of the acting tensions. Curve 2 characterizes the change in the short-term strength of PVAC coatings from the coating thickness in accordance with Eq. (21). Since the coating is in a brittle state, the steady-state long-term strength for each thickness is 0.5 of the short-term, i.e., curve 3.

An analysis of the data, (**Figure 19**), indicates that under the short-term effect of the wind load, destruction of the PVAC coatings will occur when the coating thickness is more than 850 μm, since for them $\sigma > 0{,}5\,R$ with long exposure to wind load, destruction will occur at a thickness of 300–850 μm. Coatings less than 300 microns thick are resistant to cracking, as $\sigma < R$ and $\sigma < 0{,}5\,R$. Each type of coating has its own critical pore size, exceeding which leads to cracking of the coatings.

5. Conclusion

The received results of research of a roughness of a surface of coverings confirm the assumption that quality of a substrate, namely, the degree of its uniformity, presence or absence of pollution on its surfaces, and its porosity render essential influence on quality of appearance of formed coverings that determines their stability while in service.

Studies have been conducted to evaluate the long-term strength of coatings. The values of the structurally sensitive factor are calculated. The condition for coating cracking is obtained, depending on the activation energy and operating time. Given the influence of the scale factor and the conditions for the destruction of coatings, a methodology is proposed for selecting the optimal thickness of the coatings. It is established that for each type of coating, there is a critical pore size, exceeding which leads to cracking of the coatings.

Author details

Valentina Loganina* and Yerkebulan Mazhitov
Penza State University Architecture and Construction, Penza, Russia

*Address all correspondence to: loganin@mail.ru

References

[1] Loganina VI, Orentlikher LP. Persistence of Protective and Decorative Coatings of the Exterior Walls of Buildings. Moscow: Publishing House Association of Construction Universities; 2001. p. 104

[2] Orentlicher LP, Loganina VI. Protective and Decorative Coatings of Concrete and Stone Walls. Handbook. Moscow: Stroyizdat; 1993. p. 136

[3] Andrianov KA. Organosilicon Compounds. Goskhimizdat: Moscow; 1995. p. 234

[4] Karyakina MI. Physico-Chemical Foundations of the Formation and Aging of Coatings. Moscow: Chemistry; 1980. p. 216

[5] Sukhareva LA. The Durability of Coatings. Moscow: Chemistry; 1984. p. 240

[6] Loganina VI. Durability of paint and varnish coatings depending on the quality of their appearance. IOP Conference Series: Materials Science and Engineering. 2016;**471**:022044

[7] Andryushenko EA. Lightfastness of Paint and Varnish Coatings. Moscow: Chemistry; 1986. p. 187

[8] Bartenev GM, Zuev YS. Strength and Destruction of Highly Elastic Materials. Moscow-Leningrad: Chemistry; 1984

[9] Loganina VI. The influence of surface quality of coatings on their deformation properties. Contemporary Engineering Sciences. 2014;7(36): 411241

[10] Bykhovsky AA. Distribution. Kiev: Science, Dumka; 1983. p. 191

[11] Loganina VI, Skachkov Yu P. Assessment of the stress state of the coating in depending on the porosity of the cement substrate. Key Engineering Materials. 2016;**737**:179-183

[12] GOST R 8.700–2010. State system for ensuring the uniformity of measurements. Methods of surface roughness effective height measurements by means of scanning probe atomic force microscope

[13] Zisman GA, Todes OM. The Course of General Physics. Moscow: Chemistry; 1968

[14] Loganina VI, Kislitsyna SN, Mazhitov Ye B. Development of sol-silicate composition for decoration of building walls. Case Studies in Construction Materials. 2018;**9**:e00173

[15] Gurevich MM, Itsko EF, Seredenko MM. Optical Properties of Paint Coatings. Leningrad: Chemistry; 1984. p. 120

[16] Ajzenshtadt AM, Frolova MA, Tutygin AS. Fundamentals of Thermodynamics of Highly Dispersed Systems of Rocks for Building Composites (Theory and Practice). Arkhangelsk: CPI NarFU; 2013. p. 113

[17] Loganina VI, Mazhitov YB. Research of inter-phase interaction in ZOL-silicate paints. International Journal of Engineering & Technology. 2018;7(4.5): 605-607

[18] Zhurkov SN, Narzulaev BN. Temporal dependence of solids. Journal of Technical Physics. 1953;**23**:1677

[19] Bokshitsky MN. Long-Lasting Polymer Strength. Moscow: Chemistry; 1978. p. 309